More Wire Antenna Classics

Volume 2

Compiled by Chuck Hutchinson, K8CH

Production:
Paul Lappen

Cover Design:
Sue Fagan

Published by:

ARRL *The national association for AMATEUR RADIO*

225 Main Street
Newington, CT 06111-1494

This work is publication No. 252 of the Radio Amateur's Library, published by the League.

Printed in USA

ISBN: 0-87259-770-9

First Edition
Second Printing, 2002

Contents

Foreword

Wire antennas can be fun to experiment with and rewarding to use. Many of you have shown your interest by purchasing *ARRL's Wire Antenna Classics*. As we looked through ARRL publications searching for the best articles on wire antennas for that book, it soon became obvious that there were more good items than would fit in a single book. For that reason, and because of your interest, we have published this second volume.

As in the first volume, some of the articles you'll find here are classics that were published before most of us were born. Others reflect the benefits of using the latest in computer modeling followed by actual testing.

In these pages you'll find more wire antennas and ideas. Dipoles, off-center-fed dipoles, multiband dipoles, loops, collinears, wire beams, vertically polarized, and receive antennas are all featured. Some of these antennas you may want to build and use right away. Others may provide inspiration for your own creativity.

For RF-safety information and fundamental antenna theory, be sure to refer to *The ARRL Antenna Book*. See the latest issue of *QST* for other ARRL antenna-related publications, or visit our online bookstore at: **http://www.arrl.org/catalog/**. Please take a few minutes to give us your comments and suggestions on this book. There's a handy Feedback Form for that purpose at the back.

David Sumner, K1ZZ
Executive Vice President
Newington, Connecticut
February 2000

CHAPTER ONE

DIPOLES

By Frank Gue, VE3DPC, ex-VE6BH From *QST*, June 1968

An 80-Meter Inverted V for Field Day

It's that time of year again when portable operation is a delight. Here is a full-sized low-frequency antenna that you can put up in the field or at home without any help.

A Field Day antenna for 80 meters should be light, portable, inexpensive, capable of easy erection without any special equipment, and it should not contain any fragile parts.

Of course, the antenna must perform adequately.

The popular inverted V suggests itself as an answer to these requirements. It can be constructed so that it guys itself, minimizing the inevitable exasperating tangle of rope and wire, and it requires only one mast. Like all dipoles the inverted V can be fed at the center with low impedance line, and elements for several bands can be connected in parallel at the feed point of the antenna.

The inverted V shown in the photographs and sketches is simple, easily managed (20 minutes to erect, 10 minutes to dismantle), and costs under $15 for all components. Not only is it a convenient antenna for portable operation, but it can serve as a fine permanent antenna for the home station as well. In fact, that's what has happened to the inverted V pictured; it's been in use in the back yard of VE3DPC ever since Field Day, 1967.

The inverted V as it looked this past winter in the author's backyard.

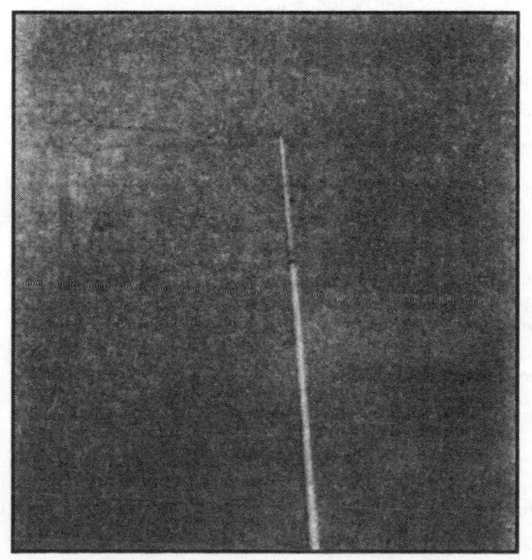

Upper portion of the 80-meter inverted V as it appeared at the Field Day site.

Assembly

As shown in Figure 1, the antenna support consists of three 10-foot sections of 1¼-inch TV mast. One end of each of the two lower sections is swaged, permitting all three sections to be force-fitted together. Hooks of music wire were installed in the top section, so that the element wire and feed line could be easily coiled for storage.

The center insulator (Figure 2) is made of maple; however, if this wood cannot be found, birch or fir can be used instead.

Figure 2 shows how the center insulator was drilled and Figure 3 shows how the feed line and antenna element were attached. Dimensions are obviously not critical, and the sketches are intended to show in a general way how the insulator was made, not to give precise detail. Just be careful where you drive the screws!

Softwood (pine) was picked for the end insulators (Figure 4) because it is adequate for the application and easier to work with than maple. Fragile ceramics should not be used in place of the wood insulators.

The 80-meter element (Figures 3 and 5) is made from small-gauge, Teflon-insulated, stranded Copperweld that was bought at a surplus outfit. This wire is worth looking for; it's tough, durable, light, and easy to handle. If operation on both 80 and 40 is desired, the builder can connect a 40 meter element to points A and B in Figure 3.[1]

The feed line can be 50-ohm coax, 72-ohm coax or 72-ohm Twin-Lead. I chose the latter because it is low cost, lightweight, very flexible, easy to store, and because it lends itself to such outrageous techniques as being tied in knots; however, since most transmitters have coax output, either a balun or an antenna coupler must be used between the feed line and the transmitter to maintain balance.[2] If elements for more than one band are to be used, the antenna coupler is preferred because it will reject harmonics of the transmitting frequency that might fall in the range of one or more of the paralleled antennas.

Raising and Lowering

Transporting the antenna to and from the Field Day site is easy. Tape the three mast sections together, wrap them with cloth to protect the car's finish, and tie the works between one of the door handles and the rear bumper. If the antenna extends past the back of the car, don't forget to tie a red flag on the end of the bundle to alert other drivers.

Upon arrival at the Field Day site, unpack the antenna and join the TV masts together. The three-section support is light enough to be "walked up" by one man, if the wind is under 10 miles per hour; however, prior to the antenna raising, you must put stakes into the ground, determine the

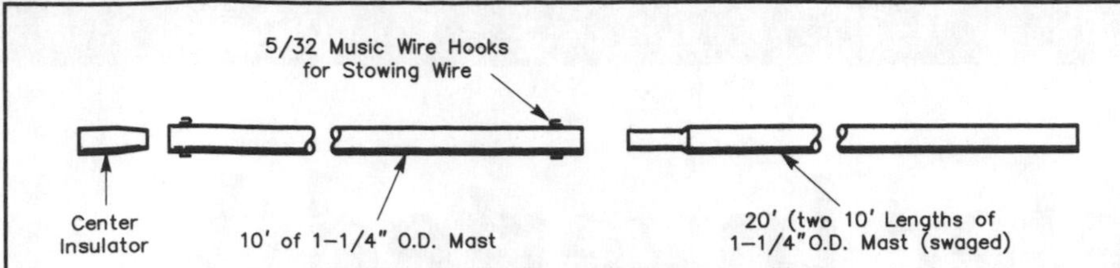

Figure 1—The makings of the inverted-V center support.

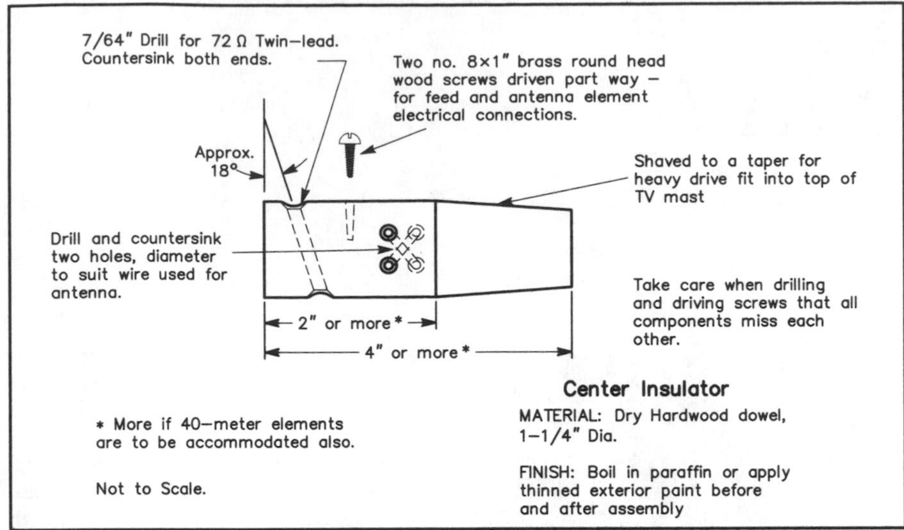

Figure 2—An X-ray view of the center insulator.

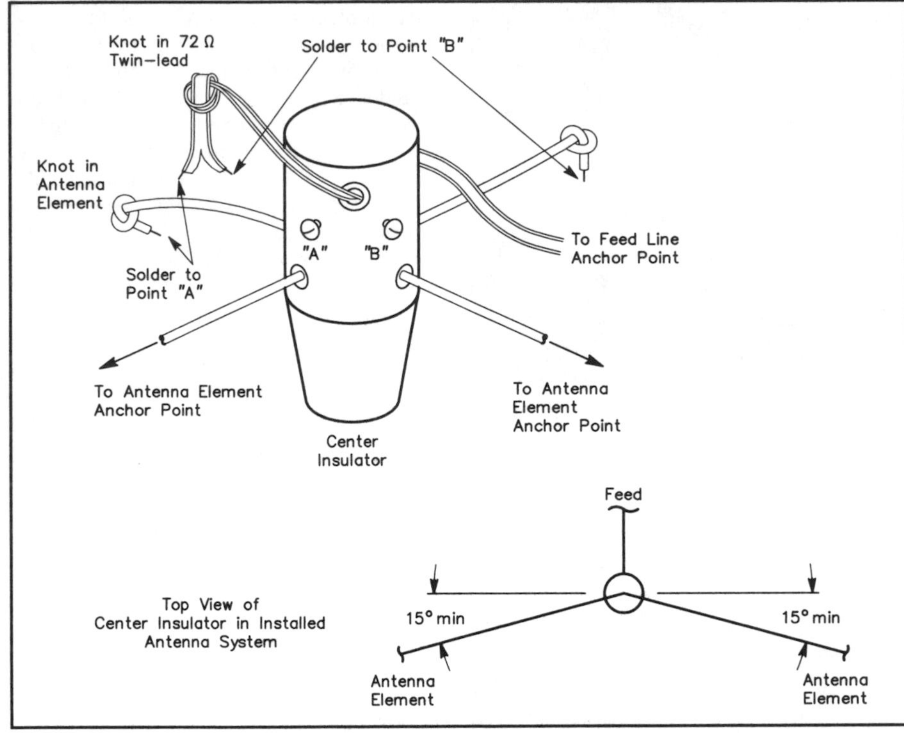

Figure 3—Details of how the feed line and element wires should be attached to the center insulator, and a top view of the installed insulator.

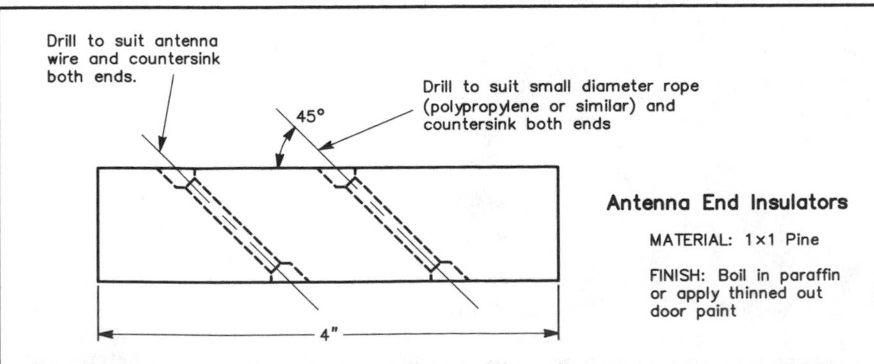

Figure 4—A sketch of one of the wooden end insulators. Two insulators are required for each dipole used.

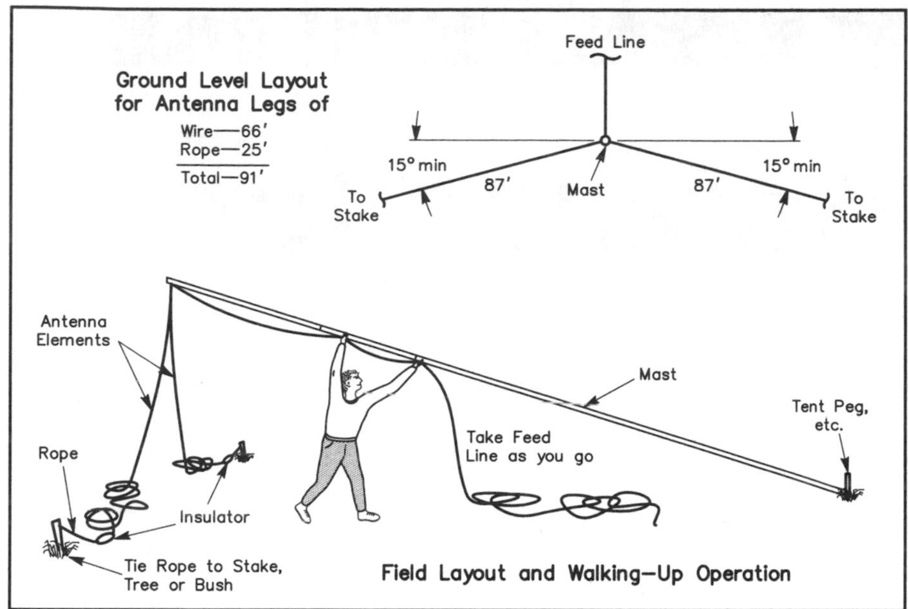

Figure 5—By using the dimensions and layout shown in the upper right corner of the sketch—so that ground stakes could be installed and proper length guys attached between them and the mast—the author was able to raise his 80-meter inverted V without any help. For other element lengths than shown, one man can easily put up the antenna by using simple geometry to determine the amount of rope required and the locations of the stakes.

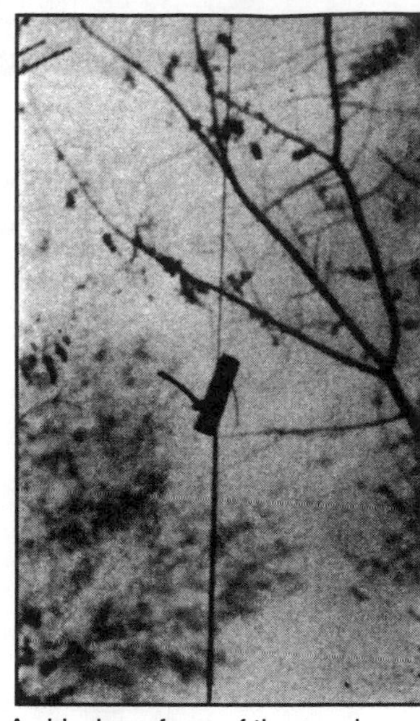

A sideview of one of the wooden end insulators.

exact length the guys will be when the antenna is in its operating position, and attach the guys to the stakes (Figure 5).

If help is available, it won't be necessary to install stakes or measure guys prior to putting up the antenna. Have the helpers pay out each half of the antenna element and keep slack out of the wire as the mast goes up. Station the men about 45 degrees off the line of the mast as it lies on the ground, and then, once the mast is up, have the helpers walk around to the final positions shown in Figure 5. The last few degrees of "walking up" is where your control is poorest.

With the rear guy (feed line) in your hand, move quickly and smoothly past top dead center and promptly and firmly apply back tension on the antenna element. Once the feed line is snubbed, you can easily correct the mast for lean by lifting the base and putting it where it suits you. If you are fortunate enough to have three helpers, there is no problem at all. Give the third man the feed line, station him as far away as possible from the mast and antenna element, and have him take up the slack as the mast goes up.

Base support can be little or nothing. The mast, which is now permanently erected in my back yard, rests directly on the grass. It was moved a couple of inches during each mowing operation, and no apparent harm came to the lawn.

Lowering the antenna is brutal but quick. Release the back guy (feed line) and let the whole system fall. If the ground is reasonably flat, this will be OK. But if the ground is very rough, the pipe may take a set as it strikes the ground. Under such conditions, it would be better to use helpers to reverse the erecting procedure.

The antenna has been left up for a period of several months, during which there have been many autumn storms and a severe ice storm. It has shown no signs of distress, and it has performed normally when wet or loaded with ice. The small size of the inverted V's parts has led to little wind and ice loading and is probably the main reason for the antenna's durability.

[1] Although dimensions for an 80-meter inverted V are given in Figure 5, these aren't necessarily the optimum dimensions for every installation: there are too many variables involved. If the antenna is not to be used with a transmatch and open-wire line, it is best to experimentally determine the length of the inverted V. This can be done by starting with an overall length equal to about 525 divided by the frequency in megahertz. The ends of the inverted V should then be trimmed until the SWR is minimum at the operating frequency. A simpler method is described in footnote 2.

[2] An easier arrangement, especially for multiband operation, can be achieved by using a transmatch and either 300-ohm Twin-Lead or 450-ohm Ladder-Line. Only one dipole—about 100 feet long (length not critical)—is all that is necessary to cover all the amateur bands between 3.5 and 30 MHz.

By John C. Allred, W5LST From *QST*, October 1961

E-Z-UP Antenna for 75 and 40

Full-size dipoles for the 75- and 40-meter bands occupy more space than is conveniently available on the 75 by 113-foot lot at W5LST. The increasing popularity of the "drooping," or inverted-vee, dipole antenna among amateurs led us to investigate it for our somewhat crowded conditions. Based on the electrical design of Glanzer,[1] this system has performed meritoriously at W5LST. Requiring only one support, it was surprisingly easy to erect, gives a satisfactory s.w.r. over the phone bands and, importantly, the cost was less than thirty dollars complete. A plan-view sketch is shown in Figure 1.

The Mast

The mast is a telescoping Channel-master, capable of 50-foot height, but extended only to 35 feet. Extending the upper sections to less than their full lengths gives rigidity to the mast, and has apparently eliminated the need for guys on each section, as recommended by the manufacturer. To date this mast has withstood gusts of 50 miles per hour without a shudder; it is yet to be tested in a real gale, however.

As shown in Figure 2, a 9-foot length of 3-inch pipe is cast in concrete with 6 feet of its length extending above ground. Three pairs of clamps, such as those used on chain-link fence, secure the mast to the upright standard. During erection, these clamps are loosened and all three lie at the base of the standard, so that the mast need only be lifted about 6 inches to be put in place. When the mast is in place, the clamps are raised and tightened.

[1] Glanzer, "The Inverted V-Shaped Dipole," *QST*, August, 1960.

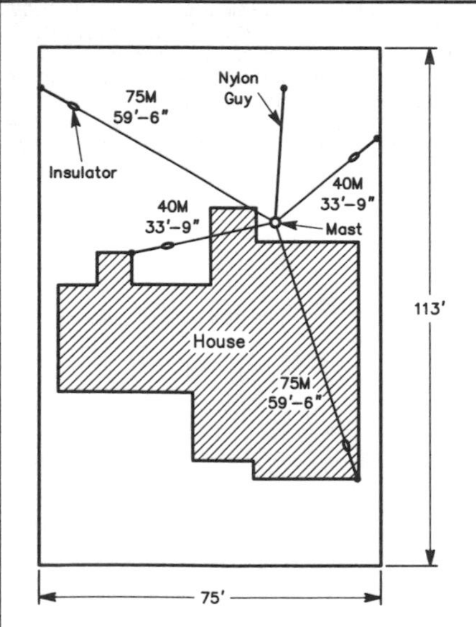

Figure 1—W5LST's layout for effective 40- and 80-meter antennas on a small lot. The two "drooping" dipoles are fed in parallel with a single coax line. Nylon-line extensions are used to reach convenient anchorages.

Figure 2—The mounting for the antenna mast. The 9-foot pipe is guyed temporarily while the concrete is poured.

Rigging

Except for the antenna conductors, all rigging is of nylon line of 500-pound test. A halyard is reeved through a pulley of suitable size which is wired securely to the top of the mast. The two ends of the halyard are made fast to a harness snap which, in turn, supports the center of the antennas. Provision of this halyard has proved to be a great convenience in permitting inspection of the antenna connections and the adjustment of tension in the wires without the necessity for lowering the mast.

Nylon line has a tendency to ravel at its ends but this problem is easily solved. Most fastenings were made with two halfhitches, followed by sewing the end of the line to itself with thread, as shown in Figure 3A, and doping with one of the quick-drying model-airplane cements.

Some weeks after the initial installation, it became apparent that some additional stabilization of the mast against occasional strong northerly winds would be desirable. Accordingly a nylon line was run from the harness snap at the top of the mast to a convenient anchor in the back yard, which happened to be the top of the children's swing set. Experience seems to show that the antenna wires, together with the additional nylon line, stabilize the mast against aerodynamically-excited vibration, without any appreciable strain on the antennas.

The Antennas

As shown in Figure 3, the two antennas are connected in parallel at the top of the mast. The lower ends are connected to convenient tie points so that the two legs of a given antenna are more or less in a straight line. To our great surprise, very little effect is produced by moving the ends of the an-

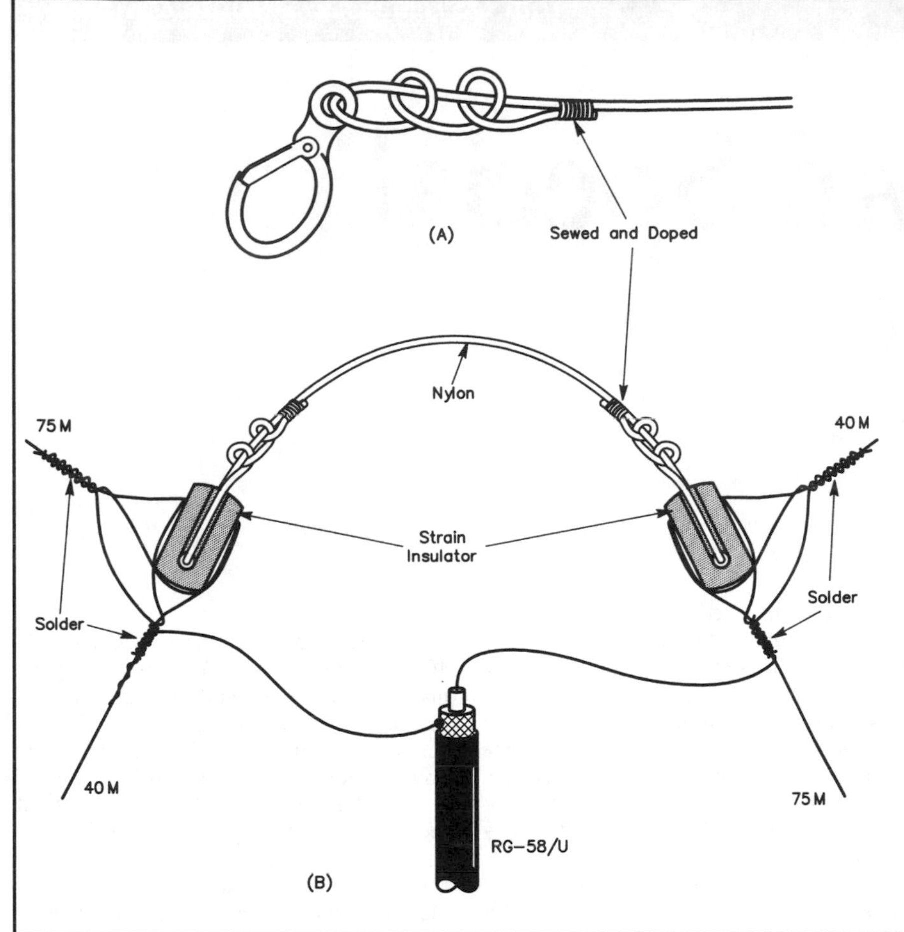

Figure 3—The halyard and feeder arrangement. The coaxial cable is sealed securely with polyethylene tape after the connections are made. The harness snap of A at the end of the hoisting halyard engages the bridle between the two insulators in B.

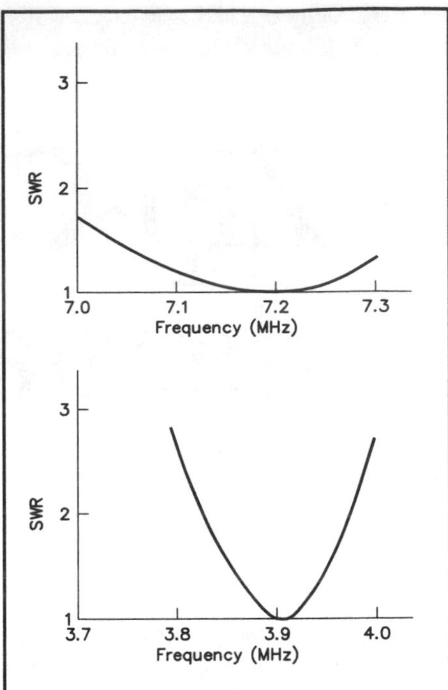

Figure 4—SWR curves as indicated by a "Monimatch"-type SWR indicator.

tennas either horizontally or vertically. There is apparently negligible electrical interaction between them as indicated by the s.w.r. bridge.

Pruning of the antennas is, as always, desirable. In our case, the optimum lengths of each leg turned out to be 33 feet 9 inches for 40 meters and 59 feet 6 inches for 75 meters. Figure 4 shows the performance of the antennas on 40 and 75 as measured by a Heath s.w.r. bridge. Although the antennas tune sharply, they are usable over the entire phone band in each case, and it would be difficult to imagine a better performer on the lower-frequency bands than this simple antenna system.

A Zip-Cord Special Antenna

Recently I found myself in need of an 80-meter antenna to help me keep in touch with some of my friends back home while operating portable in Arizona. The antenna and feed line were not readily available, but I remembered that one of my friends had recommended zipcord as a substitute feed line for a dipole, provided low power was used. Going one step further, I felt the entire antenna could be made of this low-priced material. I purchased 80 feet of zip cord at a local hardware store (at 4 cents a foot) and peeled it apart for 60 feet. This gave me the required 120 feet for a dipole, plus a 20-foot feed line which was taped to prevent further splitting. The feeder was fastened to a 2×4 which was used as a center support. The ends were extended as far as possible, then tied with heavy cord to a fence that was in the trailer park where I was staying. This made a slight inverted-V antenna.

To reach the desired frequency of 3935 kHz it was necessary to shorten the antenna somewhat. I did this by folding three feet on each end of the dipole back upon itself. This enabled me to keep skeds with many "7s" at home.

Having met with success on 80-meters I immediately began to eye the 40-meter band for possible use. I then measured out 32 feet from the center and cut the wire there, then tied a knot in the two pieces, leaving a couple of inches of bare wire dangling. Now for forty meters, I merely disconnect the wires and reconnect them for 80 meters.

As the *ARRL Antenna Book* states, a 40-meter dipole will often present a good match at 15-meters, and this one is no exception; I have worked many stations across the country on 15 meters with low power while using this simple antenna.

The *Handbook* or *Antenna Book* has not recommended this type of feed line in many years, though using a twisted pair at one time was usual practice in feeding a doublet. But, when the need is there for a quick and easy antenna, I don't think you can find an easier one as long as you remember to use low-power transmitters (100 watts or less) and trim the wire for a low value of SWR. — *Frank S. Wise, W7QYC*

Zip-Cord Antennas—Do They Work?

Basic Amateur Radio: Parallel power cord is readily available and is easy to work with. How efficient is it when used at radio frequencies? Well, that depends.

Zip cord is readily available at hardware and department stores, and it's not expensive. The nickname, zip cord, refers to that parallel-wire electrical cord with brown or white insulation used for lamps and many small appliances. The current price is about 6 cents a foot in hundred-foot rolls in most parts of the U.S. The conductors are usually no. 18 stranded copper wire, although larger sizes may also be found. Zip cord is light in weight and easy to work with.

For these reasons, zip cord is frequently used as both the transmission line and the radiator section for an emergency dipole antenna system. The radiator section is obtained simply by "unzipping" or pulling the two conductors apart for the length needed to establish resonance for the operating frequency band. The initial dipole length can be determined from the equation $\ell = 468/f$, where ℓ is the length in feet and f is the frequency in megahertz. (It would be necessary to unzip only half the length found from the formula, since each of the two wires becomes half of the dipole.) The insulation left on the wire will have some loading effect, so a bit of length trimming may be needed for exact resonance at the desired frequency.

For installation, many amateurs like to use the electrician's knot shown in Figure 1 at the dipole feed point to keep the transmission-line part of the system from unzipping itself under the tension of dipole suspension. This way, if zip cord of sufficient length for both the radiator and the feed line is obtained, a solder-free installation can be made right down to the input end of the line. Granny knots (or any other variety) can be used at the ends with cotton cord to suspend the system. You end up with a lightweight low-cost antenna system that can serve for portable or emergency use.

But just how efficient is a zip-cord antenna system? Since it is easy to locate the materials and simple to install, how about using such for a more permanent installation? Upon casual examination, zip cord looks about like high-quality 72-ohm balanced feed line. Does it work as well? Ask several amateurs these questions and you're likely to get answers ranging all the way from, "Yes, it's a very good antenna system!" to "Don't waste your time and money – it's not worth it!" Myths and hearsay seem to prevail, with little factual data. The intent of this article is to rectify that situation.

Zip Cord as a Transmission Line

In order to determine the electrical characteristics of zip cord as a radiofrequency transmission line, we purchased a 100-foot roll and subjected it to tests in the ARRL laboratory with an rf impedance bridge. Zip cord is properly called parallel power cord. The variety purchased was manufactured for GC Electronics, Rockford, IL, being 18-gauge, brown, plastic-insulated type SPT-1, GC cat. no. 14-118-2G42. Undoubtedly, minor variations in the electrical characteristics will occur among similar cords from different manufacturers, but the results presented here are probably typical.

As the first step, we checked the physical length of the wire. GC generously provided nearly 4 feet over the specified 100-ft length, and the extra length was promptly lopped off.

We wanted to avoid measurement errors that might arise from coiling the wire on the supply spool (inductance between turns) or laying it out on the vinyl-tiled cement-slab floor (capacitance to a large surface), so the second step was to suspend the wire in a long hallway, about a foot or so below the false ceiling. Cotton twine was used for the supporting material. Undaunted by snide comments from fellow staff members about cheating the telephone company out of installation work and facetious suggestions

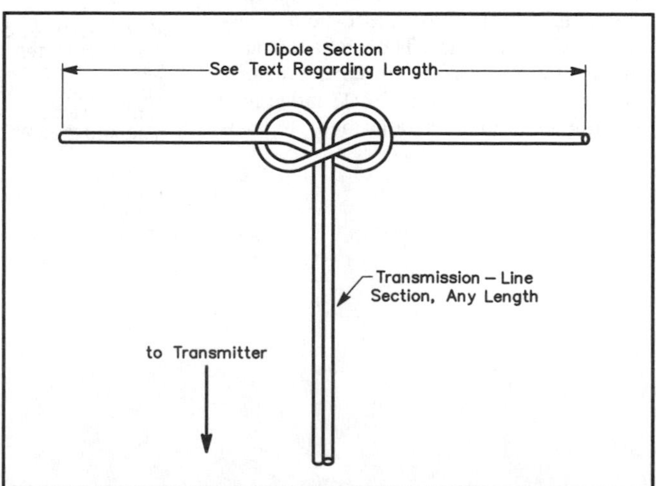

Figure 1—This electrician's knot, often used inside lamp bases and appliances in lieu of a plastic grip, can also serve to prevent the transmission-line section of a zip-cord antenna from unzipping itself under the tension of dipole suspension. To tie the knot, first use the right-hand conductor to form a loop, passing the wire behind the unseparated zip cord and off to the left. Then pass the left-hand wire of the pair behind the wire extending off to the left, in front of the unseparated pair, and thread it through the loop already formed. Adjust the knot for symmetry while pulling on the two dipole wires.

that we use two paper cups and a string for our intercom, we continued with preparation for the tests.

With a General Radio 1606A rf impedance-measuring bridge, we made measurements at the input end of the line on 10, 15 and 29 MHz, while the far end of the line was terminated first in a short and then in an open circuit. We did notice during measurements that our readings changed somewhat as people walked in the hallway, especially during the 29-MHz tests. After a seemingly endless period of shooing away would-be helpers who disrupted measurements by grabbing the line, hanging objects on it, and so forth, we finally arrived at a set of readings considered to be satisfactory for our purposes. From there, a bit of work with an electronic calculator and a Smith Chart told us what we wanted to know — the electrical characteristics at radio frequencies.

The results? Well, as the saying goes, I have some good news and some bad news.

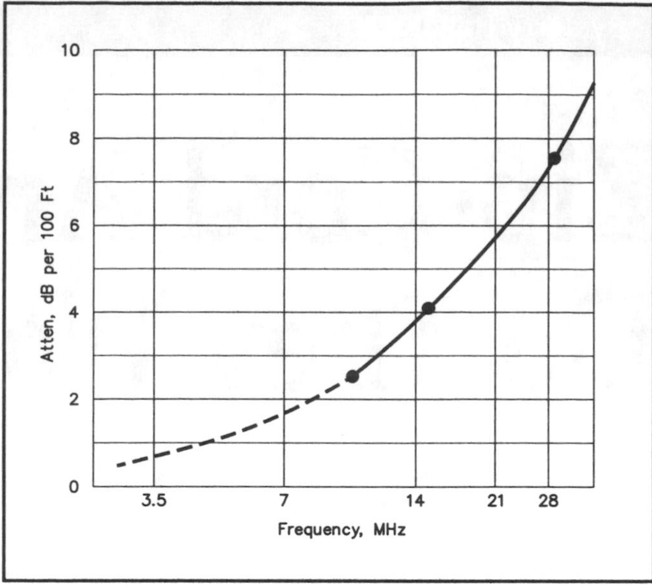

Figure 2—Attenuation of zip cord in decibels per hundred feet when used as a transmission line at radio frequencies. Measurements were made only at the three frequencies where plot points are shown, but the curve has been extrapolated to cover all high-frequency amateur bands.

First, the Good News

If ever you need a balanced 105-ohm transmission line, then zip cord is the stuff for you! Its characteristic impedance was determined to be 107 ohms at 10 MHz, dropping in value to 105 ohms at 15 MHz and to a slightly lower value at 29 MHz. The nominal value is 105 ohms at hf. The velocity factor of the line was determined to be 69.5 percent.

This reported change in impedance with frequency may raise a few eyebrows, but it is a fact that most lines do not exhibit a constant characteristic impedance across a range of frequencies. However, this writer has not previously encountered any line which was not "flat" between 3 and about 50 MHz. Zip cord may be exceptional from this standpoint, especially considering that the insulating material was not chosen for its qualities at rf.

And Now the Bad News

Who needs a 105-ohm line, especially to feed a dipole? Most of us know that a dipole in free space exhibits a feed-point resistance of 73 ohms, and at heights above ground of less than 1/4 wavelength the resistance is even lower. An 80-meter dipole at 35 feet, for example, will exhibit a feed-point resistance of about 40 ohms. Thus, for a resonant antenna, the SWR in the zip-cord transmission line can be 105/40 or 2.6:1, and maybe even higher in some installations. Depending on the type of transmitter in use, the rig may not like working into the load presented by the zip-cord antenna system.

But the really bad news is still to come–line loss! Figure 2 is a plot of line attenuation in decibels per hundred feet of line versus frequency. Chart values are based on the assumption that the line is perfectly matched (sees a 105-ohm load as its terminating impedance). For lengths other than 100 feet, multiply the figure by the actual length in feet and then divide by 100.

In a feed-line, losses up to about one decibel or so can be tolerated, because at the receiver a 1-dB difference in signal strength is just barely detectable. But for losses above about 1-dB, beware. Remember that if the total losses are 3 dB, half of your power will be used up just to heat the transmission line.[1]

Based on this information, we can see that a hundred feet or so of zip-cord transmission line on 80 meters might be acceptable, as might 50 feet on 40 meters. But for longer lengths and higher frequencies, look out! The losses become appreciable.

[1]Additional losses over those charted in Figure 2 will occur when standing waves are present. See Gibilisco, "What Does Your SWR Cost You?" January 1979 *QST*. Trouble is, you can't use a 50- or 75-ohm SWR instrument to measure the SWR in zip-cord line accucately.

What About Zip Cord Wire as the Radiator?

For years, amateurs have been using ordinary copper house wire as the radiator section of an antenna, erecting it without bothering to strip the plastic insulation. Other than the loading effects of the insulation mentioned earlier, no noticeable change in performance has been noted with the insulation present. And the insulation does offer a measure of protection against the weather. These same statements can be applied to single conductors of zip cord.

The situation in a radiating wire covered with insulation is not quite the same as in two parallel conductors, where there may be a leaky dielectric path between the two conductors. In the parallel line, it is this current leakage which contributes to line losses. The current flowing through the insulation on a single radiating wire is quite small by comparison, and so as a radiator the efficiency is high.

Now back to the original question: How efficient *is* a zip-cord antenna system? Well, that *does* depend, on the length of the wire used for the feed-line section and on the frequency. In a pinch on 160, 80, 40 and perhaps 20 meters, communications can certainly be established with this kind of antenna. For higher frequencies, especially with long line lengths for the feeder, you're on your own.

Piggyback Antenna For The Ten-Meter Band

Most amateurs from Novices on up use dipole antennas for the eighty- and forty-meter bands. In many cases, these two antennas were all the Novice needed since the fifteen-meter band could also be covered with the forty-meter dipole.

However, with the advent of the new ten-meter subband for Novices, a simple and inexpensive method of getting the needed coverage is to use a "piggyback" antenna. It can be easily added to an existing dipole, eliminating the need for additional supports or feed lines. Also, it will have no effect on either an eighty- or forty-meter dipole.

Of course, performance will not be as good as that obtained with a beam, but it should not be sold short either. A KH6 was worked during a recent contest with a power output of 1.8 watts. — *Robert M. May, II, WA4DEG*

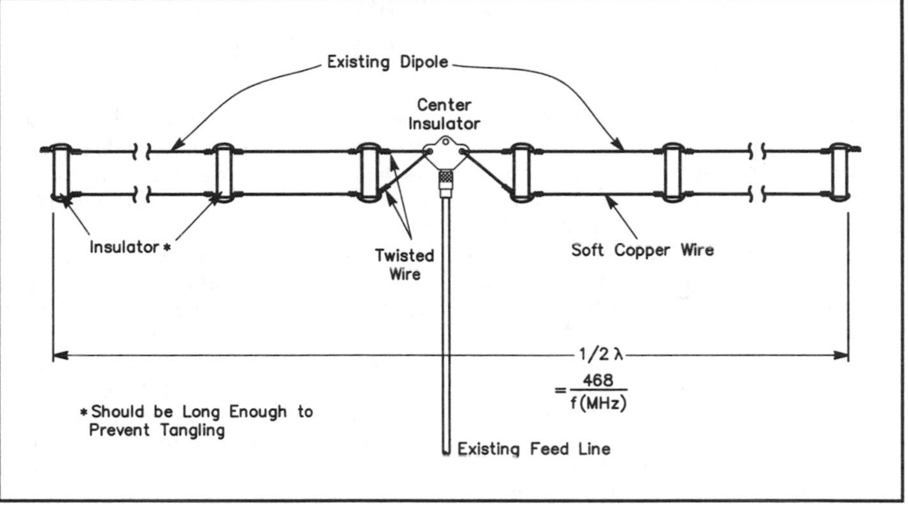

RFD-1 and RFD-2: Resonant Feed-Line Dipoles

This unique design offers simplicity and general utility. In short, these are "reel" great antennas for the HF bands.

During several decades of operation on the 80-meter band, I have encountered a number of hams who need an antenna system that is efficient, simple to construct, and yet can be deployed easily in difficult locations. The resonant feed-line dipoles described here provide an excellent match to the transceiver without a separate antenna tuner. They are easily transported on plastic cord reels. The basic design is extensible to other bands—all of this without a dangling feed line to contend with!

Consider some realistic situations encountered by fellow hams in their efforts to get on 80 meters:

Scenario 1: You live on a small suburban lot with trees, but there is no way to stretch a straight 120-foot-long dipole with a feed line.

Scenario 2: Your small backyard ends at the edge of a steep decline. A radiator wire could be snaked through the trees, but running a separate feed line is out of the question.

Scenario 3: You live in a high-rise building and could secretly stretch a wire upward or downward, but there is no way to install a feed line.

Scenario 4: You're going on vacation to the mountains and you know that your cottage is nestled among hundred-foot-high pines You can take along a bow and arrow and cord, but then what would you do?

In response to these problems, I began to search for a universal solution under the following ground rules. (1) No separate antenna tuner would be required, (2) the antenna could be deployed with no more difficulty than stringing up a length of coax and (3) the antenna could be easily stored and unwound from a cord reel without a tangle of cable and wire.

Several years ago I used a 10-meter vertical linear coaxial sleeve antenna, as shown in Figure 1. A vertical dipole is constructed from a quarter-wave whip and a quarter wavelength of shielding braid. Its feed line passes through the braid, yielding a simplified geometry. Although this concept could be adapted to 80 meters, who wants to deal with 60 feet of shielding braid? The important lesson I learned was that the RF current has no trouble traveling up the inside of the coax and making a 180° turn to travel back on the outside of the braid!

Because this is true, perhaps we don't need the separate outer braid. Why not just use the outside of the coax itself? If we do this, however, how do we let the RF "know" when it should stop flowing and reflect back toward the center of the radiator, as it did when it came to the end of the added braid in the 10-meter vertical? The current on the outside of the coax shield is called "common-mode" current—there is no counteracting equal and opposite current as there is inside the coax. One design approach utilizes the primary function of a balun transformer: to place in unbalanced reactance in the path of this common-mode current without affecting the desired balanced transmission line currents. The development of this concept is discussed in the Appendix.

The common-mode current on the coax shield is transformed by a quarter-wavelength stub to maximum near the transceiver. It was not surprising to note experimentally that a coil of a few turns placed at the fed end of the dipole decreases the resonant frequency of the system substantially (because of the added inductive reactance). However, an unexpected dividend of this is that, at the resonant frequency, an almost perfect impedance match to the 50-ohm source is realized! Now we have the design for our simplified antenna system! We can increase the resonant frequency by moving the coil along the cable away from the current maximum while retaining the perfect match. And at the same time we can decrease the common-mode current on this part of the line, because of the coil's inductive reactance. For the values chosen, the coil is near self-resonance from the distributed capacitance of the coil windings. The equivalent parallel-resonant circuit serves to increase the reactance at this point in the antenna, which assures a reflection when the RF reaches this virtual end of the dipole. To my knowledge, this configuration and method of resonating is unique and novel, and I refer to it as a T choke.

Constructon, Installation and Adjustment

The simple arrangement of the resonant coaxial linear dipole is shown in Figure 2. The dimensions and test results are for a nominal

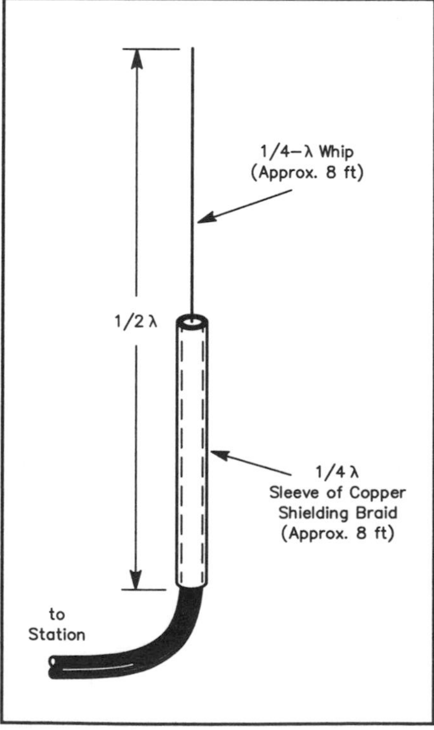

Figure 1—A vertical linear coaxial sleeve antenna for 10 meters.

frequency of 3.95 MHz. To dramatize the simplicity of the antenna, I list all required parts in Table 1.

The coaxial cable connector is assembled at the input end of the coax for connecting to the transceiver. The center conductor of the far end of the cable is connected to the antenna wire to form a hook and eye, and is securely soldered. This junction is potted in the center of a short length of PVC pipe to form a robust center insulator assembly. At the outset, 13 turns of cable can be wound on the cord reel at a point approximately 59 feet plus 25 inches from the center of the dipole. The 13 turns can then be taped in place on the reel with duct tape or equivalent. The remaining coax and wire can now be wound on the reel for ease of transporting. To avoid kinks when unwinding, secure the end of the wire and rotate the reel, keeping the wire and coax taut.

Ground losses are a great enemy of an HF antenna, especially at the lower frequencies, so place the dipole for 80 meters as far above the ground as possible! After installation, check the resonant frequency with a noise bridge or an SWR meter. If these construction details are followed, the resonant frequency should be approximately 3.95 MHz. If the resonance indication is indefinite or if the resistance is not close to 50 ohms, adjust the self-resonance of the coil by moving the turns slightly on the reel. This alters the inter-turn capacitance, permitting adjustment of the reflection of RF at the end of the dipole. During this adjustment, remember that the greatest effect is between the input and output turns, where the voltage difference is greatest. If you want to lower the resonant frequency of the antenna, remove the tape and rotate the reel to increase the 25-inch distance, thereby increasing the length of the dipole. This is done in a manner to retain the 13-turn coil—you are winding and unwinding equal lengths of cable. When the desired resonant frequency is attained, you are ready to operate! This is a broadband antenna, and not much adjustment is required.

The 13 turns is mentioned only as a nominal coil size. I have used both 11-turn and 13-turn coils. Self-resonance can be adjusted as mentioned earlier. As the number of turns is increased, the initial 25-inch distance will be decreased. For example, in my test installation a change from 11 turns to 13 turns altered this distance required for resonance from 68 inches to 25 inches. A greater impedance at this point serves to improve the isolation of the dipole from ground.

The RFD-2 for Two-Band Operation

The preceding construction information is for an antenna dedicated to a single amateur band, the RFD-1. With a slightly different configuration, referred to as the RFD-2, the RFD-1 design can be extended to cover the 40-meter band without severing the coaxial cable. Conversion from 80 to 40 meters and back requires only a few minutes. The RFD-2 design allows you to change operation to 40 meters for a few days without permanently altering the 80-meter lengths. In other words, the total dimensions of the 80-meter coaxial feed-line dipole are retained but operation is adapted to 40 meters. Tbis is done simply by winding coaxial cable on reels. This is readily achieved by altering the winding of the coax on the original reel and then adding a second coil near the far end of the dipole. Once the values have been established, this band change is accomplished simply.

The RFD-2, as it has evolved at W2OZH, is shown in Figure 3. Total dimensions are given in Figure 3A. The length of the feed line has been increased from that of the RFD-1 to 143 feet to make allowance for the cable used in the 13-turn coil. This assures an isolating stub which is approximately 1/4 λ on 80 meters, and at the same time permits a stub length that is close to 1/4 λ on 40 meters, thereby achieving the desired high impedance for isolation on both bands. Also, the no. 12 terminating antenna wire has been replaced by a length of coaxial cable—the outer jacket provides insulation so the coax can now be coiled to provide the desired dipole length on 40 meters.

As pointed out earlier, a self-resonant coil is used to assure a high coefficient of reflection at the fed end of the 80-meter dipole. For other bands it is necessary to calculate the approximate number of turns required to approach self-resonance. A suitable winding for 80 meters was determined to be 13 turns on a 6-inch diameter reel, so the numbers for other bands can be obtained by a simple scaling calculation. Details are given in the Appendix. For guidance in adapting the design for the other HF bands, the approximate number of turns for the terminating coils is shown in Table 2.

For wavelengths shorter than 40 meters it may be desirable to use a radiating section longer then 1/2 λ. The length either side of center can be any odd number of quarter wavelengths. For example, on 20 meters the radiator could be 3/2 λ (three

Table 1
RFD-1 Parts List

- 1–118-foot length RG-8X (minifoam) coaxial cable
- 1–59-foot length no. 12 stranded or Copperweld wire
- 1–PL-259 male coaxial cable connector
- 1–10-inch-diameter cord reel (Doscocil model no. 32500 or equiv.)
- 1–3-inch length of 1/2-inch OD PVC pipe (for center insulator)
- 1–Epoxy potting compound (sufficient to fill pipe)

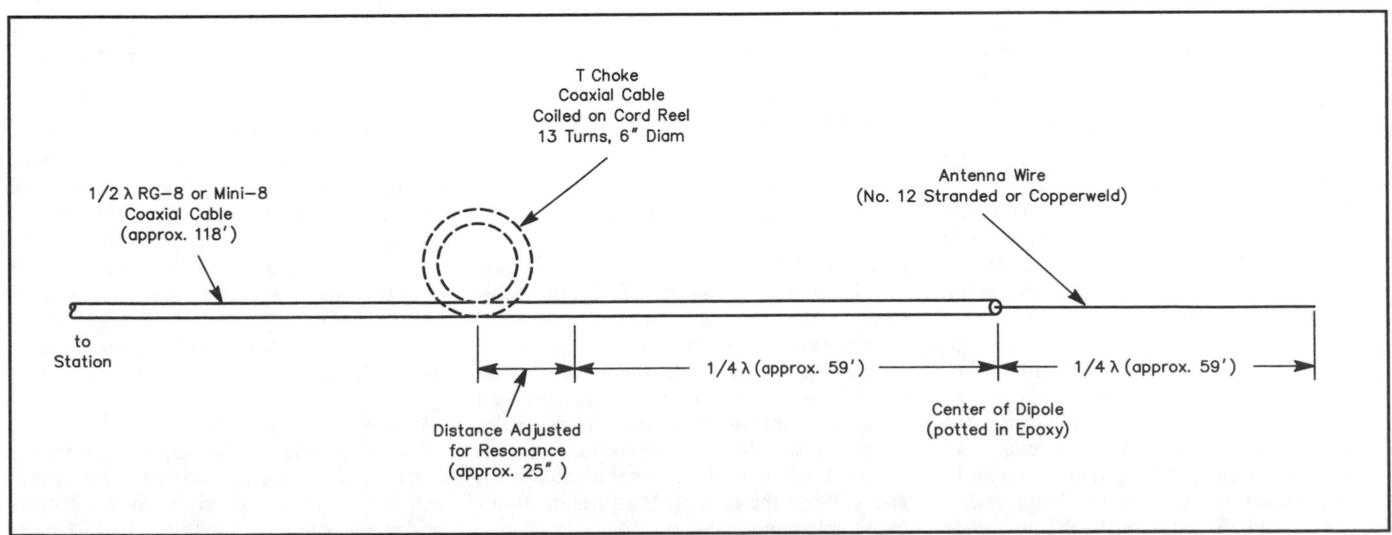

Figure 2—The RFD-1 (resonant feed-line dipole) antenna for 80 meters.

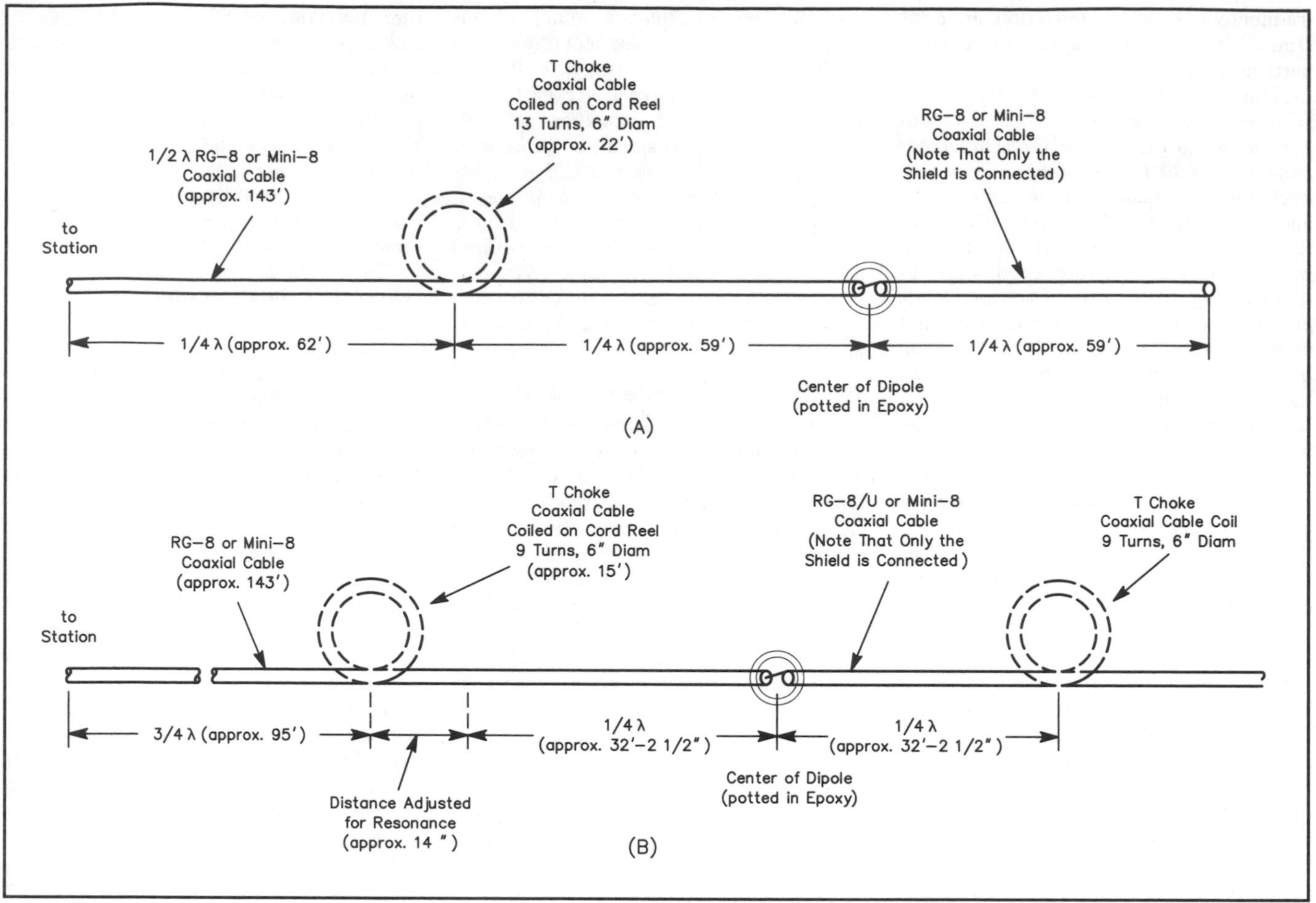

Figure 3—The RFD-2 antenna can be configured for either of two bands. Shown at A is the setup for 80 meters, and at B for 40 meters.

quarter waves either side of center). Depending upon the antenna orientation, the additional radiation lobes may be advantageous. Also, the unused conductor lengths at the ends are less apt to cause trouble because of parasitic excitation from the main radiator.

Arrangement and Construction

FromTable 2, for the 40-meter band we need to reduce the resonant coil at the input end of the dipole to nine turns, and we need to move the coil along the coax to a point slightly more than 32 feet, 2³/₄ inches from the center. This is half the dipole length calculated for 7.26 MHz with an allowance for tuning adjustment, using the dipole-length approximation equation $L = 468/f_{MHz}$. As with the RFD-1, this coil establishes the desired, high impedance at the fed end of the dipole.

In addition, we need to establish a high impedance at the far end of the dipole because, in this case, the desired end of the 40-meter dipole does not coincide with the end of the conductor. This is because we want to retain the full length of the conductor measured for the 80-meter band, rather than cutting it. The desired high impedance reflecting termination is achieved by use of a second, 9-turn, self-resonant coil of coax.

Table 2
Scaling of RFD-1 to Other Bands

Band	*Frequency*	*Turns*
160 m	1.90	19
80 m	3.95	13
40 m	7.26	9
30 m	10.12	8
20 m	14.29	7
17 m	18.14	6
15 m	21.38	6
12 m	24.96	5
10 m	28.65	5

Figure 3B shows the RFD-2 configuration for 40 meters. A cord reel can be used for the second coil, although any suitable insulating 6-inch diameter form can be used. I wound the coil in two layers using a 1-gallon windshield-washer-fluid bottle. After the windings had been properly adjusted, I taped them in place with duct tape, and slipped the coil off the form so it became self-supporting.

As with the RFD-1, it is desirable to place the antenna as high above ground as possible to reduce ground losses and to belp achieve low-angle radiation, if that is desired. The dipole does not have to be installed in a straight line, and if you can gain appreciably greater height by bending it, this may be a desirable compromise.

After the antenna with its two coils is erected, measure the resonant frequency using a noise bridge or an SWR meter. If the construction details have been followed, the resonant frequency should not be far from the nominal value of 7.26 MHz. The resonance point and the input resistance can be shifted moderately by both changing the 14-inch offset distance and changing the inter-turn spacing of the coils. The greatest effect is produced by changing the spacing between the input and the output turns (that is, between coil turns 1 and 9). This is because the voltage difference, and therefore the change in capacitive influence, is greatest there.

Results

My experience in developing antennas is that the good concepts really "want to work," and this antenna was no exception. Noise-bridge measurements with the RFD-1 indicated an input resistance at resonance of

49 ohms, and from a practical standpoint the SWR is 1:1. An H-field antenna probe was used to evaluate the power radiated at the center of the dipole compared with that at the current loop near the feed-point. The current ratio was 5.5 to 1, which corresponds to a power ratio of 30 to 1, or 15 decibels. This indicates that the coil is very effective in attenuating the common-mode current flowing back toward the feed point.

For the RFD-2 with the 40-meter dimensions shown in Figure 3B, the initial resonance point was within the 40-meter phone band and the SWR was essentially 1:1. Slight adjustment brought the resonant frequency up to the desired value, and the input resistance was very close to 50 ohms. A salient characteristic of the RFD antennas is the ease with which an impedance match is attained. It appears always to be easy to get a reflected power indication of zero. This is probably because the common-mode current on the shield of the coax is indistinguishable from the desired radiating current—in other words, the common mode is used rather than avoided.

Appendix

When the separate outer braid of the antenna shown in Figure 1 is removed, there must be some way to let the RF "know" when it should stop flowing on the shield of the coax and reflect back toward the center of the radiator. A balun box that I made up for another purpose is one approach: 30 turns of bifilar winding on an Amidon T-200-2 iron-powder toroidal core. The turns formula for such a balun transformer is

$$T=\sqrt{\frac{\text{desired } L(\mu H)}{A_L}}$$

where

T = no. of turns

A_L = inductance index (microhenries per 100 turns)

From this, the inductive reactance for unbalanced current is

$$X_L=2\pi fL=\frac{2\pi f T^2 A_L}{10^4}$$

At a frequency of 4 MHz, this 30-turn coil has an unbalanced reactance of only 270 ohms. We would need about ten of these in series to support the RF field at the end of a dipole antenna!

An alternative method involves placing a toroidal isolation transformer in the coaxial line at the fed end of the linear dipole. I actually wound such a transformer on two stacked T-200-2 cores, to provide a 1-kW power capability. This configuration worked, after a fashion, but the impedance match was less than desired, probably because of excessive capacitance between secondary and primary windings. However, before I took steps to control this, a simpler approach came to mind.

We need to isolate the transceiver from the fed end of the dipole, so why not cut the coaxial feed line to be a quarter wavelength long? (This length is measured in free-space, not in the line.) This serves to tranform the high impedance of the fed end of the dipole to the low impedance at the grounded transceiver. Any unused portion of this approximate 60-foot length of coax can be wound in a coil and used for further isolation and, as it turns out, for tuning the system to resonance.

Scaling for a Second Band

For coverage of a second band, we would like to have the same inductive reactance that we had for the 80-meter coil.

From handbooks, the inductance of a coil of assumed dimensions is

$$L = AN^2$$

where

A = a constant determined by the coil geometry

N = number of turns

Thus, we have for the reactance

$$X_L = 2\pi fL = 2\pi fAN^2$$

For equal reactances we can calculate N_2, the new number of turns, at frequency f_2, from N_1, the known number of turns at frequency f_1, by the equation

$$N_2=N_1\sqrt{\frac{f_1}{f_2}}$$

Off-Center-Loaded Dipole Antennas

In these times when much of our amateur population lives in urban areas, the subject of shortened antennas for the lower frequency amateur bands is a very popular one. Physically short ground-mounted vertical antennas with lumped-constant loading to make them resonant can be quite efficient radiators, if a good radial system has been installed. This has certainly been evidenced in Sevick's series of recent *QST* articles.[1] To many amateurs, however, the "hitch" in constructing such a system is the installation of a *good* radial system. It must be admitted that for the "top" amateur bands, 160 and 80/75 meters, an efficient system of buried radials requires a sizable amount of real estate, even for a physically short radiator. On the average city-size lot, 50 or 75 by 120 to 150 feet, it's almost impossible to install a highly efficient radial system for 80/75 meters, much less for 160 meters, when structures like a house and perhaps a separate garage exist. Or to some amateurs, just the thought of burying hundreds or maybe thousands of feet of wire is enough to turn off any enthusiasm for the project. What's the alternative? A dipole type of antenna with lumped constant loading. At modest heights, 30 or 40 feet, such an antenna will prove to be quite satisfactory if it is physically longer than about 0.2 wavelength. Shorter lengths may also be used, at reduced efficiency. Such an antenna can be fed directly with 50-Ω coaxial line, and it can be operated with no earth ground. (Of course the chassis of the transmitter and/or receiver should be grounded adequately for protection against shock hazard.)

Nearly all of us are familiar with the concept behind the use of inductive loading. A vertical antenna which is shorter than a quarter wave (or a dipole antenna which is shorter than a half wave) will exhibit capacitive reactance at its base (or center) feed point. To cancel such capacitive reactance, a coil having the proper inductive reactance may be connected in series with the base feed point of the vertical. The same result will be obtained through the use of two such coils for a dipole, one coil connected in series with each half. It is not necessary for the inductor to be installed at the feed point, however. In fact, greater radiating efficiency results through improved current distribution if the inductor is located along the radiator some distance away from the low-impedance feed point, *viz*, in the manner of a center-loaded mobile whip antenna. Figure 1 shows this concept extended to a dipole element, with off-center loading. The inductors resonate the antenna to the operating frequency, but do little actual radiating themselves. (This is in contrast to helically wound or continuously loaded elements, where a long thin inductor is the radiator as well as the loading element.)

In the antenna represented by Figure 1, there are many variable factors to be considered when a practical antenna for a given frequency is being constructed. Of primary consideration from an efficiency standpoint is the overall length, shown as dimension A. Another consideration for efficiency is the distance of the coils from center, dimension B. The longer the overall length (A), up to a half wave, and the farther the loading coils are placed from the center (B), the greater is the efficiency of the antenna. However, the greater is distance B (for a fixed overall antenna size), the larger the inductors must be to maintain resonance. Theoretically, if the coils were placed at the outer ends of the dipole, they should be infinite in value to maintain resonance. Capacitive loading of the ends, either through proximity of the antenna to other objects or through the addition of capacitance hats, will reduce this requirement to a more practical value.

What Inductance Values?

As a matter of personal interest, this writer has been doing experimental work for a number of years with off-center-loaded antennas. One big drawback to such experimentation was the ever-present need for a large amount of cut-and-try work to arrive at resonance whenever a new set of dimensions was to be used. Probably the number of pruned-off turns from coil stock from such experiments, if straightened out and soldered end to end, would make up several full-sized half-wave antennas for the 160-meter band. Therefore, most of the writer's work of late in this area has been in going through paperwork exercises, looking for a way whereby at least "ball-park" values of inductance needed for a particular system could be calculated.

The equation contained in the Mobile chapter of *The ARRL Antenna Book* for determining the capacitance of a vertical antenna shorter than a quarter wavelength looked promising in early computations,

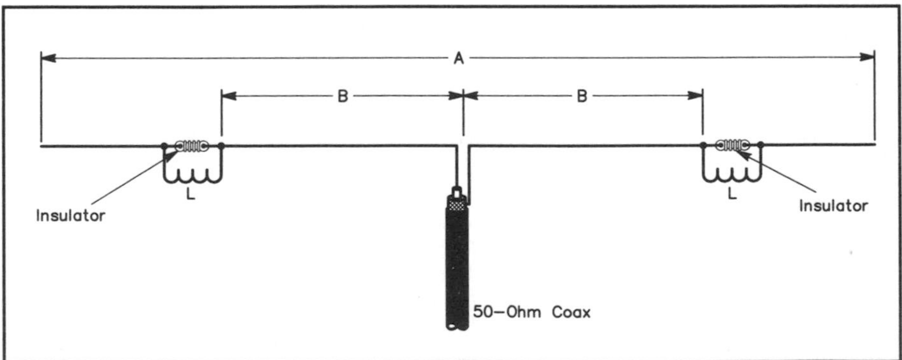

Figure 1—A dipole antenna lengthened electrically with off-center loading coils. For a fixed dimension A, greater efficiency will be realized with greater distance B, but as B is increased, L must be larger in value to maintain resonance.

and, indeed, it became the basis for the calculation procedure which finally resulted.

This procedure has been found to produce results much closer than mere "ball-park" values for the necessary inductance – for wire antennas "in the clear" at moderate heights, the final inductance values found by cut-and-try pruning for lowest SWR at the desired frequency have been so close to the value from calculations that a laboratory bridge was necessary to measure the difference. The results are equally good for elements using tubing. Once the needed inductance value is determined by calculations, it is generally found sufficient to obtain coil dimensions from an ARRL L/C/F Calculator (see LaPlaca*) or by equation. Any significant pruning which has been found necessary could always be attributed to objects in proximity to the ends of the antenna.

The complete set of calculations is expressed in the mathematical relationship below as Eq. 1, presented here primarily for mathematics buffs or those having access to electronic computers.

This equation yields the inductance required, in microhenrys, for single-band resonance of a shortened antenna of a particular physical size at a given frequency, for a specific position of the loading coils from the center of the antenna. To spare the reader the task of performing some rather tedious calculations, Figure 2 has been prepared from Eq. 1. The curves of the chart have been normalized, and may be used for any frequency of resonance. The chart is based on a halfwavelength/diameter ratio of the radiator of approximately 24,000. (This corresponds to No. 14 wire on 80 meters or No. 8 wire on 160 meters.) For "thinner" conductors, the required inductance will be somewhat greater than that determined from Figure 2, and less inductance will be required for "thicker" conductors.

The use of the chart is as follows: At the intersection of the appropriate curve from the body for dimension A and the proper value for the coil position from the horizontal scale at the bottom of the chart, read the required inductive reactance for resonance from the scale at the left. Dimensions A and B are shown in Figure 1, and for use with the chart are expressed as percentages. Dimension A is taken as percent length of the shortened antenna with respect to the length of a resonant half-wave dipole of the same conductor material. Dimension B is taken as the percent of coil distance from the feed point to the end of the shortened antenna. For example, resonating an antenna which is 50% or half the size of a half-wave dipole (one-quarter wavelength overall), with loading coils positioned midway between the feed point and each end (50% out), would require loading coils having an inductive reactance of approximately 950 ohms at the operating frequency. If the antenna is hung "in the clear," and if the length/diameter ratio of the conductor is near 24,000, inductance values as determined from the chart will be very close to actual values required. (Eq. 1 above takes the diameter of the radiator into account, and thus may be used for any length/diameter ratio.) For practical purposes, dimension B may be taken as that distance from the center of the feed-point insulator to the inside eye of the loading-coil insulator, and dimension A as the eye-to-eye distance inside the end insulators (which are not drawn in Figure 1).

Proximity of surrounding objects in individual installations may require some pruning of the coils, and the exact amount of final inductance required should be determined experimentally. If the antenna is hung in inverted-V style, with the ends brought near the earth, the required inductance will almost always be somewhat less than that determined from the chart or equation. A grid-dip meter, Macromatcher, (see Hall and Kaufmann*) or SWR indicator may be used, during the final adjustment procedure.

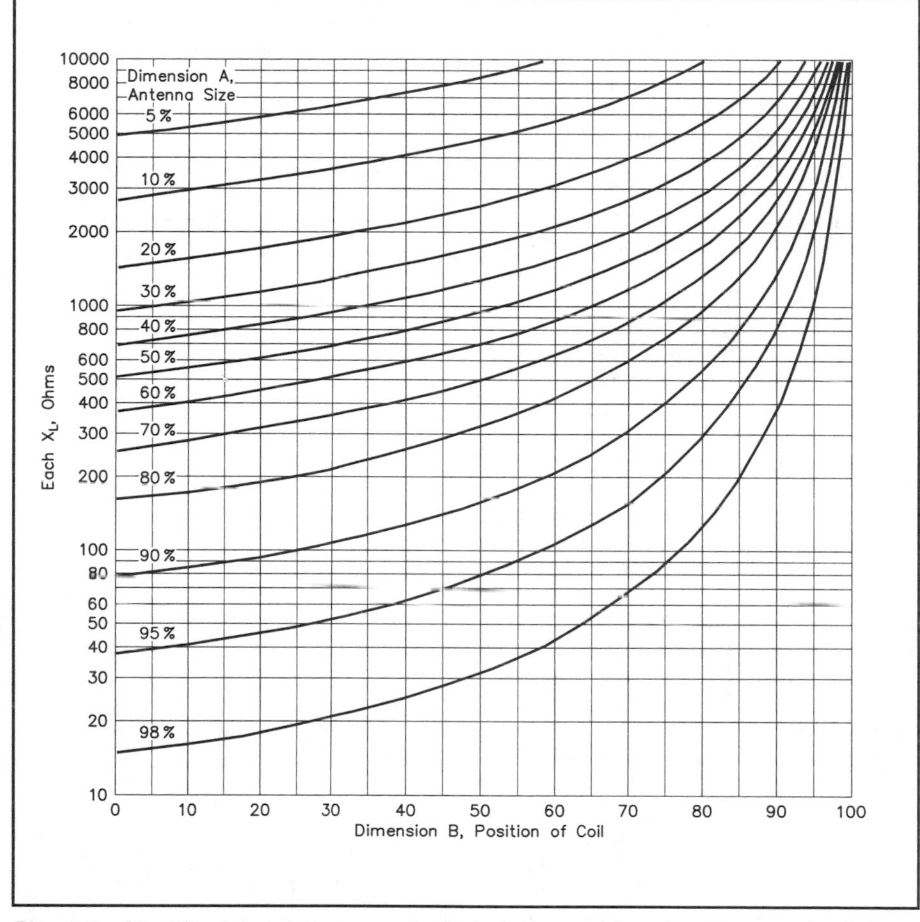

Figure 2—Chart for determining approximate inductance values for off-center-loaded dipoles. At the intersection of the appropriate curve from the body of the chart for dimension A and the proper value for the coil position from the horizontal scale at the bottom of the chart, read the required inductive reactance for resonance from the scale at left. See Figure 1 regarding dimensions A and B.

(Eq. 1):

$$L_{\mu H} = \frac{10^6}{68\pi^2 f^2}\left\{\frac{\left[\ln\frac{24\left(\frac{234}{f}-B\right)}{D}-1\right]\left[\left(1-\frac{fB}{234}\right)^2-1\right]}{\frac{234}{f}-B} - \frac{\left[\ln\frac{24\left(\frac{A}{2}-B\right)}{D}-1\right]\left[\left(\frac{\frac{fA}{2}-fB}{234}\right)^2-1\right]}{\frac{A}{2}-B}\right\}$$

where
$L_{\mu H}$ = inductance required for resonance
ln = natural log
f = frequency, megahertz
A = overall antenna length, feet
B = distance from center to each loading coil, feet
D = diameter of radiator, inches

Practical Antennas

Although one might erect an inductively loaded antenna that is cut for a single amateur band, it is possible to use the antenna itself for two, three, or more bands of operation, if provision is made to lower the antenna for band changes. A simple rope halyard and pulley arrangement at one of the supports will do the trick. Figure 3A shows a 3-band antenna of this nature, for 160, 80, and 20 meters. If the insulators shown are left open, with nothing bridging them, the antenna is a simple half-wave dipole cut for 14.18 MHz. (The 48.5-foot lengths act merely as support wires, and have negligible effect on operation of the antenna.) If the insulators are bridged with short lengths of antenna wire, the antenna becomes a center-fed 80-meter dipole, resonant at about 3.6 MHz. For 160-m operation the 20-meter insulators may be bridged with loading coils to resonate the antenna at 1.8 MHz, as shown in Figure 3A. Burndy or other manufacturers' "Servit" type of electrical connectors may be used, for ease in making band changes quickly, as shown in Figure 4.

The calculation procedure for determining loading-coil values for the antenna of Figure 3A, using the chart of Figure 2, goes like this. If operation is desired on 1.8 MHz, the length of a full-sized half-wave dipole is found from the relationship $468/f$ to be 260 feet. The 130-foot length of Figure 3A represents 50% of this size, meaning that the dimension-A curve marked "50%" in Figure 2 is to be used. The position of the coils is $16.5/(16.5 + 48.5) \times 100$ or 25% of the distance out from center, dimension B. From the intersection of 25 (horizontal scale at bottom), and the 50% curve, the required inductive reactance is read from the scale at the left of Figure 2 to be 650 ohms. The inductance, L, is $650/2\pi f$ or 57.5 microhenrys, if No. 8 wire is to be used. For smaller diameter wire, the inductance should be somewhat larger. (Calculations from Eq. 1 for No. 12 wire indicate the required inductance is 60.99 µH.)

The radiation resistance of a shortened antenna loaded to resonance is less than that of a full-sized antenna. Further, the shortened antenna is "sharper," meaning that the change in reactance versus frequency is greater. In other words, the shortened antenna acts as a tuned circuit having a higher Q than a full-sized antenna. To check these characteristics, the line input impedances for the antenna of Figure 3A were measured with a laboratory bridge, and the electrical line length at the measurement frequency was then taken into account to determine the impedance at the antenna feed point. The antenna was constructed of No. 12 wire and hung at a height of 50 feet as a "flat-top" radiator.

The solid curve of Figure 3B is a plot of the feed-point impedance versus frequency for this antenna. The plot on Smith Chart coordinates is more meaningful than a simple SWR-vs.-frequency curve because the magnitudes of the resistive and reactive components are shown, as well as the sign of the reactance. (Capacitive reactance is negative, plotted to the left of the vertical center line, and inductive reactance is positive, plotted to the right.) In this presentation, a 50-ohm nonreactive impedance will appear at the exact center of the chart. The SWR in 50-ohm line for a given frequency may be determined by first noting the distance from the center of the chart to the particular impedance plot on the curve, and next measuring this same distance down the vertical center line from chart center (a drawing compass is helpful for this task), and finally dividing 50 into the value read at that point on the center line. For example, the SWR at 1.8 MHz equals 120/50 or 2.4, as indicated by the segment of the 2.4 SWR circle in Figure 3B. It may be seen that resonance (zero reactance) occurs at approximately 1810 kHz, where the resistance is about 22 ohms. The SWR at resonance is 2.33:1, and climbs to 3:1 at 1825 kHz. At 1850 kHz, the SWR is 10:1. Without any matching provisions the antenna is relatively sharp, as mentioned earlier. If one sets the usable bandwidth as the frequency range where the SWR is 3:1 or less, it is approximately 35 kHz, or 1.9% of the resonant frequency. As far as efficiency is concerned, ohmic losses are low, and the antenna is a good performer on 160 meters. Because of its horizontal polarization, it has proved to be most effective at night, and stations several hundred miles away have been worked with S-9 reports received for the 50-watt signal.

For a comparison of impedances, the broken curve of Figure 3B is a plot of measured impedances of a full-size half-wave dipole, 260 feet long overall, hung in place of the shortened antenna. From this curve it may be seen that resonance occurs at 1810 kHz, where the resistance is 59 ohms. The 3:1-SWR bandwidth for the half-wave antenna is in the order of 60 kHz, or 3.3% of the reso-

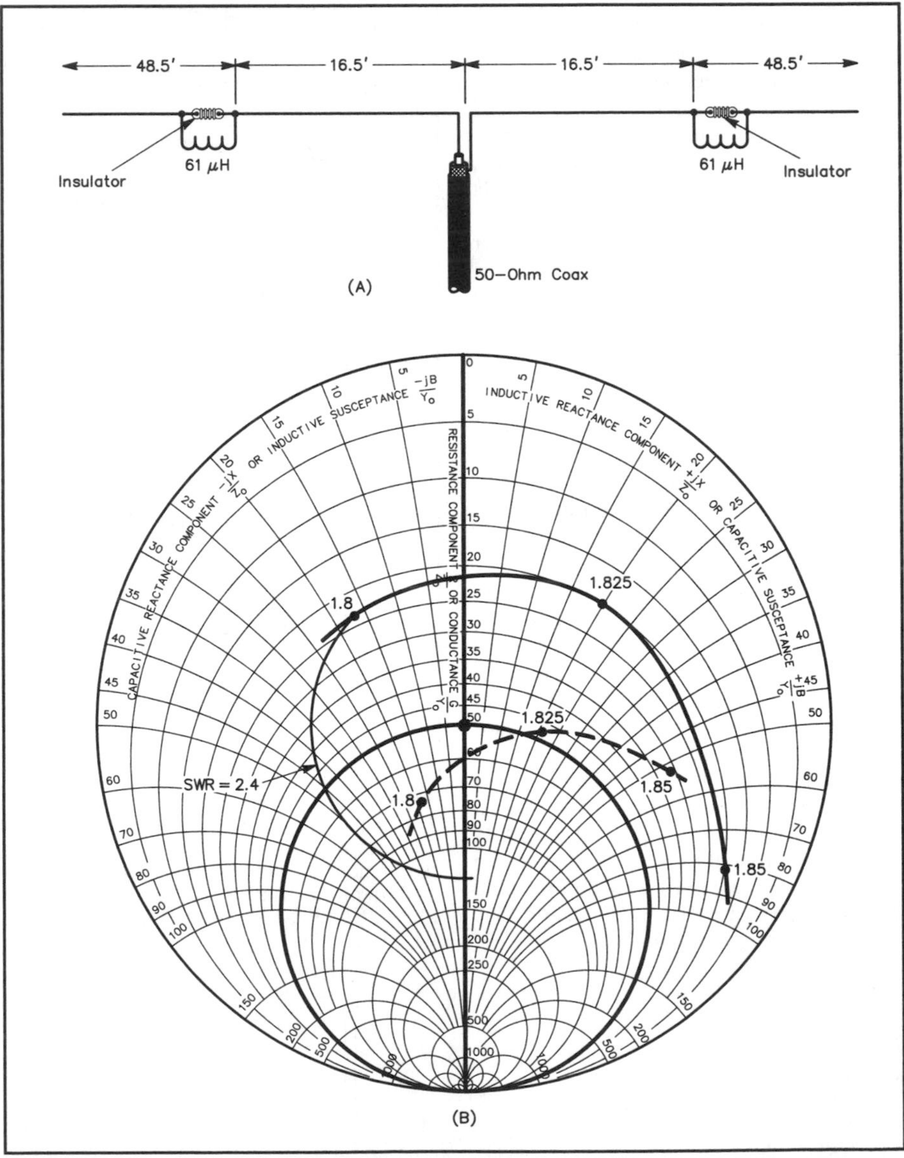

Figure 3—At A, an 80-meter dipole loaded for 160-meter operation. The inductors are 37 turns of coil stock having a 3-inch diameter and 10 turns per inch (B&W 3035). At B, the impedance plot of this antenna installed at a height of 50 feet (solid curve), and the plot of a 160-m half-wave dipole (broken curve).

nant frequency. It is interesting to note on this curve that the SWR at resonance is 1.18:1, and that it is a somewhat lower value, 1.15:1, at a frequency a few kilohertz above resonance. (Measurements were made every 5 kHz across this band, but plot points are shown only for 25-kHz increments to avoid crowding of the data.) This evidence refutes the oft-heard statement that the SWR-vs.-frequency curve is *always* lowest at antenna resonance. Points to remember are that the SWR in a transmission line is completely dependent upon the characteristic-impedance value of the line in use. Using a line of different impedance may shift the position of the SWR curve along the frequency axis in a simple SWR-vs.-frequency plot. This is definitely true in this case — if the 160-meter half-wave dipole were to be fed with 75-ohm line, the lowest SWR would occur at a frequency about 5 kHz *below* antenna resonance, whereas with 50-ohm line the lowest SWR is at a frequency slightly *above* resonance. The reason this happens is that the resistive component of the impedance, which consists of the radiation resistance plus any loss resistance, is not constant with frequency, even over a rather narrow frequency range. It must be acknowledged that the differences here are very slight, however, and for practical purposes the frequency of lowest SWR is (within a few kilohertz) the resonant frequency of the antenna.

Another point concerning the SWR values bears noting. The values as determined from the plots in the manner described above are quite accurate, having been determined by measurements with laboratory equipment. In contrast, measurements with simple SWR indicators usually cannot be relied upon for anywhere near the equivalent accuracy.

For example, the author owns a commercially manufactured SWR indicator of the Monimatch type (see McCoy*) which, under a particular set of conditions, indicates a 2.5:1 SWR in a line where laboratory measuring equipment shows the *true SWR to be 4:1*. A significant difference! Herein lies another reason why impedance plots on Smith Chart coordinates are more meaningful than a simple SWR-vs.-frequency curve — greater accuracy may generally be expected.

A Half-Size 80-Meter Antenna

Figure 5A shows the 3-band concept described earlier as it can be applied to 80, 40, and 20 meters. Its overall length is 66 feet, not a difficult length to use on a small lot. This antenna was constructed for 80-m operation with a design-center frequency of 3.55 MHz, using No. 12 antenna wire and 40-μH loading coils—27 turns of stock having a diameter of 3 inches and a pitch of 10 turns per inch (tpi). Feed-point impedances versus 80-m frequency for the antenna, hung at a height of 50 feet, are shown by the solid curve at B of Figure 5. Actual resonance occurred at 3.54 MHz, where the resistance was about 26 ohms. The bandwidth within which the SWR is 3:1 is

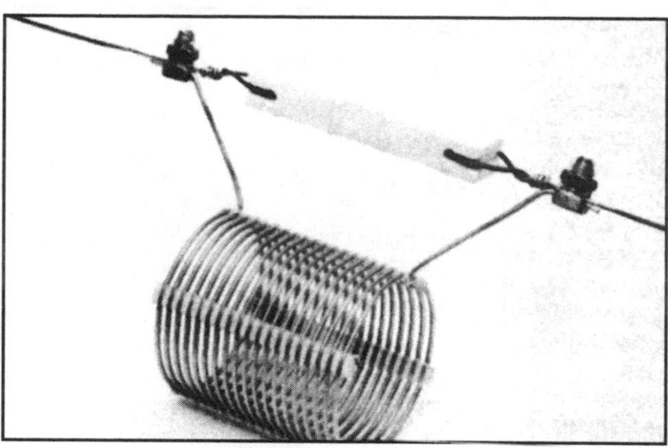

Figure 4—Copper electrical service connectors, sold under one trade name of Servit, provide a simple means of installing the loading coils. The antenna wire and the ends of the coil wires should be tinned to prevent corrosion. In addition, a protective coating of acrylic spray may be used at each connection.

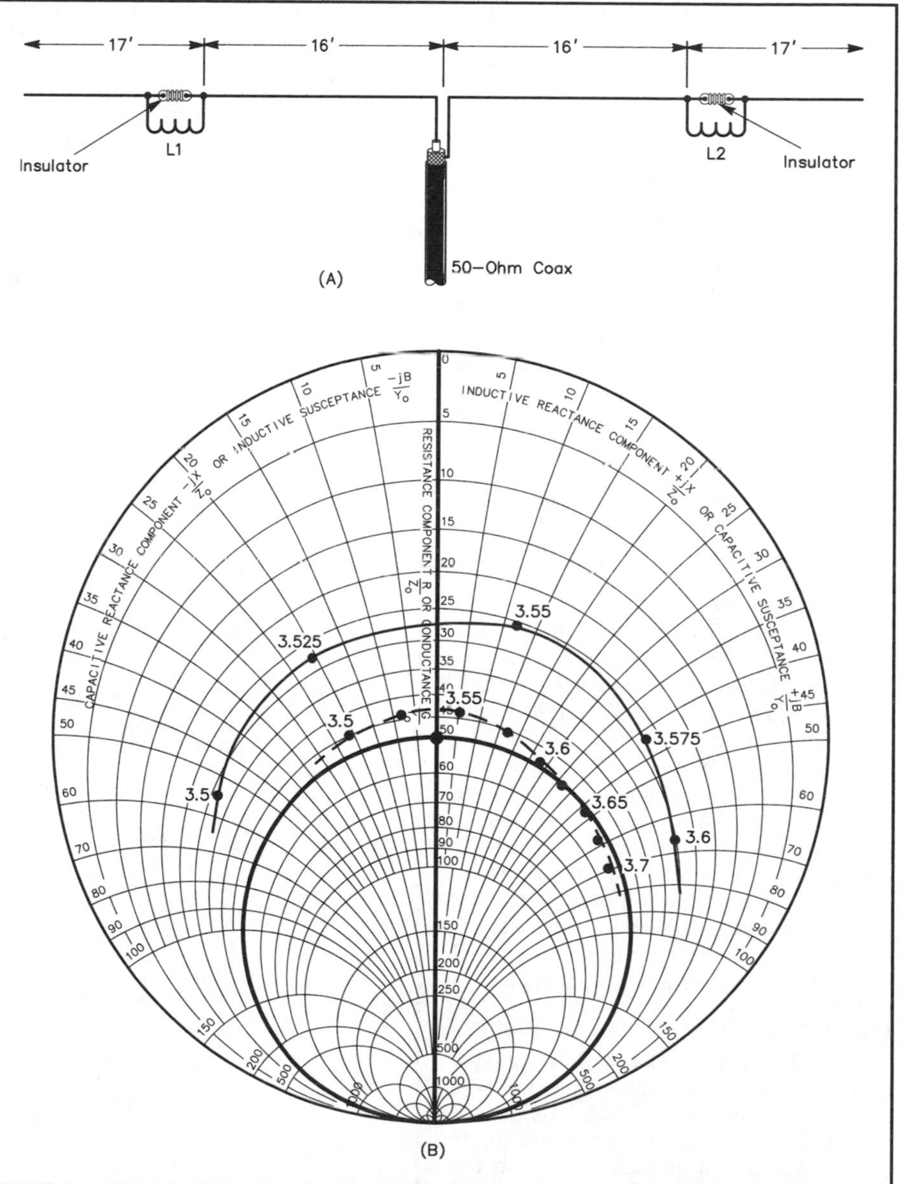

Figure 5—At A, a 40-meter dipole loaded for 80-meter operation. For resonance at 3.55 MHz the coils should be approximately 40 μH (27 turns of stock); 3.75 MHz, 35 μH (24 turns); 3.9 MHz, 31 μH (22 turns); and 4.0 MHz, 29 μH (21 turns). These are circulated inductance values for No. 12 antenna wire. Coil stock referenced above is 3-inch diameter, 10 turns per inch (B&W 3035). At B, the impedance plot of the 3.55-MHz version (solid curve) and of an 80-m half-wave dipole (broken curve).

60 kHz, or 1.69% of the resonant frequency.

Also shown in Figure 5B, by the broken curve, are the feed-point impedances of a half-wave dipole, 132 feet overall length, hung in place of the shortened antenna. Resonance occurs at 3.54 MHz, where the resistance is 43.5 ohms and the SWR is 1.15:1. The broader nature of the half-wave antenna is exhibited by the "tighter" curve which swings closer to the 50-ohm center point of the chart than the shorter, loaded antenna. The SWR at 3.5 MHz is 1.6:1 and remains below 3:1 to 3.67 MHz.

Capacitive and Inductive Loading

One would assume that a combination of capacitive and inductive loading might provide a different feed-point impedance than would inductive loading alone, because of different current distributions in the radiators. To check out this assumption, the antenna of Figure 5A was used as a "test bed" for comparative measurements. Capacitance hats were attached at different points along the 17-foot lengths of wire outside the coils, and the coils were pruned to reresonate the antenna at about the same frequency as before. The impedance measurements were then repeated.

Dangling End Sections:

First, "hats" consisting of 18 inches of No. 12 wire were affixed to the antenna ends and permitted to dangle. This lowered the resonant frequency to 3435 kHz. By calculations, this was approximately the same effect as that of extending the 17-foot portions of the antenna by the same amount as the dangling lengths, so it would seem to make little difference whether short sections of extra length are added inside the supporting insulators or are at the ends, suspended at right angles to the main antenna wire.

The inductors were reduced from 40 to 36.5 µH (25-turn coils replaced the original 27-turn coils), and resonance occurred at about 3575 kHz. At this frequency the resistance was 26 ohms and the SWR 1.90:1. The 3:1-SWR bandwidth, 64 kHz, is 1.79% of the frequency of resonance. The impedance plot for this arrangement is shown as Curve A in Figure 6. The resistance at resonance for this antenna is identical to that with the coils alone, and the bandwidth is only 4 kHz greater, 64 kHz vs. 60. From these results, one would conclude that the main advantage offered by the "danglers" is a small saving of space over a flat-top antenna.

Capacitance Hats Near Loading Coils:

Next the dangling end sections were removed and a pair of capacitance hats was formed, each from two 36-inch lengths of No. 12 solid wire. The two wires for a single hat were attached at their centers to the antenna wire at a point just outside one of the loading coils. The hat wires were then bent radially to form an X at right angles to the antenna wire, like four spokes of a wheel with the main antenna wire at the hub. The diameter of the X-shaped hat was thus 36 inches. The second hat was placed in a like manner just outside the second coil. Burndy connectors were used to affix the hat wires. The resonant frequency of this configuration with the original 40-µH loading coils was found to be 3290 kHz. The effect of adding the hats was about the same as that of extending the 17-foot lengths to 19 feet.

When the inductors were replaced with 23-turn coils (32.5 µH), the antenna resonated at about 3.575 MHz, the resistance being 23 ohms. The SWR at resonance is 2.15:1, and the 3:1-SWR bandwidth for this configuration is 60 kHz, 1.68% of the resonant frequency. The impedance of this arrangement versus frequency is shown by Curve B of Figure 6.

It is surprising to note that, by the standards of most amateurs, the characteristics of this antenna are not as good as those of the same length antenna with loading coils alone. The SWR at resonance for the antenna with combination capacitive and in-

Table 1
Characteristics of various loading techniques, 66-foot 80-m dipole.

Loading	*Approx. feed-point resistance, resonance*	*SWR at resonance*	*3:1-SWR bandwidth, % of resonant freq.*
40-µH coils only	26 ohms	1.92:1	1.69
36.5-µH coils, 18″ dangling ends	26	1.90:1	1.79
36″ hats outside 32.5-µH coils	23	2.15:1	1.68
30-µH coils, 36″ hats at ends	25	1.98:1	2.05
None (λ/2 dipole)	43.5	1.15	Greater than 3.6

Coil positions for each loaded antenna were 16 feet from antenna center. All antennas were constructed of No. 12 wire and installed at a height of 50 feet.

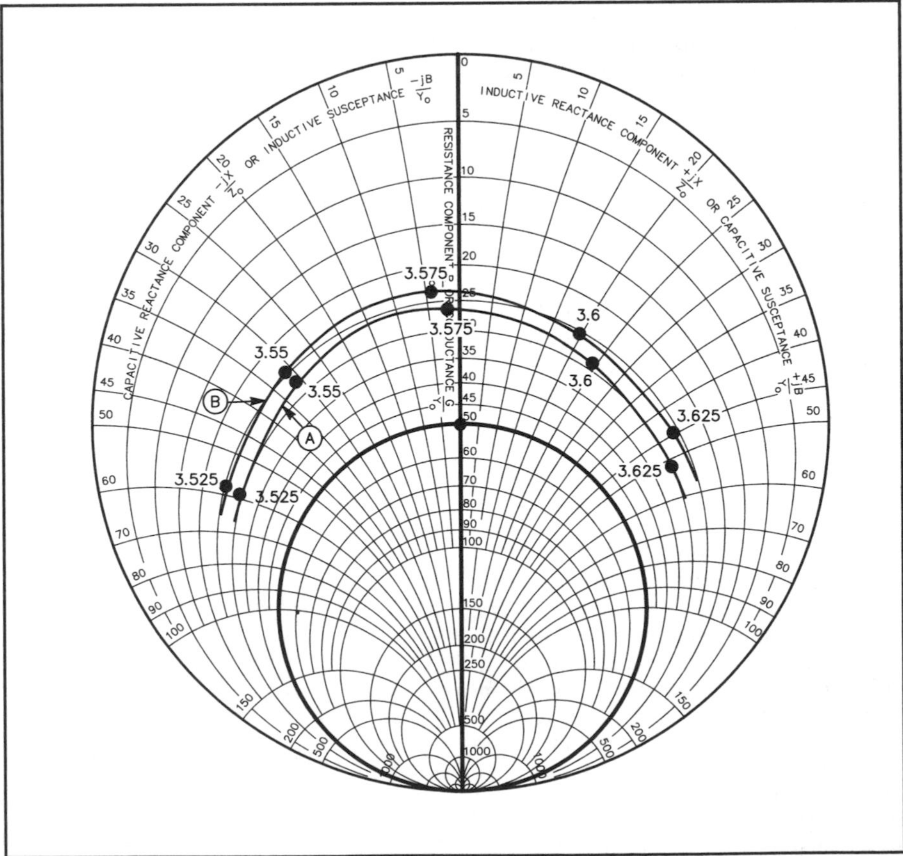

Figure 6—Curve A is the impedance plot of the antenna of Figure 5A with 1.5-foot dangling end sections added and the coil trimmed to restore resonance near the original frequency. Curve B is a plot of the same antenna with X-shaped capacitance hats added at a point just outside the loading coils (dangling sections removed and coils trimmed to reestablish resonance).

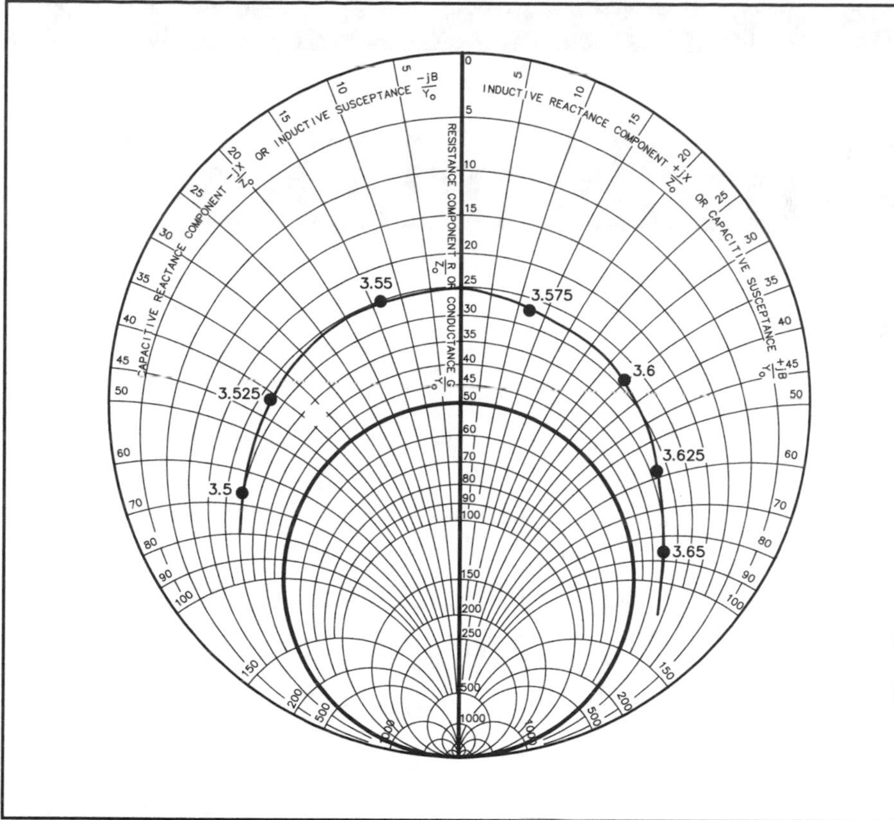

Figure 7—Impedance plot of a 66-foot dipole using a combination of off-center inductive loading and capacitive end loading. Of all the shortened configurations tried, this arrangement offered the greatest bandwidth.

ductive loading is higher (2.15 vs. 1.92), and the 3:1-SWR bandwidths are the same, 60 kHz. Perhaps a significant factor here, though, is that the diameter of the capacitance hats used for these measurements was small, only .011 wavelength. Supporting much larger hats presents mechanical problems with wire antennas, however, as even these were a bit flimsy and would require reshaping after gusty weather.

Capacitance Hats at Antenna Ends:

Finally, the X-shaped capacitance hats were moved to the outside ends of the antenna, just inside the end insulators. With the original 40-µH coils, resonance appeared at 3215 kHz. From calculations, it was as if the 17-foot end sections were actually 21 feet long. With 30-µH coils (22 turns) in place, the resonant frequency was 3560 kHz. At this frequency the resistance was 25 ohms and the SWR 1.98:1. The 3:1-SWR bandwidth is 73 kHz, or 2.05% of the resonant frequency. The impedance plot of this antenna is given in Figure 7.

It is interesting to note that the position and shape of the plot for this antenna on Smith Chart coordinates is nearly identical to that for the same length antenna with loading coils only, the solid curve of Figure 5B. For this antenna, however, the plot points for 25-kHz frequency increments appear closer together, which accounts for the increased bandwidth.

Conclusions:

The measured characteristics of these various configurations of loading for the 80-meter antennas are tabulated in Table 1. Remember that the overall "flat-top" length of each antenna arrangement, is 66 feet, and that the loading coils are always positioned 16 feet each side of the center of the antenna, being pruned for resonance at approximately 3550 kHz. For comparison, information for a half-wave dipole is also included.

Of the various arrangements, capacitive end loading decidedly provides the greatest bandwidth, excepting the full-size half-wave antenna, of course. Although there are slight differences in the resistance value at resonance, all are of the same order of magnitude. These values, as well as those for the 160-m antenna discussed earlier, tend to confirm a *broad* rule of thumb that the writer has formulated for this type of antenna: The feed-point impedance value at resonance is roughly proportional to the length of the antenna. That is, a loaded antenna which is half the size of a half-wave dipole will have approximately half the radiation resistance of the full-sized antenna.

Eq. 1 given earlier or the chart of Figure 2 allows one to calculate loading-coil values for antennas with loading coils only. Additional capacitive loading is not taken into account. Calculating the effects of various capacitive loading arrangements appears to be difficult, and work remains to be done in this area.

Multiband Antennas with Loading Coils

All of the foregoing material has been devoted to the loading of an antenna for resonance at a single frequency. Resonated as described, the antenna is electrically a half wave in length. It will, however, operate well on higher frequencies — frequencies at which it is an odd multiple of half waves, in *electrical* length . . . three half waves, five half waves, etc. Because of the lumped loading of the shortened antenna, these higher frequencies will likely not be closely related to odd-order harmonics of the fundamental frequency, as the case would be for a nonloaded radiator. (For example, it is a well-known fact that a 7-MHz half-wave dipole operates well on its third harmonic, 21 MHz.)

A loaded dipole will become an electrical 3/2-λ antenna at some frequency *below* that which is three times the fundamental resonant frequency. Depending upon the overall antenna length, coil value, and coil position, it is possible for an 80-meter loaded dipole to become a 3/2-λ performer on 40 meters. With such an arrangement, one would have a dual-band antenna without requiring the use of traps. The idea can be expanded upon to arrive at a loaded antenna without traps which will operate on more than two bands. This scheme offers considerable constructional simplification as compared with trap arrangements.

The multiband loading-coil concept has been recognized for better than half a century, but little use of the technique has been made by amateurs. Some years ago a very good article on the subject a was published by William Lattin, W4JRW.* That article is recommended reading for anyone interested in more details on the concept. Supplemental information has been published by Buchanan.* Attempts by this writer to calculate antenna sizes and coil values for dual-band antennas have met with some success. From calculations and experiments to date, it appears that with only two loading coils (one each side of center), the antenna must always be greater than a half wave in physical length for the higher of the two frequency bands. In other words: any 80/40 meter arrangement, for example, apparently would need to be longer than 66 feet from tip to tip. However, much work also remains to be done in this area.

References

Buchanan, "An Inexpensive 40- and 80-Meter Antenna," Hints and Kinks, *QST*, September, 1962, p. 62.

Hall and Kaufmann, "The Macromatcher, An RF Impedance Bridge for Coax Lines," *QST*, January, 1972, p. 14. Information also contained in the 50th and subsequent editions of *The Radio Amateur's Handbook*, and the 13th edition of *The ARRL Antenna Book*.

LaPlaca, "Using the ARRL L/C/F Calculator," *QST*, December, 1973, p.26.

Lattin, "Multiband Antennas Using Loading Coils," *QST*, April, 1961, p.43.

McCoy, "Monimatch, Mark II," *QST*, February, 1957, p. 38. Also, "An Etched-Circuit Monimatch For Checking Your Antenna System," *QST*, October, 1969, p.29.

Sevick, "The Ground-Image Vertical Antenna," *QST*, July, 1971, p. 16. Also, "The W2FMI Ground-Mounted Short Vertical," *QST*, March, 1973; "A High Performance 20- 40- and 80-Meter Vertical System," *QST*, December, 1973, p. 30; and "The Constant-Impedance Trap Vertical," *QST*, March,1974,p.29.

By Frank Witt, AI1H From *QST*, September 1993

A Simple Broadband Dipole for 80 Meters

Turn your existing 80-meter dipole into a broadband antenna by simply modifying the feed line. Multiband operation is an option.

A conventional coax-fed, half-wave dipole doesn't provide a low SWR over the entire 80-meter band—an inconvenience for those of us who like to operate phone and CW on that band. Several approaches to overcoming this limitation, short of an antenna tuner in the station, have been described.[1,2] The antenna system described here is simpler than any of its predecessors and has the following features:

- A 2:1 SWR or better is achieved over all or most of the 80-meter band.
- Antenna length and appearance are the same as those of a conventional half-wave dipole. Consequently, it's lightweight and has small wind and ice loading.
- The antenna configuration permits multiband operation with a single feed line.
- The losses due to broadband matching are acceptable.
- The cost is about the same as a conventional half-wave dipole.

All the SWR data given in this article were measured at the transmitter end of the feed line. The reference impedance is 50 Ω, since most equipment is designed for this impedance. The term *antenna system* as used throughout this article includes not only the radiating wire, but also the feed line, balun (if used), any lightning-protection measures, antenna tuner and so forth.

The dipole antenna itself is not broadband; the system uses a broadband *match*. The key broadbanding element of this antenna system is the *transmission-line resonator*: Part of the transmission line compensates for the reactance presented by the dipole away from its resonant frequency. This part of the line is a multiple of an electrical half wavelength. Another part of the line presents an appropriate source impedance to the transmission-line resonator (TLR).

First I'll describe a version of the broadband antenna system, along with some practical results. Then I'll cover the important matter of antenna-system loss. Following that are some variations to suit specific requirements, and a method for using the antenna for several bands. I'll also compare transmission-line-resonator broadbanding to other broadbanding methods.

The 80-Meter Broadband Antenna System

Figure 1 shows the simple broadband antenna system as used at my station. The antenna proper is a center-fed half-wavelength dipole. The transmission line is segmented into one electrical wavelength of 50-Ω coax and an electrical quarter wavelength of 75-Ω coax. The calculated and actual lengths are shown in Table 1. Lengths were calculated using the formulas given later in this article, using a center frequency (F_0) of 3.75 MHz and VF (velocity factor) of 0.66. The actual lengths resulted after I performed the tuning procedure described later. Manufacturing variations from the published cable velocity factors, and some stretching of the coax, contributed to the differences between actual and measured values. (The actual lengths were measured on untensioned cable.) The antenna is installed as an inverted V with a 140° included angle and an apex height of 60 feet. The wire size is #14, but is not critical.

Table 1
Calculated and Actual Lengths of the Broadband Dipole Antenna at AI1H

	Calculated	*Actual*
1/4-λ Coax	43.3 feet	43.3 feet
1-λ Coax	173.1 feet	170.5 feet*
Dipole	124.5 feet	122.7 feet

*Includes 11 inches for balun.

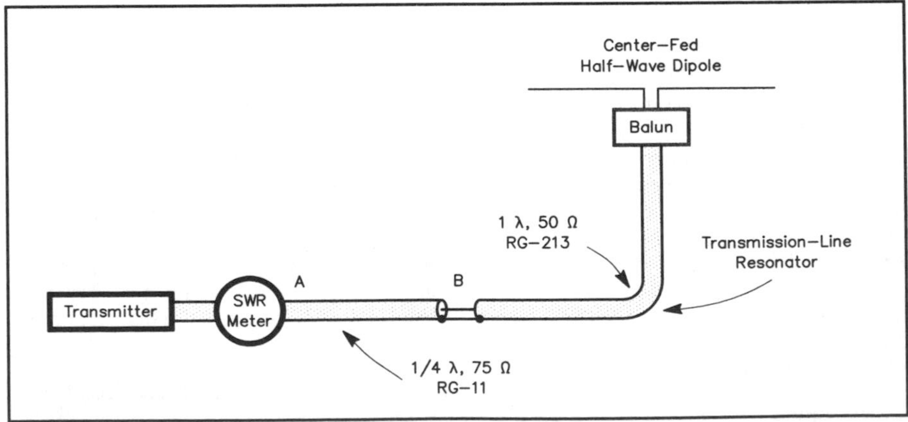

Figure 1—One form of the simple broadband antenna system. It resembles a conventional dipole except for the 1/4-wavelength, 75-Ω segment. Points A and B are discussed in the text.

This system's SWR (at the transmitter) as a function of frequency is shown in Figure 2. For comparison, the SWR for the same dipole fed with about 5/4 wavelengths (214 feet) of RG-213 coax is also shown. (This is the same total length as the RG-213 and RG-11 segments used in the broadband system.) The broadband system's 2:1 SWR bandwidth is 2.2 times that of the conventional system—and the *only* difference is the feedline configuration!

The radiating properties of the broadband antenna over the 80-meter band are essentially identical to those of a dipole cut for any specific frequency in the band. Also, since the antenna system is designed for a 50-Ω transmitter, the feed-line length may be extended by adding the required length of 50-Ω coax between the transmitter and the quarter-wave segment (point A in Figure 1).

A 1:1 current balun should be installed at the antenna's feed point. I use the balun on general principles. Often, it provides no visible difference in operation, but the balun does minimize feed-line radiation. You can determine whether your antenna needs a balun by measuring the SWR versus frequency with and without a balun installed. If the balun is not needed, the two sets of data will be identical.

Antenna-System Losses

It's important to know the losses in any antenna system. This is especially true for broadband antennas, because loss alone can broadband an antenna system. As the next section shows, the configurations presented in this article do not yield a significant loss penalty. Although other loss contributors exist in antenna systems, we will focus on the primary ones: *feed-line loss* and *mismatch loss*. Other losses, such as ohmic loss in the antenna wire, are the same for both the conventional and broadband systems described here.

Feed-line loss is the easiest to understand. It is unavoidable, and is lowest when the feed line is *flat* (when the line SWR is close to 1:1). At HF, feed-line loss results primarily from ohmic losses in the copper conductors.

Mismatch loss occurs when the impedance seen by the transmitter is not the complex conjugate of the transmitter's impedance (when the line SWR at the transmitter is not 1:1). For a 50-Ω transmitter, the mismatch loss is 0 dB when the load impedance is 50 Ω. When the load impedance is not 50 Ω, the mismatch loss can be made to be 0 dB if a transmitter with a tunable output stage (such as a conventional tube-type linear amplifier) is tuned for a conjugate match. An antenna tuner can also provide this match. In this case, however, the antenna-tuner loss (perhaps as much as 1 dB) replaces the mismatch loss in the total-loss equation. That subject isn't discussed here.

If you don't use an antenna tuner and the transmitter has a fixed-tuned 50-Ω output, loads that present the transmitter with an SWR under 2:1 are highly desirable. The impact of high SWR on mismatch loss will become clear in the next section.

Loss must be kept in perspective. All of the broadband antenna systems described here have a worst-case total loss of less than 3 dB—not enough to notice in many 80-meter QSOs. (If the loss is 3 dB, half of the transmitter's output power is radiated and half is lost elsewhere.) The main effect of loss is stress on system components: that on the transmitter due to the mismatched load, and that on the transmission line due to heating.

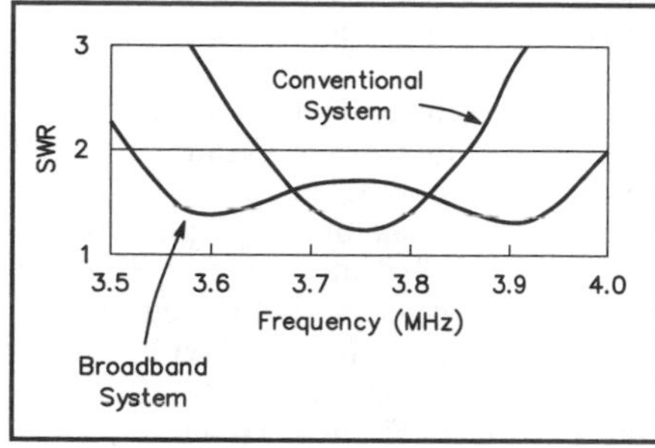

Figure 2—Measured SWR versus frequency for the broadband and conventional antenna systems.

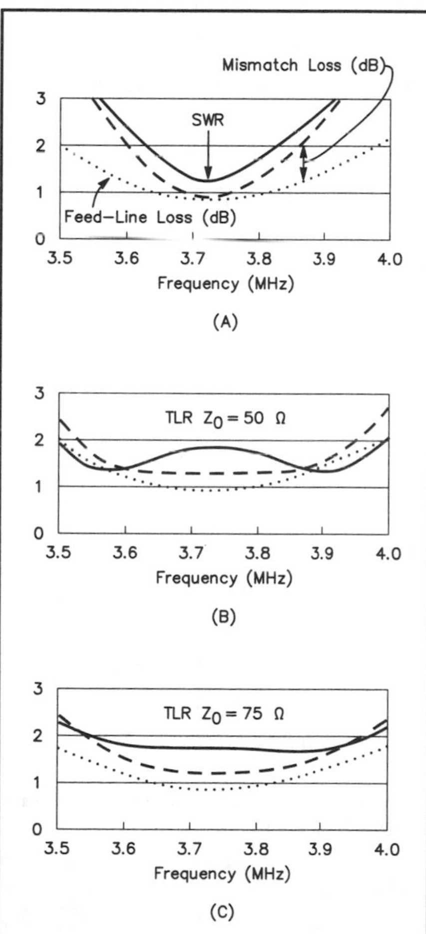

Figure 3—Antenna-system configurations for long feed-line runs. The solid lines are SWR; the dotted lines are feed-line loss; and the dashed lines are feed-line loss plus mismatch loss. At A, a conventional system using an RG-213 feed line; at B, the feed line is a 1/4-λ section of RG-11 (75 Ω) followed by 1 λ of RG-213; and at C, a 1/4-λ segment of RG-11 is followed by 1 λ of RG-11 (one 5/4-λ piece of RG-11). The total feed-line length in each case is 216.4 feet.

Variations

The broadband antenna system described above is well-suited for the installation at my station, where the distance between the shack and the antenna is rela-

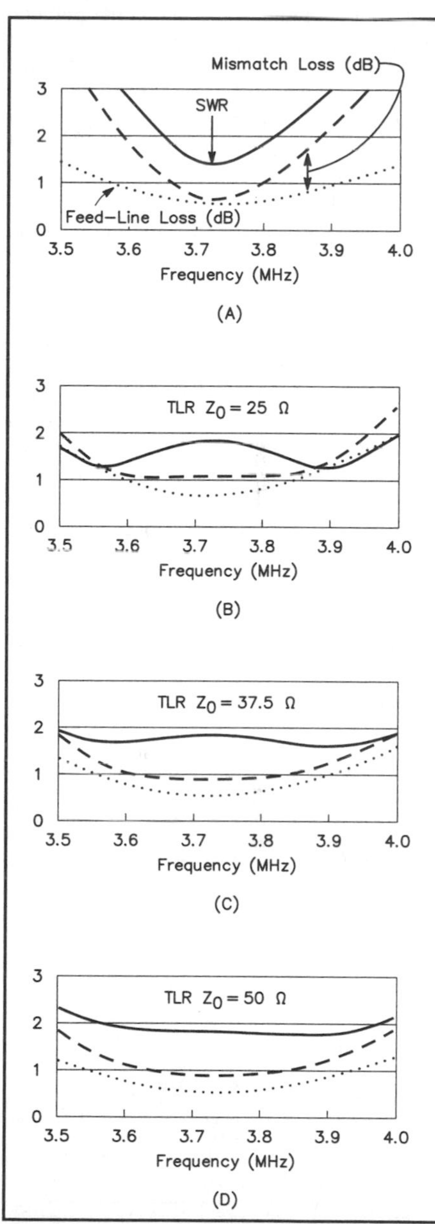

Figure 4—Antenna systems for high power and shorter feed-line runs. The solid lines show SWR; the dotted lines represent feedline loss; and the dashed lines show feed-line loss plus mismatch loss. At A, the feed line is RG-213; at B, it's a 1/4-λ section of RG-11 followed by two paralleled 1/2-λ lengths of RG-213; at C, a 1/4-λ segment of RG-11 is followed by two paralleled 1/2-λ lengths of RG- 11, and at D, 1/4-λ of RG-11 is followed by 1/2-λ of RG-213. The total feed-line length in each case is 129.8 feet.

tively long (more than 200 feet) and because I use a 1-kW amplifier. Other feed-line combinations are better suited to other installations. Some of these are shown in Figures 3 through 5, along with calculated SWR and loss data. From this information, you can select an appropriate feed-line combination for your needs.

The figures also show the characteristics of conventional dipole antenna systems. If you compare them, you'll see that the transmission-line resonator provides broadbanding without a significant loss penalty. I haven't tried all these combinations, but based on my experience, they should perform as predicted in most situations if the radiator doesn't deviate significantly from the model I used in my calculations: a dipole 125 feet long, 40 feet high, and made of #14 wire. This model is based on data provided by Walt Maxwell, W2DU, in his book, *Reflections*.[3] I chose his data since it is typical of many 80-meter installations.

All of the broadband antenna systems use a 1/4-wave section and either a 1/2- or 1-wavelength section. Figure 3 illustrates a system for long feed-line runs. It uses RG- 11 and RG-213 cable and should be considered for all power levels. Figure 3B covers the case shown in Figure 1 and used at my station. The feed line of Figure 3C is a continuous length of RG-11 cable 5/4 wavelengths long. The transmission-line resonator is the 1-wavelength section of the cable nearest the antenna.

This approach would also work with surplus 75-Ω CATV Hardline. A 3/4, 5/4 or 7/4-λ section of 1/2-inch Hardline yields less than 2 dB feed-line loss plus mismatch loss over the entire band, and less than 1 dB total loss over any 300 kHz of the band. This configuration is particularly attractive to contesters and DXers, because even a fairly long line7/4 λ is 372 feet of 1/2-inch CATV Hardline—gives low loss and a very good match over, say, the 3.5- to 3.8-MHz range.

Three broadband antenna systems are shown in Figure 4. All of these are candidates for applications requiring shorter feedline lengths. Figures 4B and 4C show the performance realized when coax cables are paralleled to achieve a low equivalent characteristic impedance. Figure 3B, which results from a 1-wavelength RG-213 transmission-line resonator, and Figure 4B, are very similar. The latter system uses the same amount of cable, but it's cut in half and parallel-connected. This will become clear in the sidebar, "How It Works." The configuration in Figure 4D is attractive because of its simplicity.

Lower-power applications without long feed-line runs can use RG-58 and RG-59 coax. Figure 5B shows how excellent broadbanding is achieved with a remarkably simple feed line. Again, no loss penalty results from the broadbanding.

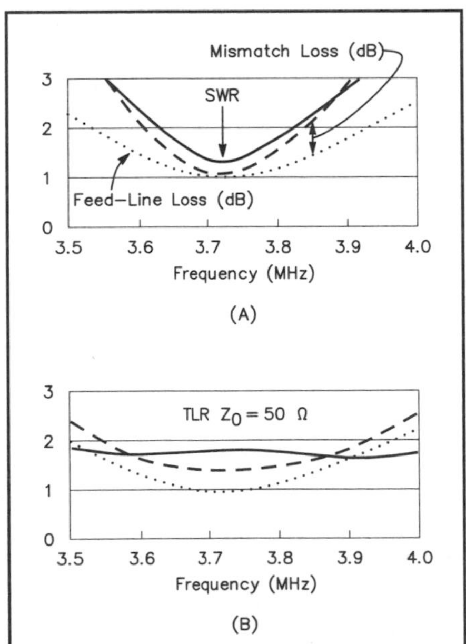

Figure 5—Antenna systems for low power and shorter feed-line runs. The solid lines show SWR; the dotted lines represent feedline loss; and the dashed lines show feed-line loss plus mismatch loss. At A, the feed line is RG-58; at B, it's a 1/4-λ section of RG-59 followed by 1/2 λ of RG-58. The total feedline length in each case is 129.8 feet.

Adjusting the Broadband Antenna System

The antenna system is easy to build and adjust. First calculate the lengths (in feet) of the transmission-line segments:

$$L_{quarter} = \frac{245.9\,VF}{F_0} \quad \text{(Eq 1)}$$

$$L_{half} = \frac{491.8\,VF}{F_0} \quad \text{(Eq 2)}$$

$$L_{full} = \frac{983.6\,VF}{F_0} \quad \text{(Eq 3)}$$

where

$L_{quarter}$=length of quarter-wave segment

L_{half} = length of half-wave segment

L_{full} = length of full-wave segment

VF = velocity factor

F_0 = center frequency in MHz

A good starting point for the dipole wire length (in feet) is:

$$L_{dipole} = \frac{467}{F_0} \quad \text{(Eq 4)}$$

For the 80-meter application, I suggest using an F_0 of 3.75 MHz. It's a good idea to cut the wires so that the overall length is 4 feet longer than necessary, in case you need to lengthen the wire during tuning. Pass 2 feet of the extra wire through each end insulator and wrap it back around the antenna wire.

To tune the antenna system, you'll change only the dipole and transmission-line-resonator lengths. The best approach is to build the antenna system as I have outlined here and to measure the SWR at the transmitter end of the system. Any tilt or frequency offset in the SWR characteristic can be removed by increasing or decreasing the dipole or transmission-line resonator length. Start by changing the length of the dipole. To improve the SWR at the high end of the band, the dipole must be shortened; to improve the SWR at the low end of the band, the dipole must be lengthened. Progressively add or subtract 6 inches from both legs of the dipole until the SWR curve is symmetrical about the center frequency.

Frequency offset may be required to center the SWR characteristic in the 80-meter band. You can move the entire curve along the frequency axis without causing asymmetry by changing both the dipole and transmission-line resonator lengths using the following equation:

$$L_{New} = L_{Old}\left(\frac{3750 - \Delta F}{3750}\right) \quad \text{(Eq 5)}$$

ΔF is the required frequency offset in kilohertz. Shortening the dipole and resonator moves the curve center up in frequency, and lengthening them moves the center down. The length of the quarter-wave segment need not be changed, since the SWR characteristic is not very sensitive to its length.

Lightning Protection

Every antenna system should be designed to minimize the likelihood of a lightning strike. One part of this is keeping all parts of the antenna proper at ground potential. The grounding should be done *outside the shack*, by means of a good ground rod.

I recommend that you install a coaxial lightning protector, which bleeds any static charge from the center conductor, at point B of Figure 1. The protector (and therefore the feed-line shield) should be connected to a high-quality ground rod (the kind electricians use) driven 8 feet into the ground.

Conversion of Existing 80-Meter Dipoles

A study of the cases shown in Figures 3B, 4D and 5B suggests that it's possible to easily convert many existing 80-meter half-wave dipole antennas. Because the most popular way to feed an 80-meter dipole is with a 50-Ω coaxial feed line, the conversion to a broadband antenna system is straightforward. First trim the dipole for resonance at about 3.75 MHz. Then cut the 50-Ω feed line at a multiple of an electrical half-wavelength (at 3.75 MHz) from the antenna. Calculate this length using Eq 2 or Eq 3. Add the 75-Ω quarter-wave section, then complete the run to the shack (if necessary) with 50-Ω coax. Then use the tuning procedure described earlier to optimize the system.

Multiband Operation

Most broadband 80-meter antenna systems are usable only on the 80-meter band, because the broadbanding elements do not allow efficient power transfer on other bands. This is not true with the approach described here, since the structure consists only of a center-fed dipole and a transmission line. Moreover, the transmission-line segments are close to multiples of an electrical half-wavelength near 40 meters and other bands. This opens the possibility for paralleling other half-wave dipoles with the 80-meter dipole and sharing the feed line.

To minimize their interaction, the various dipoles should be spaced from each other away from the feed point. Of course, some interaction will occur and you must tune the multiband system to meet your requirements. I recommend first tuning the 80-meter broadband system and then the next-highest-frequency dipole, and so forth. Only the 80-meter antenna will be broadband, but such broadbanding is not required on the other bands. Figure 6 shows the result of adding a 40-meter dipole to the Figure 1 antenna. Each dipole leg is 34.4 feet long. Note that the SWR on 80 meters changes very little compared to Figure 2. No change was made to the 80-meter dipole or the transmission line.

The multiple-dipole approach described above achieves resonance on several bands and eliminates the need for an antenna tuner on those bands. Of course, if you use an antenna tuner, operation on all HF bands should be possible, but this arrangement is usually not as effective as the multiple-resonance antenna system described here because the feed-line loss is much higher.

Comparison with the Coaxial-Resonator Match

How does the simple broadband dipole described here stack up against other approaches for achieving a good match over the entire 80-meter band? The coaxial resonator match broadband dipole[4,5] represents one of the more efficient designs published to date. It achieves broadband matching at the antenna by the integration of 1/4 wavelength of coaxial cable as a part of the antenna.

Since the coaxial-resonator match achieves a good match at the antenna, the SWR on the feed line is low and the feedline loss is about the same as its matched loss. However, the coaxial cable in the match itself increases the system loss. The net result is that the total loss is about the same with the coaxial-resonator match, but the SWR at the transmitter is lower, never exceeding about 1.6:1 between 3.5 and 4 MHz. Once the SWR is less than 2:1, however, a lower SWR has little value unless you're using a transmitter that significantly reduces power at such SWRs.

Note that the approach described in this article uses a thin wire for the antenna. Most other broadbanding approaches use additional wires or radiators made partly from coaxial cable and are vulnerable to damage from wind and ice loading. Their additional weight and complexity are also limitations.

From the above comparison, the simple broadband antenna system has, by its very simplicity, an edge over the coaxial resonator match, at least in applications where the simpler approach is feasible. Because of the limitations of available coaxial cables, the opportunity for a satisfactory design is constrained. On the other hand, the coaxial resonator match has more adjustment parameters, is useful over a much broader range of applications and yields the lowest SWR over the band.

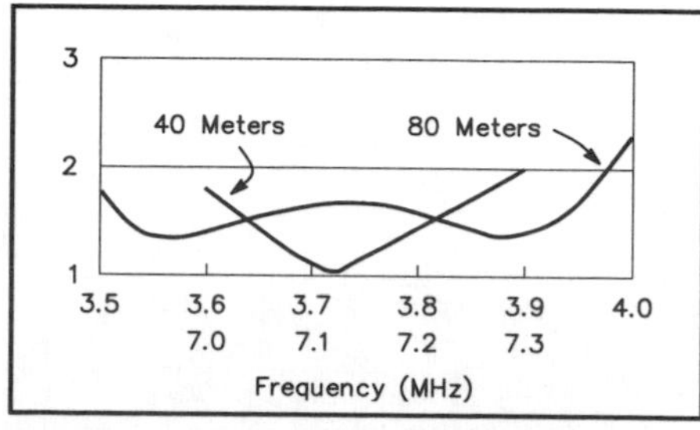

Figure 6—Measured SWR for the 80- and 40-meter multiband antenna system.

How It Works

A fundamental way of achieving a broadband match to a resonant dipole antenna involves a parallel-tuned LC network and an appropriate source resistance. In an *RF Design* article,* I described the method for designing such networks, even with lossy resonators. The top of Figure A shows the equivalent circuits of the antenna and matching network. The bottom of Figure A illustrates the corresponding elements in the antenna system.

The role of the resonator is played by the transmission-line segment nearest the antenna. It must be a multiple of an electrical half-wavelength. The quarter-wavelength "Q"-section, made from 75-Ω coax, transforms the 50-Ω transmitter reactance to 112.5 Ω ($75^2/50=112.5$). I won't go into the design details here; they're the subject of another article, "Broadband Matching Using the Transmission-Line Resonator," in preparation for *The ARRL Antenna Compendium, Volume 4.*

For the structure of Figure A to yield a broadband match, the characteristic impedance of the transmission-line resonator and the transmitter resistance must be within a range of values. Fortunately, commonly used transmission lines, which are available in 50- and 75-Ω characteristic impedances, work well in this application. The broadband systems of Figures 3 through 5 show the usefulness of this approach.

Figure B makes another significant point. For this application, the network parameters of a one-wavelength transmission-line resonator (top) are similar to those of a half-wavelength resonator (bottom) with half the characteristic impedance of the upper resonator. Parallel-connecting two identical cables is a convenient way of achieving lower characteristic impedances. This explains the similarity of Figures 3B and 4B and the similarity of Figures 3C and 4C.—*AI1H*

*F. Witt, "Optimum Lossy Broadband Matching Networks for Resonant Antennas," *RF Design*, Apr 1990, pp 44-51 and Jul 1990, p 10.

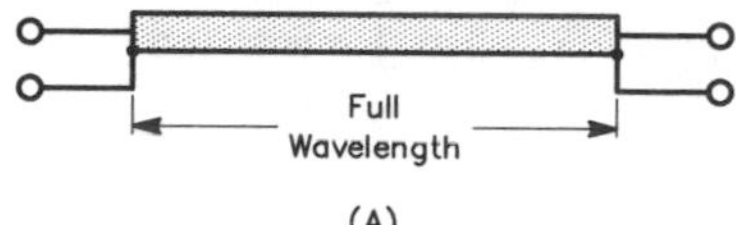

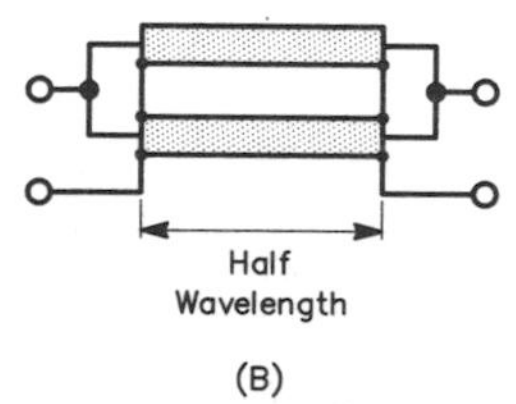

Figure A–Lossy broadband-matching-network equivalent circuit (top), and corresponding simple broadband antenna system elements (bottom).

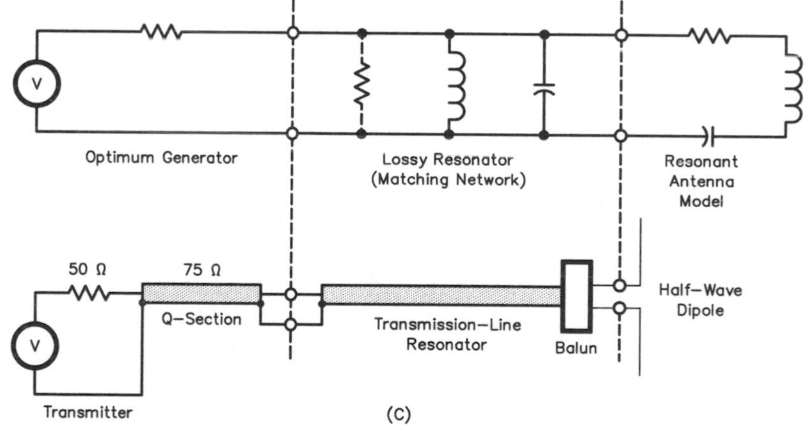

Figure B–These two transmission-line resonators behave essentially the same in this application. The characteristic impedance of each cable segment is the same, making the characteristic impedance of the lower resonator half that of the upper one.

Summary

The simple broadbanding technique I've described here capitalizes on the common availability of coaxial cables that fit the application. It overcomes the narrow bandwidth limitations of a conventional 80-meter, half-wave dipole without significant disadvantages. Even parallel dipoles for other bands may be fed with the same feed line.

The limitation of available coaxial cable parameters can be overcome by using the transmission-line resonator as a resonant transformer. Applying this technique is described in an upcoming *ARRL Antenna Compendium* article, "Broadband Matching Using the Transmission-Line Resonator."

This work has benefited from the support and encouragement of my wife, Barbara, N1DIS. Also, I must credit Andrew Griffith, W4ULD, for helping to turn my attention to the approach described here. After reading my *QST* article on match bandwidth of resonant antenna systems,[6] Andy noted that antenna systems should be viewed from their match to a 50-Ω transmitter, even if the feed line does not have a 50-Ω characteristic impedance. He showed examples of the narrowing of match bandwidth to make his point. In my response, published with Andy's letter in *QST*,[7] I pointed out that match bandwidth of an antenna system may actually be *increased* by selecting the right cable length and characteristic impedance. As an example, I showed in Figure 3 of that correspondence the large match bandwidth of a dipole fed with a 5/4-wavelength, 75-Ω RG-11 cable. Note that this is the same case shown in Figure 3C of this article. Thank you, Andy!

Notes

[1]G. Hall, ed, *The ARRL Antenna Book*, 16th ed (Newington: ARRL, 1991), pp 9-1 through 9-12.

[2]M.W. Maxwell, *Reflections: Transmission Lines and Antennas* (Newington: ARRL, 1990), pp 18-1 through 18-6.

[3]See Note 2, p 15-19. I frequency-scaled Walt's data, which is equivalent to changing the wire length. This antenna has a Q of 13 and a resonant resistance of 65-Ω. I took into account the fact that the antenna's radiation resistance increases with frequency.

[4]F. Witt, "The Coaxial Resonator Match and the Broadband Dipole," *QST*, Apr 1989, pp 22-27.

[5]F. Witt, "The Coaxial Resonator Match," *The ARRL Antenna Compendium, Volume 2* (Newington: ARRL, 1989), pp 110-118.

[6]F. Witt, "Match Bandwidth of Resonant Antenna Systems," *QST*, Oct 1991, pp 21-25.

[7]A. Griffith and F. Witt, "Match Bandwidth Revisited," *QST,* Technical Correspondence, Jun 1992, pp 71-72.

Once More With the 80-Meter Broadband Dipole

With reference to earlier items by K1TD and W7ZOI in *QST*, it *is* possible to design a simple, adjustment-free matching network that will make an 80-meter dipole have a reasonable SWR across the band.[1,2] When using Hayward's RLC model of the dipole, a simple parallel-tuned circuit across the antenna feed point gives less than 2:1 SWR across all but the bottom 10- to 20-kHz of the band, as shown in Figure 1.

This method *does* require a source impedance of 140 ohms, so a broadband matching transformer will be needed at the input. I see no reason why you couldn't just tap the input down on L1 and use L1 as an autotransformer to accomplish the impedance matching function. In fact, although I have not tried it, it may be possible to include the balanced-to-unbalanced transformation in the same coil, as shown in Figure 2. To tune this network, use a dip meter with no antenna connected. Then install the network on the antenna and adjust the input tap for the best compromise SWR across the band.

By choosing a lower characteristic impedance, you can get much better SWR over a narrower bandwidth (Figure 1). Retune both the antenna and the network to center the response on either the phone or CW portion of the band.

You can get better SWR in both the fullband and narrow-band case by adding a series-tuned circuit to the input (Figure 3). The improvement is marginal, however, and complicates the impedance step-up arrangement, so it is probably not worth the effort. It would be better to take Hayward's suggestion and make the antenna itself inherently more broadband. One method is to "fatten" the elements by using a wire cage arrangement or similar. Of course this would change the values of Zo, L1 and C1.

The original graphs for Figures 1 and 3 were drawn with an HP-87. I would be happy to send a copy of the rogram to anyone who sends me an s.a.s.e.—*Alan Bloom, N1AL*

[1] J. Hall, "The Search for a Simple, Broadband: 80-Meter Dipole," *QST*, April 1983, p. 22.
[2] W. Hayward, "Limitations to Broadband Impedance Matching," Technical Correspondence, *QST*, July 1984, p. 45.

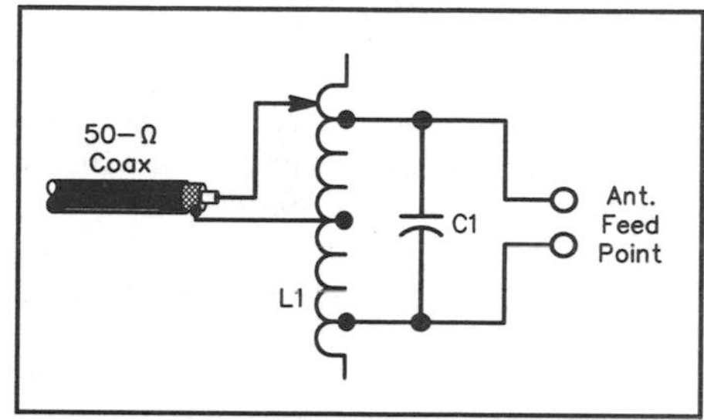

Figure 2—Proposed method of feeding a balanced antenna with unbalanced line. The shielded conductor of the coax line must be connected at the exact center of the coil, which is an electrically neutral point in the antenna system.

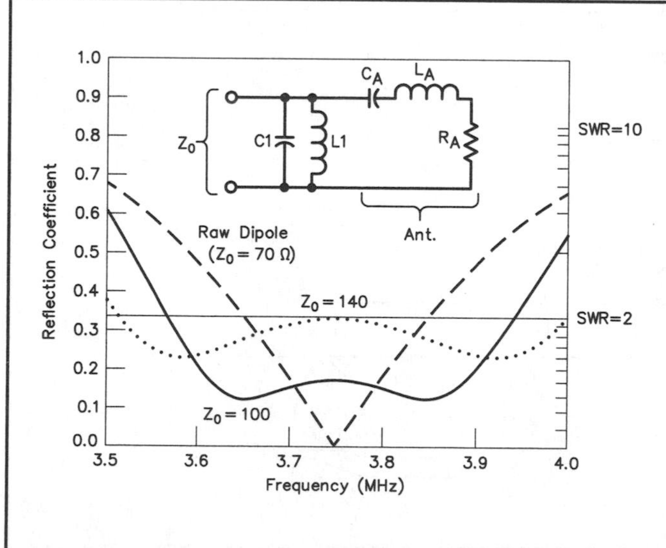

Figure 1—Calculated reflection coefficient and standing-wave ratio vs. frequency for a single-wire dipole and dipoles with lumped constants (C1 and L1) at the feed point.

For the curve where Zo is 140 ohms:
C1 = 2000 pF
L1 = 0.9006 μH

For the curve where Zo is 100 ohms:
C1 = 3500 pF
L1 = 0.5146 μH.

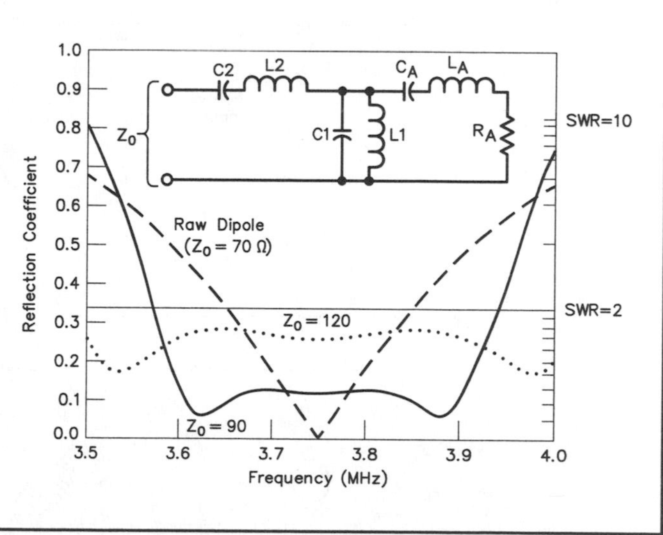

Figure 3—Calculated reflection coefficient and standing-wave ratio vs. frequency for an antenna with shunt and series lumped constants at the feed point (C1-L1 and C2-L2, respectively).

For the curve where Zo equals 120 ohms:
C1–3500 pF
C2–60 pF
L1–0.5146μH
L2–30.0211 μH

For the curve where Zo equals 90 ohms:
C1–7000 pF
C2–60 pF
L1–0.2573 μH
L2–30.0211 μH

Fat Dipoles

Antennas without problems make radio communications enjoyable. I design overseas radio stations for a living, so I'd rather not have to fight my own ham station when just relaxing and rag chewing. Fat dipoles do the things I want. They match the coax line well over a wide band, and they launch the signal remarkably well.

Theory

Making a dipole conductor thicker than normal with respect to wavelength will increase the bandwidth and modify the working impedance of the antenna. The trick is to make a dipole "fat" in such a way that it may be easily constructed from cheap materials, be highly efficient and at the same time arrange things so that it will match the transmission line from the lower band edge to the upper band edge.

I started with the assumption that my band of interest would be the 80/75-meter band. From end to end, this requires a 13% bandwidth to the 2:1 SWR points for my broadband-solid-state final. I also assumed that this antenna was going to be at a nominal height of 30 feet or 0.11 wavelength above ground. The calculations indicated that a dipole built of four quarter-wavelength no. 14 wires (0.064 inch) with a spacing of 0.0114 wavelength would produce the necessary results. The correct length would have to be 0.45 wavelength to match a 50-ohm line.

Length = 442.5/f feet
Width = 11.25/f feet
Height = 112.5/f feet
where f = center frequency in MHz

Construction

Very few problems will be encountered in building this simple fat dipole if you follow the drawing (Figure 1). First you will need five good insulators. I prefer egg type

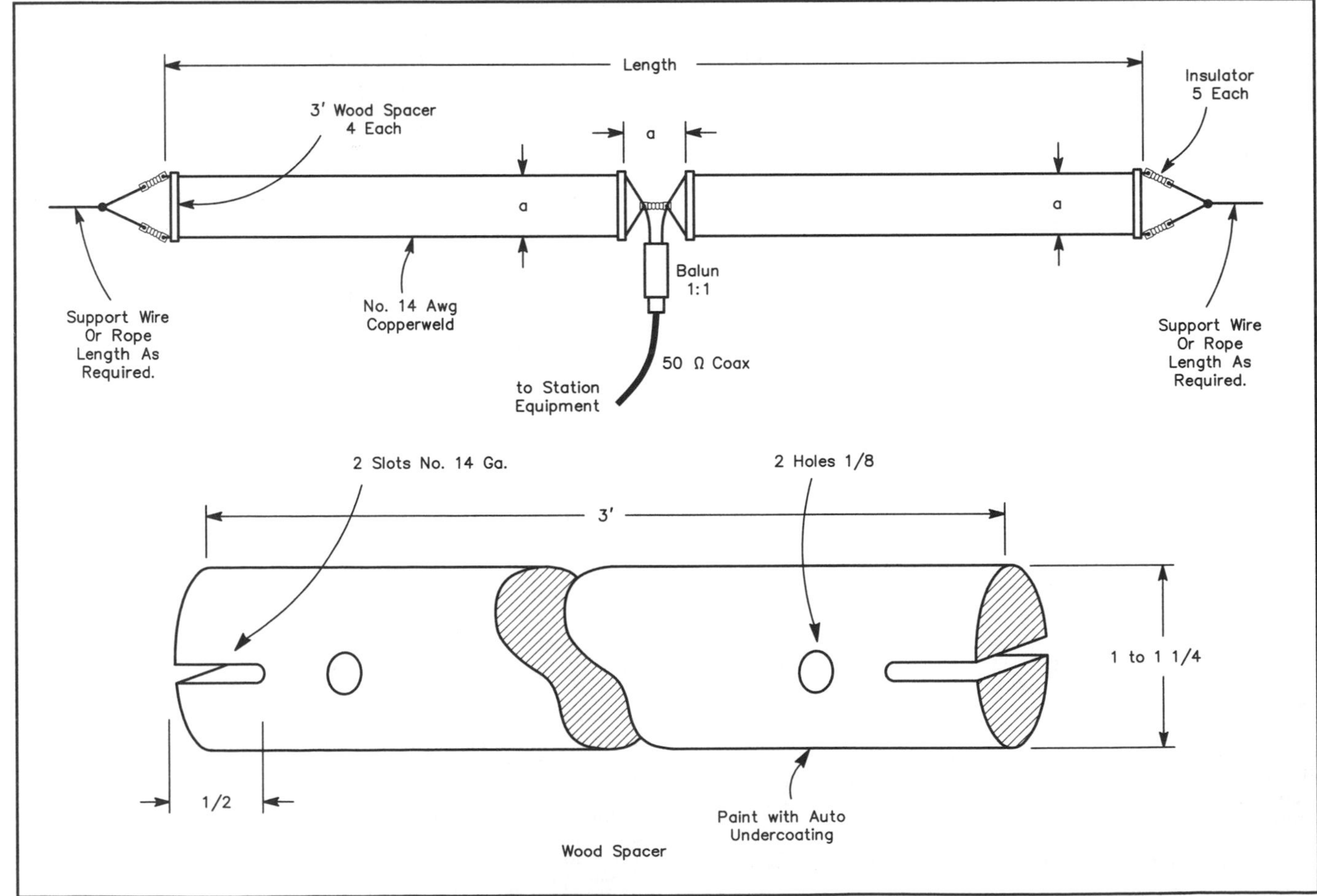

Figure 1—The fat dipole with construction details of the spreaders. See Table 1 for the length and spreader dimensions (A).

insulators but there is no critical problem here. One insulator is for the center of the dipole and the others are for the four ends. You will need four 3-foot-long broom sticks or 1-inch wood or plastic rods with good weatherproofing. I painted mine with auto undercoating but outdoor paint or varnish should also work. Copperweld wire is very desirable because it won't stretch and change the tuning of your antenna. This type of wire is available through advertisements in *QST*. The same source may also be able to supply the essential wide-band balun transformer and coax. Either the RG-58 or RG-8 types of coax cable are satisfactory but the latter requires more support because of its greater weight.

Measure your wire carefully and leave enough extra so that the insulators can be attached. The final length will need to be the calculated value, from insulator wire end to insulator wire end. After building the four-wire section, attach dowel rods as shown to act as spreaders. Fasten each rod in place with pieces of wire threaded through the holes and then wrapped around the antenna wire. Wrap these spreader wires tight enough so that the rods will not slip out of place. Snip off projecting wire ends wherever they occur to prevent RF corona power loss. Then, using either wire or rope, make a bridle to hold the ends of the antenna.

Table 1

Dimensions and Bandwidths on Various Bands

Frequency MHz	*Length, ft*	*Spacing (A), ft*	*Bandwidth, MHz*
1.9	233	5.9	0.252
3.75	118	3	0.5
7.15	62	1.6	0.951
14.175	31.2	0.8	1.885
21.225	20.85	0.53	2.823
28.860	15.34	0.39	3.837

Last, solder the balanced end of the balun transformer to the dipole. Each wire from the balun should go to the pair of wires on the same side of the dipole. The solder job should be of the best quality and permanent because it is hard to repair later. The coax needs to be connected to the unbalanced side of the balun. If you use large-diameter coax (3/8 inch) then think about ways to support the weight. Perhaps a piece of nylon rope from the dipole center insulator to the coax will help take the load, but I'll leave the details of the problem up to you. After this final construction step, haul the antenna up in the trees, using care that no twists are allowed.

Operation

For once I had a 75-meter antenna that worked better than predicted. The SWR was 1.6:1 or better from 3.5 to 4.0 MHz. Better yet, reports received were excellent with my old 100-watt solid-state transceiver. Moving up and down the band gave no loading problems from the broadband final. The fat dipole is just what I needed for a good, relaxing rag chew after a hard day with the 500-kW rig at the office.

CHAPTER TWO

OFF-CENTER-FED-DIPOLES

By John Belrose, VE2CV and Peter Bouliane, VE3KLO From *QST*, August 1990

The Off-Center-Fed Dipole Revisited: A Broadband, Multiband Antenna

The search for a simple broadband, multiband antenna with acceptable SWR over the entire 80/75meter band continued with the publication of the article by Witt, AI1H, in April 1989 *QST*. This inspired author Belrose, VE2CV, to reopen his files on the "Windom antenna," or, more properly, the off-center-fed dipole. Particularly since the opening of the 10, 18 and 24-MHz bands, this time-honored, somewhat controversial antenna is attracting a revival in interest and usage because of its multiband characteristics. Thus, the off-center-fed dipole deserves an update. Certainly, most North American radio amateurs have not heard about the German version of this antenna.

Even though it is one of the simplest antennas, the drooping dipole (a dipole with drooping ends, popularly referred to as an inverted V) is effective on 80 meters. For good DX performance, the apex of an 80-meter drooping-dipole antenna should be as high as possible above ground (at least 15 meters [50 feet]), because the antenna's vertical radiation angle decreases with height above ground. The arms of an 80-meter drooping dipole should not be too close to the ground (say, greater than 3 m [10 ft]). The input impedance ($R_a \pm jX_a$) of a drooping dipole depends on the operating frequency, the length of the dipole, the angle between its arms, the dipole's height above ground, and–particularly the resistive component–on the conductivity of the earth beneath the antenna. The included angle between the arms of an 80-meter drooping dipole should be about 127°, since, in this configuration, the antenna's pattern is dipole-like, and its input impedance is about 50 Ω for antenna heights typically employed by radio amateurs. The principal disadvantage of such an antenna is that its SWR bandwidth–less than 200 kHz–is too narrow to cover the 80/75-meter band.

The broadband performance of an antenna can be improved through the use of a matching network at the feed point. This network can comprise discrete components (see Hall,[1] Hayward,[2] Bloom,[3] Hately,[4] and Li, et al[5]); or transmission-line stubs (Snyder,[6] Hansen,[7] and Witt,[8]). Authors Li, et al give a microcomputer program for the design of LCR networks for broadband matching.

Another way to improve broadband performance is to use two dipoles fed in parallel: one dimensioned for the middle of the lower half of the band, and the other for the middle of the upper half of the band. The drooping-dipole configuration is ideal for this arrangement, since the ends of the two dipoles can be fanned (for angular separation) in either the vertical or horizontal plane. An alternative is to use an off-center-fed dipole, which, in addition to broadbandedness, has multiband performance characteristics. This article is concerned with such an antenna.

The "Windom Antenna" and Single-Wire Feed

The original Windom antenna (devised in 1928-29), named after Loren Windom, W8GZ, the amateur who wrote a comprehensive article about it,[9] employed single wire feed at a point of 1/6 to somewhat over 1/7 of the antenna length from the center (see Figure 1A). Windom reported on a detailed experimental study by colleagues John Byrne, W8DKZ, Edward Brooke, W2QZ, Jack Ryder, W8DKJ, and Prof W. L. Everitt of the Ohio State University Dept of Electrical Engineering. They found that if ammeters were placed on the antenna, with the single-wire feeder at the position just described, the current distribution on the dipole was sinusoidal and symmetrical, with no discontinuity in the vicinity of the feed point, and no standing wave on the feeder. Clearly, the feeder was terminated in its characteristic impedance.

This article did not discuss pattern, and it was many years before the computational tools for predicting pattern were available. Parfitt and Griffin[10] have recently analyzed the single-wire-fed dipole (the Windom antenna), and their results show, as anticipated, that radiation from the feed wire does

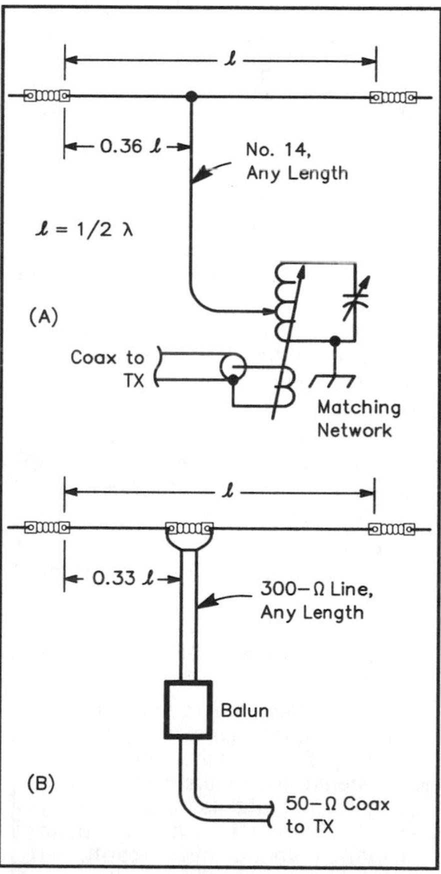

Figure 1—Two versions of the off-center-fed dipole. At A, the original "Windom antenna"–actually a Hertz (1/2 λ) element excited via a single-wire feeder; at B, a dipole off-center-fed via balanced line. The single-wire-fed version dates from the late 1920s; the balanced-line-fed version (sans balun) dates from the 1940s, and the balanced-line-and-balun version appeared in ARRL publications beginning in the late 1950s.

modify the antenna pattern, primarily at low elevation angles with the horizon. The feed line for their modeled antenna was vertical.[11] The pattern was basically dipole-like, with a squint away from the broadside. Current on a wire, whether uniform (a traveling wave) or sinusoidal (a standing wave) leads to radiation, unless the radiation that results from this feedwire current is canceled by an equal amplitude, out-of-phase current on an adjacent parallel wire (as in a balanced transmission line). For the Windom antenna, feeder radiation could be significant, depending on the arrangement and the length of the transmission line.

The Windom antenna was widely used during the 1930s and into the 1940s. For example, in author Belrose's experience, the British Columbia Forest Service were using the single-wire, off-center-fed dipole in the mid-1940s. It was a simple antenna that was easy to tune and match with the π network employed in the output circuit of their transmitter.

The Off-Center-Fed Dipole and Balanced Transmission Line

A later version (1940s) of an off-center fed dipole (*mis*called Windom) employed 300-Ω ribbon feed at a point 1/3 of the antenna length from one end (see Figure 1B). Such an arrangement was the first antenna used by author Belrose for operation on the 80, 40 and 20-meter bands in the 1940s. Scholle, DJ7SH, and Steins, DL1BBC, have more recently modified this antenna, devising a double-dipole version that provides a good match to 50-Ω line on all Amateur Radio bands from 3.5 through 28 MHz.[12] [See the Appendix for translations of this article and that cited in Note 13–*Ed.*] Note that their double-dipole version provides a good impedance match to coax on the 21-MHz band, which the single-dipole 80-meter antenna did not. A double-size version provides 1.8-MHz coverage as well.[13]

Scholle and Steins added a shorter off-center-fed dipole in parallel with the longer element, and employed a 50-Ω feed coaxial cable with a 300- to 50-Ω (6:1) balun located at the feed point. Table 1 summarizes information available on the element lengths for single and double off-center-fed dipoles. Since the dipole must be resonant on harmonic frequencies, its length must be somewhat longer than optimum for the 75-meter band, as noted by Scholle and Steins. It is apparent from their SWR curves that the resonant frequency of their antenna is less than 3.5 MHz.

Table 1
Summary of Element Lengths for Published Off-Center-Fed-Dipole Designs

Figure 1B Version	13.5 m (44.29 ft)	27.94 m (91.67 ft)
cq-DL double-dipole version	13.8 m (45.28 ft) 4.69 m (15.39 ft)	27.7 m (90.88 ft) 9.38 m (30.77 ft)
cq-DL double-sized, double-dipole version	25.88 m (84.91 ft) 4.69 m (15.39 ft)	51.77 m (169.85 ft) 9.38 m (30.77 ft)
Author's double-dipole version, (balanced coaxial-line feed, Figure 4)	23.3 m (76.44 ft) 6.78 m (22.24 ft)	48.2 m (158.13 ft) 13.58 m (44.55 ft)

In the arrangement employed by Scholle and Steins, the longer elements were horizontal and the shorter elements sloped (the included angle between the arms of which was Λ≈100°). The authors' version was a drooping-dipole configuration (Λ≈127°) with the angle between the arms of the two dipoles approximately 45°. This antenna was fed with two 15-meter (50-foot) lengths of RG-62A (foam dielectric) coaxial cable configured as a balanced transmission line.[14]

We fabricated a *single* off-center-fed dipole (dimensionally identical with the longer element of the Scholle and Steins "double dipole") and used it as a Field Day antenna (in 1985) for 75/80 in. The stimulus for this, and for carrying out a detailed study, was a search for a simple all-band antenna conducted for the Canadian National Institute for the Blind.[15] We experimented with 4:1 and 6:1 baluns, primarily because 4:1 baluns are easier to come by (more on baluns later). The antenna was operated in a drooping-dipole configuration (Λ≈127°), with the apex (which, in our case, was the feed point) at about 12 meters (40 feet). (In retrospect, we should have positioned the center of the antenna at apex height, since in this configuration the current maximum on the antenna [for 80-meter operation] would be at maximum height.) Extensive impedance and SWR measurements were made. To facilitate this, we employed a low-loss (foam-dielectric) coaxial feeder consisting of a 32.3-meter (106-foot) length of Belden 8214 (velocity factor, 0.8.) This feeder was 1/2 λ long at 3.5 MHz, 1 λ at 7 MHz, and so on. High-quality commercial baluns (made by Antenna Engineering, Australia[16]) were used between the feed line and the antenna.

Figure 2 shows our SWR measurements for the 80/75-meter band. With the 6:1 balun in line, the 2:1 SWR bandwidth was 3.47 to 3.93 MHz—broadbanded, indeed. The SWR was even lower on 40 and 20 meters, and less than 2:1 for the lower half of the 10-meter band; see Figure 3. Clearly, these bandwidths are not in accord with simple

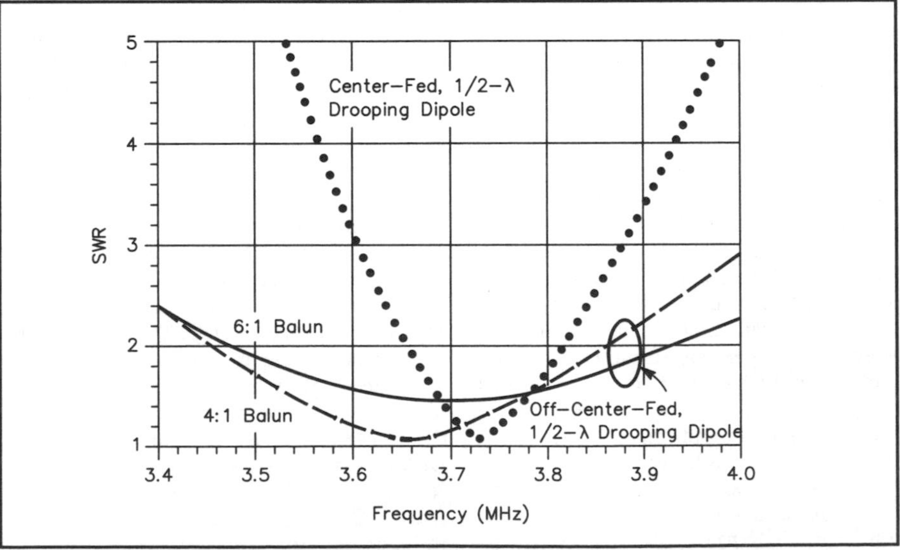

Figure 2—SWR vs frequency for a single off-center-fed dipole employing different baluns (4:1 and 6:1) at the feed point, compared with a center-fed 1/2-λ dipole. Both antennas were operated in a drooping-dipole configuration.

Center-Fed versus Off-Center-Fed Dipoles

A center-fed dipole is a resonant antenna at its fundamental frequency* (f_0) and at $3f_0$, $5f_0$, $7f_0$ and so on, whereas an off-center-fed dipole is resonant at f_0, $2f_0$, $4f_0$, $8f_0$, and its bandwidth (apparently) is greater. As a low-band, multiband antenna, this off-center-fed dipole has advantages over the center-fed dipole, but a disadvantage (or advantage, depending on the antenna configuration and one's point of view) is that feeder radiation can contribute to the antenna's resultant radiation pattern. The use of a current balun minimizes feeder radiation.—*VE2CV and VE3KLO*

*The frequency at which it is 1/2 λ long.

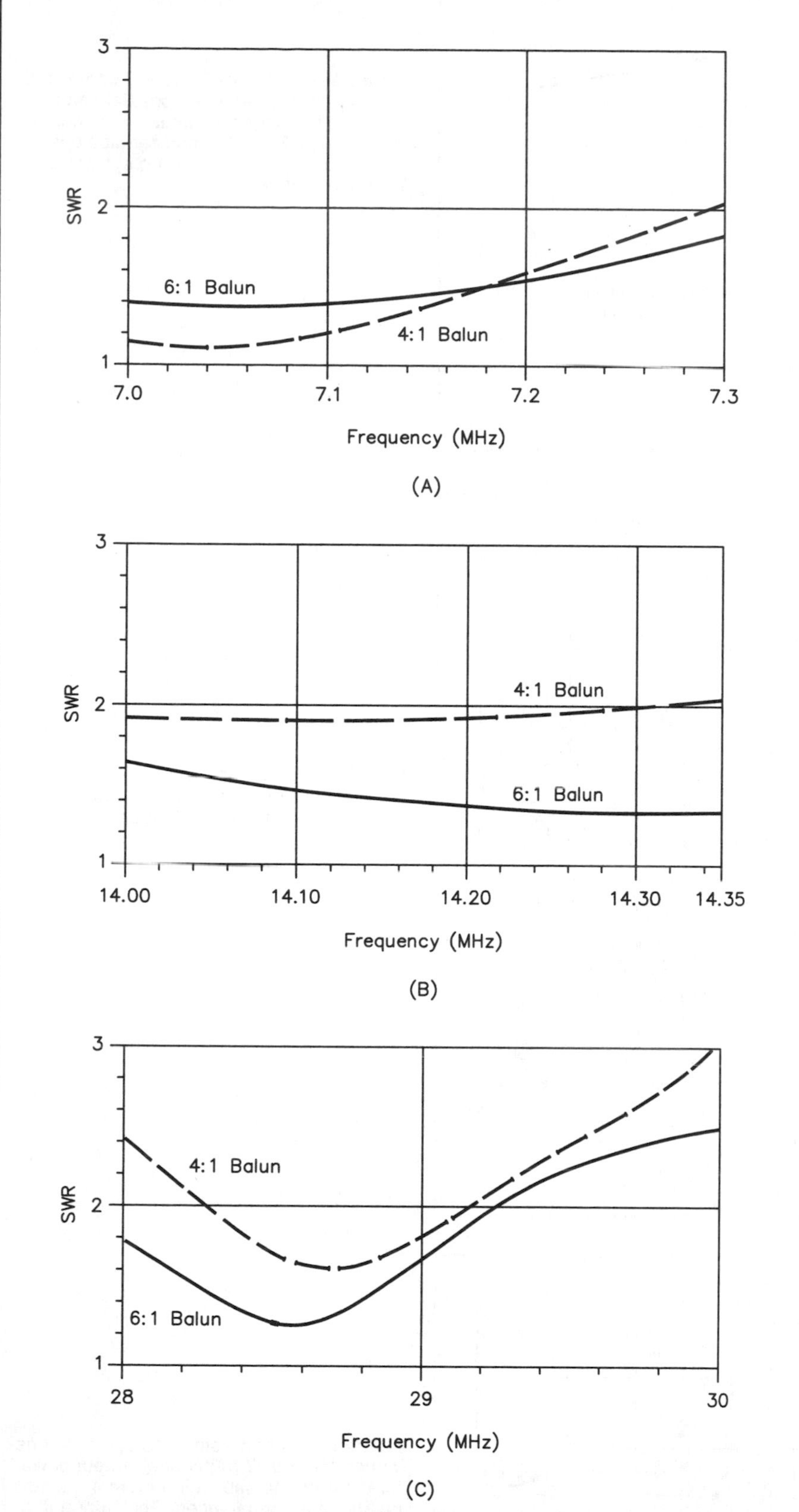

Figure 3—SWR vs frequency for the off-center-fed dipole of Figure 2 at 40, 20 and 10 meters.

theory.[17] Fifteen-meter data are not shown because this single off-center-fed dipole was not resonant on the 15-meter band.

The 4:1 balun provided somewhat sharper resonances and lower SWR at resonance in the 80/75- and 40-meter bands, but a low SWR was not obtainable in the 20-meter band. The measured impedance at resonance in the 80/75-meter band with this balun was exactly 50 Ω; hence the antenna's effective feed-point impedance was 200 Ω. This explains why the minimum SWR with the 6:1 balun was 1.5:1; the 6:1 balun, however, seems to be a better compromise if operation on other bands is wanted.

As noted, we employed high-quality commercial baluns (from the AEA Model 250 series). Radio amateurs have several alternatives in this connection; these include using a 4:1 balun and 75-Ω coax; or purchasing or winding your own 6:1 balun (see Orr, W6SAI[18]),and using 50-Ω coax.

Aside from impedance matching, however, there are other factors to consider in feeding an off-center-fed dipole and selecting a balun for this service. Next, we address some of these aspects.

Balun and Feeder Considerations

In the autumn of 1989, we decided to have another look at off-center-fed dipoles. Because of the multiple-resonant-frequency response of the double off-center-fed dipole, we decided to explore the potential of this antenna system for broadband frequency coverage. If used with a suitable antenna-system tuning unit (ASTU), a double off-center-fed-dipole system could perhaps be used on any frequency from 1.8 to 30 MHz. Since conventional baluns do not perform satisfactorily into reactive loads, we decided to eliminate the balun, at least insofar as the antenna and its feeder were concerned. We fed our antenna with a balanced 190-Ω transmission line consisting of two 15-meter (50-foot) lengths of RG-62A (foam-dielectric) coaxial cable as shown in Figure 4 and described in the work cited in Note 15. Teflon-dielectric cable would have been preferable because of Teflon's superior insulating characteristics but 95-Ω, Teflon-dielectric coax is unavailable.

We started out with a single dipole element dimensioned to 1/2 λ at 2 MHz. This antenna exhibited an antiresonant response (very high input impedance, measured at the transmitter end of the transmission line) at 7.6 MHz. Next, we installed the shorter dipole and dimensioned it to *minimize* the system's impedance at 7.6 MHz. Figure 5 shows the system's impedance v frequency. Except for a narrow band of frequencies near 6 MHz, the input resistance fell in the range of 20 to 400 Ω; and the input reactance fell in the range $+j100$ to $-j200$ Ω. This double off-center-fed dipole system is rather easily tuned and matched, since its reactance is low.

Figure 6 graphs antenna currents for a transmitter output power of 100 W. The antenna system was tuned using an unbalanced T network, with a ferrite-bead current balun of the type described by Maxwell[19] between the ASTU and the antenna. A balanced T network patterned after that described by Belrose was initially used since this is the type of ASTU that, in principle, should be used to tune and match a balanced antenna system. Unfortunately, however, this tuner employed a standard toroidal-core balun—a *voltage* balun, which applies almost-equal *voltages* to the wires of a balanced feeder. Even though the balun was on the "tuned side" of the ASTU—a practice we recommend—and the ASTU it-

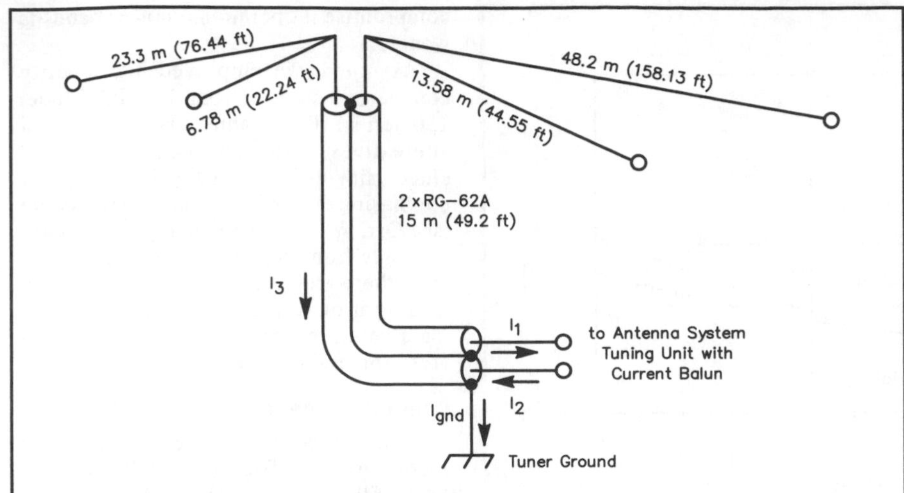

Figure 4—A double-dipole, off-center-fed antenna with a balanced coaxial-line feeder. The longer dipole is dimensioned to 1/2 λ at 2 MHz; the shorter dipole is dimensioned to minimize the system at 7.6 MHz. See text.

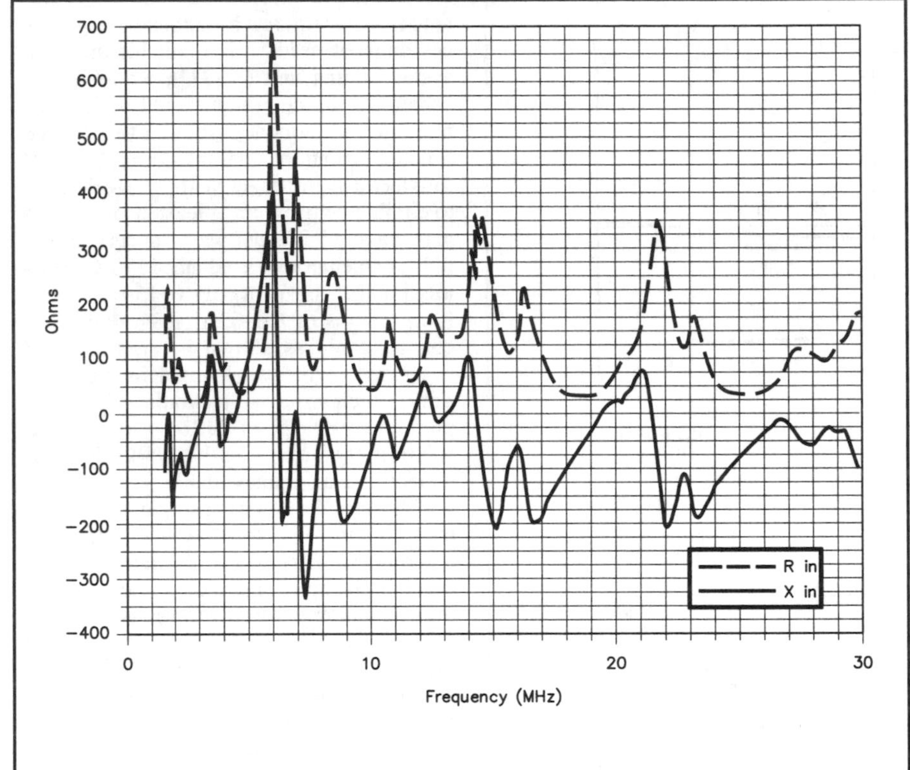

Figure 5—Impedance (input resistance [R in] and input reactance [X in]) vs frequency for the antenna shown in Figure 4.

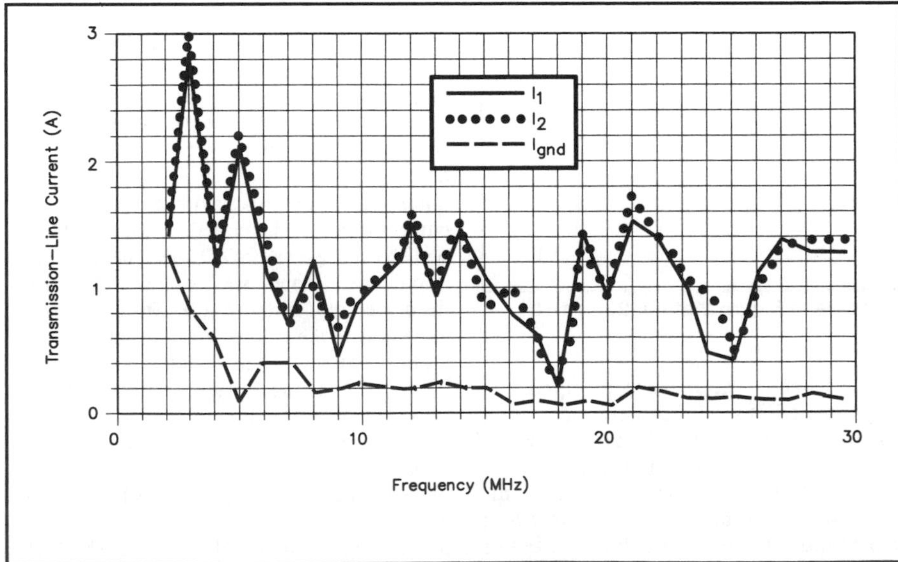

Figure 6—Transmission-line currents (peak values for 100-W transmitter output power) for the antenna shown in Figure 4, tuned by means of an unbalanced T network and a ferrite-bead-choke (W2DU) current balun.

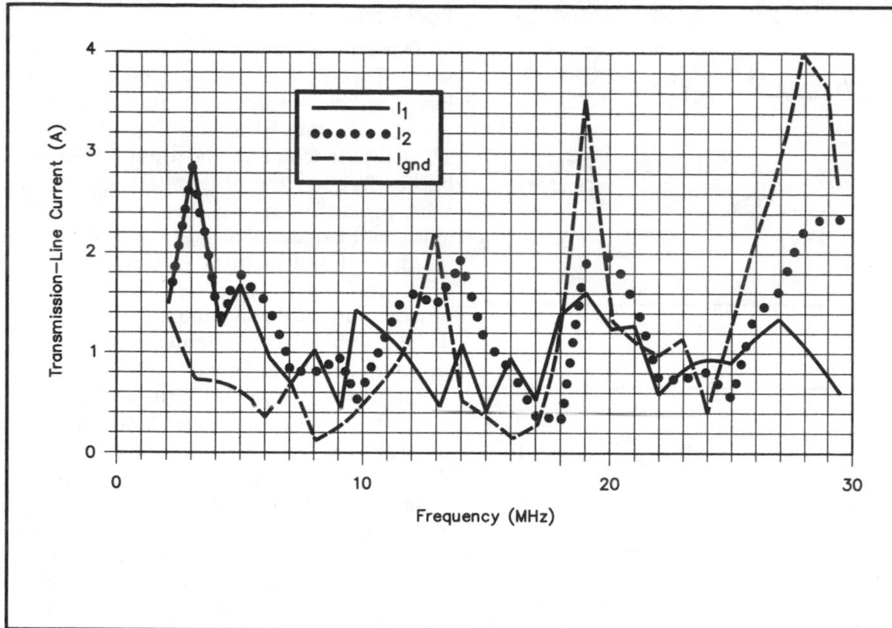

Figure 7—Transmission-line currents (peak values for 100-W transmitter output power) for the Figure 4 antenna, tuned by VE2CV's balanced tuner, which employed a voltage balun.

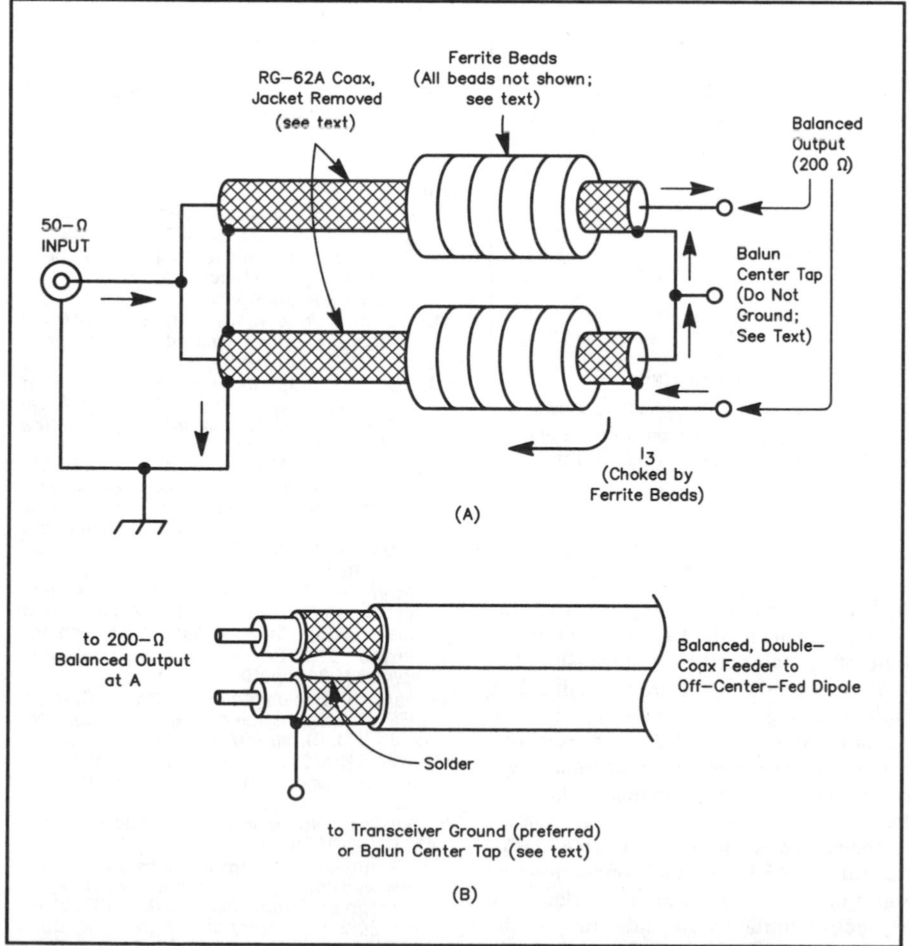

Figure 8—At A, a 4:1, ferrite-bead current balun; at B, connections to the balun end of the dual-coax-balanced feed line.

self was balanced, we found that the transmission-line currents (I_1 and I_2 in Figure 4) differed greatly in amplitude at some frequencies. The difference may surprise you; it did us (see Figure 7).

For an antenna that is asymmetrical with respect to its feeder–such as an off-center-fed dipole–a *current* balun must be used, since this type of balun forces almost equal *currents* into each conductor of the balanced line. This is necessary if transmission-line radiation is to be minimized. If these currents are exactly equal, there will be no difference in current flowing in the ground lead (I_{gnd}), which connects the braids of the coax to the tuner ground. Any current on this lead is then due to radiation-coupled current (I_3) induced on the *outside* surface of the coaxial shield. Clearly, we want this current to be small (to minimize radiation from the transmission line), and indeed it *is* small except at frequencies below 3 MHz (see Figure 6). (Our antenna was suspended from a bracket at the 12-meter [40-foot] level on a 21-meter [70-foot] aluminum-lattice mast. In the final analysis, reradiation by this mast should be considered because current may have been induced on the mast surface.)

Off-Center-Fed Dipoles for the Amateur Radio Experimenter

The off-center-fed dipole used by author Belrose in the 1940s was fed with 300-Ω twin lead via a balanced ASTU that was link coupled to a balanced (push-pull) power amplifier. There was no concern about balun losses and what type of balun to use because the system contained no balun. [20]

The authors' double-dipole, off-center-fed antenna employed a balanced coaxial feed line. This antenna was attractive for the authors because it could be used throughout the HF range if fed via a suitable matching network. If you decide to fabricate and try an off-center-fed dipole system, we suggested that you dimension it in accord with the Scholle and Steins versions, which are optimized for the amateur bands. Furthermore, we suggest that you feed such an antenna with a balanced transmission line. We have used dual RG-62A cables to make a balanced 190-Ω line, but one could use paralleled RG-63 (125-Ω) coaxial cable, which would make a balanced transmission line more in accord (Z_o ≈250 Ω) with the traditional 300-Ω-twin-lead feeder. In such a system, the balun, and the ASTU (if required), can be in the shack to allow experimentation in achieving balanced current feed and reducing losses in the balun. Whatever method you use, feed your off-center-fed antenna via a current balun.

A shield-choke current balun can be constructed by slipping ferrite beads (–43 or –73 material) over a length of coaxial cable. (Depending on the beads obtainable and the diameter of the cable you use, you may need to remove the cable's outer jacket and install the beads directly around the

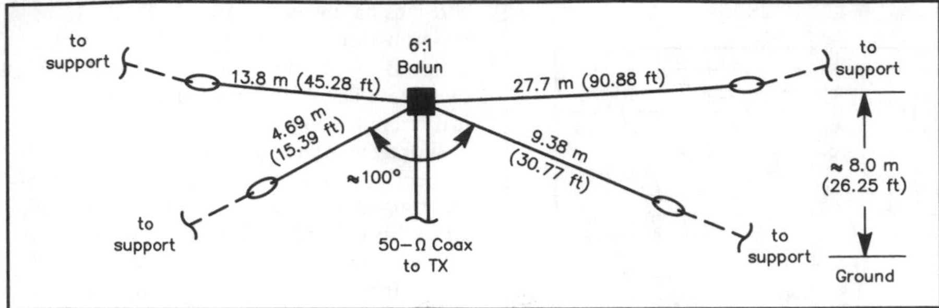

Figure A—Double Windom.

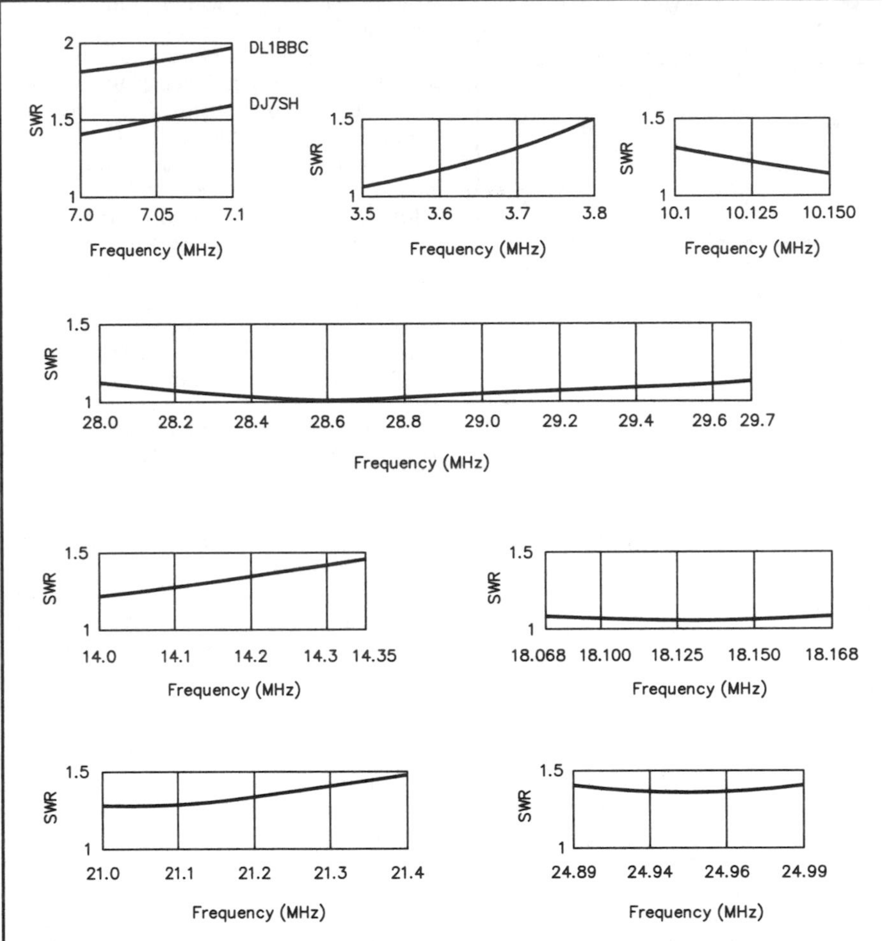

Figure B—SWR curves for the eight-band double Windom. [These curves do not cover all US amateur frequencies because allocations in the FRG differ from those in the US—*Ed.*]

shield.) As described in the work cited at Note 19, Walt Maxwell, W2DU, used 50 beads of no. 73 ferrite (Amidon no. FB-73-2401, Palomar FB-24-73 or equiv) on a piece of Teflon-dielectric cable to make a practical, low-loss, 1.8-to-30-MHz balun about 12 inches long.[21] Such a balun, however, is a 1:1 transformer—not very useful for the present application, where impedance transformation is also required. A 4:1 balun of similar type (see Note 20) can be constructed by using two equal lengths of RG-62 (95-Ω) coaxial cable, each fitted with ferrite beads. The inputs to these two coaxial-cable baluns are connected in parallel, and the outputs in series (see Figure 8). This makes a center-tapped 4:1 balun.

Our antenna feeder has a center tap: the braid of the balanced coaxial transmission line. The braids of the transmission-line cables can be grounded to the center tap of the balun, or to the ASTU (or transceiver) ground. Connecting them to the transceiver ground is the better arrangement because the transmission-line currents are better balanced, and the braid current is less. (For the braid-to-ASTU-ground connection, the braid current is measured in the wire connection to the balun center tap; in the braid-to-transceiver-ground case, the braid current is measured in the connection to the equipment ground.)

Performance

Our single-element Field Day antenna worked well for us; as previously noted, however, we employed it only on 80/75 meters. On this band, it is essentially a 1/2-λ dipole with unconventional feed. At higher frequencies, the pattern develops lobes because the dipole acts as a long-wire antenna. Provided that this directivity coincides with directions of interest, the off-center-fed dipole is a good, simple broadband/multiband antenna.

The authors' 1989 (double-dipole) version, while not designed specifically for Amateur Radio communication, has been used on various amateur bands (during the winter of 1989-90). It works, but how well? We checked into various nets on 75 and 40 meters. On 75, for instance, we checked into the ONTARS (Ontario Amateur Radio Service) net, and the Newfoundland Phone Net (1730 EST), two regions at quite different distances from the Ottawa area. The reports received were comparable with those given to other stations by net control. On the 160-meter band, we found that we could work stations we could hear, provided that they were running comparable power (100 W) and their local noise levels were reasonable. We have not yet determined the antenna's gain and pattern.

Acknowledgment

The authors work at the Communications Research Centre, Shirleys Bay, Ontario. The measurements for the double-dipole off-center-fed dipole system were made at the Laboratory as part of a study of broadband HF wire antennas.

Notes

[1]J. Hall, "The Search for a Simple, Broadband 80-Meter Dipole," *QST*, Apr 1983, pp 22-27.

[2]W. Hayward, "Limitation to Broadband Impedance Matching," Technical Correspondence, *QST*, Jul 1984, pp 45-47.

[3]A. Bloom, "Once More with the 80-Meter Broadband Dipole," Technical Correspondence, *QST*, Jun 1985, p 42.

[4]M. Hately, "Multiband Dipole and Ground Plane Antennas," *Third IEE International Conference on HF Communication Systems and Techniques*, Conf Series Pub 245, pp 102-106,1985. Maurice Hately, GM3HAT, has patented his method of feeding and matching HF dipoles (UK Patent GB 2112579), and he manufactures and markets his "dipoles of delight."

[5]S. Li, J. Rockwell, J. Logan, and D. Tan, *Microcomputer Tools for Communications* (610 Washington St, Dedham, MA 02026: Artech House, 1983), Ch 14: Broadband Matching.

[6]R. Snyder, "The Snyder Antenna," *rf design*, Sep/Oct 1984, pp 49-51.

[7]R. Hansen, "Evaluation of the Snyder Dipole," *IEEE Trans on Antennas and Propagation*, AP-35 (No. 2), pp 207-210, Feb 1987.

[8]F. Witt, "The Coaxial Resonator Match and the Broadband Dipole," *QST*, Apr 1989, pp 22-27.

[9]L. Windom, "Notes on Ethereal Adornments," *QST*, Sep 1929, pp 19-22, 84.

[10]A. Parfitt and D. Griffin, "Analysis of the singlewire-fed dipole antenna," *IEEE AP-S International Symposium, Vol III* (IEEE cat no. CH2654-2/89, Library of Congress no. 89—84327, Jun 1989, pp 1344-1347.

[11]Such an arrangement is an extreme case because the feeder is perpendicular to the

ground—a configuration probably atypical of singlewire feed in Amateur Radio installations, and one that affords practically zero cancellation between feeder and ground currents. At the other extreme, one of the amateur systems described in Windom's "Notes on Ethereal Adornments" (see Note 9) employed a *1200-foot-long* (365.8-m) feed wire that likely more nearly approached parallelism with the ground. Clearly, "modeling a Windom" is a fuzzy procedure.–*Ed.*

[12]H. Scholle and R. Steins, "Eine Doppel-Windom Antenne für Acht Bänder," *cq-DL*, Sep 1983, p 427.

[13]H. Scholle, "Eine Doppel-Windom Antenne für Neun Bänder," *cq-DL*, Jul 1984, pp 332-333.

[14]J. Belrose, "Tuning and Constructing Balanced Lines," Technical Correspondence, *QST*, May 1981, p 43.

[15]J. Swail, VE3KF, and W. Loucks, VE3AR, private communications, Apr 1985.

[16]Antenna Engineering, Australia Pty Ltd, PO Box 191, Croydon, Victoria 3136, Australia.

[17]W. Wrigley, "Impedance Characteristics of Harmonic Antennas," *QST*, Feb 1954, pp 10-14.

[18]W. Orr, Ham Radio Techniques, *ham radio*, Jan 1983, pp 68-69.

[19]M. Maxwell, "Some Aspects of the Balun Problem," *QST*, Mar 1983, pp 38-40.

[20]J. Belrose, "The Balun Saga Continued," *QST*, in preparation.

[21]As a less-costly alternative, 10≈1-inch-long mix-77 beads—with an inner diameter to pass the coax used—*may* serve as a substitute. (The authors do not have experience with beads of this material, however.) Consult your ferrite-bead supplier on the availability of beads of this type.–*Ed.*

APPENDIX

A Double Windom Antenna for Eight or Nine Bands

This work, which originally appeared as two articles in *cq-DL*, was translated from the German by Dr. George Elliott Tucker, WA5NVI.

Part 1: A Double Windom Antenna for Eight Bands

By Hubert Scholle, DJ7SH, and Rolf Steins, DL1BBC

The asymmetrical dipole antenna developed and described by Windom (W8GZ) in 1929 has been used by many amateurs for many years as the FD4. This has also been the case in Germany.

We discovered in an older periodical (*QRV*) the explanation by F. Spillner (DJ2KY) that this antenna, with the addition of a small one-band Windom for 15 m, can be used as a five-band Windom. After the installation of the additional elements, this antenna worked very well for two years at DL1BBC.

With the opening of new bands (10, 18 and 24 MHz), the thought occurred to try out a new extension of the FD4 to eight bands (3.5 to 29.7 MHz).

What worked for 21 MHz must also be possible for 10 MHz.

So we took off the 21-MHz extension to my antenna and hung two elements of 4.69 and 9.38 m (15.39 and 30.77 ft), respectively, on the FD4 and stretched these downwards from insulators as an inverted V (Figure A).

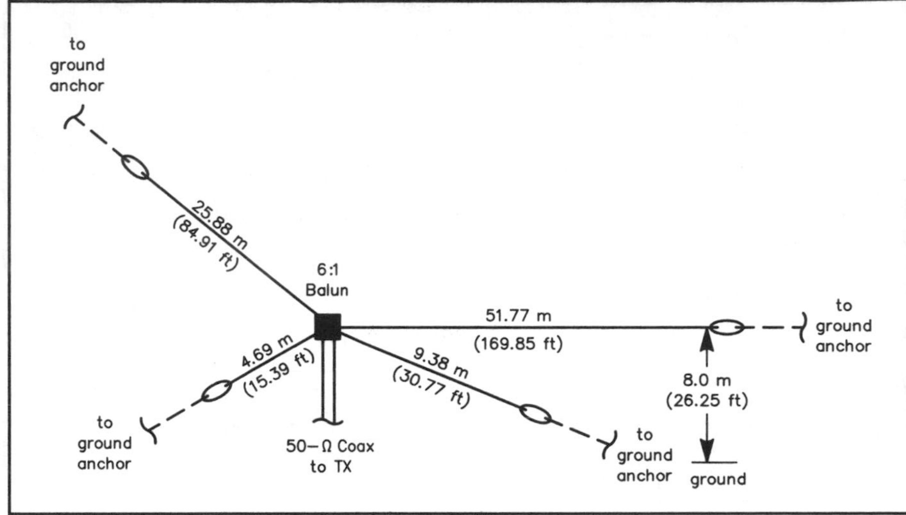

Figure C—A double Windom antenna for nine bands.

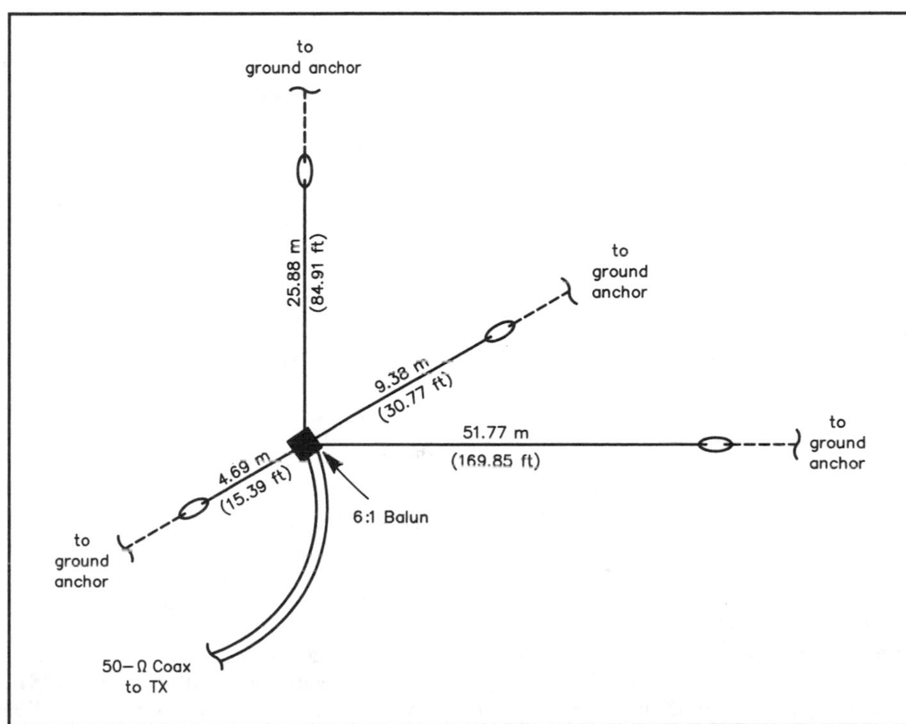

Figure D—Top view of the nine-band double Windom. [The elements are positioned to reduce coupling between the antenna's two off-center-fed dipoles.—*Ed.*]

To calculate the length we used the formula:

$$L/2 = 142.5 \div f \qquad \text{(Eq A)}$$

Whatever would work for 30 m should also work on 15 m.

As suspected, it worked.

As a by-product, it turned out in the measurements that this double Windom resonated just as well on 18 MHz and 24 MHz. So our eight-band Windom came into being with really simple means.

Construction

Thanks to our neighbors, we were able to extend the basic antenna (FD4) to its full length.

At DL1BBC it was installed about 6.9 m (22.63 ft) above the ground, rising to about 8 m (26.25 ft) at each support point. At DJ7SH it hung about 5 m (16.4 ft) above the ground and partly ran over a garage roof. Both extension legs were stretched downwards as an inverted V with an angle of about 100°. Changing this angle allows the whole antenna to be easily tuned during final adjustments.

After construction, the first measurements showed that because of the length of the 30-m elements, the 80, 40 and 20-m bands each had a resonance point that was shifted towards the low end of the band. This effect was eliminated by lengthening slightly the 30-m section, so the resonance points fell more in the middle of the bands.

With this adjustment, the resonance

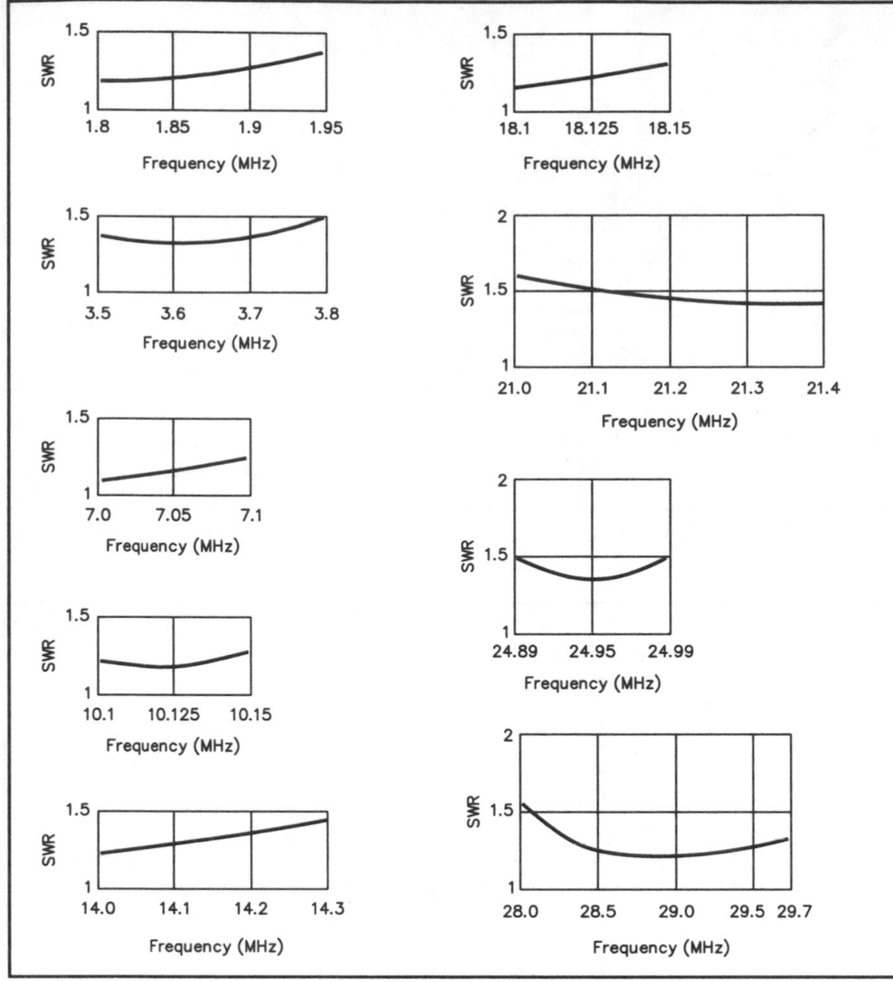

Figure E—SWR curves for the nine-band Double Windom.

point on 30 m shifted slightly towards the end of the band, but this can be tolerated.

With all measurements of the Windom, it was very clear that how the feed line ran played a decisive role.

According to our results, it must be stressed that the feed line must run first vertically downwards from the feed point to the ground and only then to the shack, as otherwise the entire antenna may be detuned. This is especially the case when the height of the antenna is under 10 m (32.8 ft). The 50-ohm-coax feed at DL1BBC was pulled through an old garden hose and then buried under the lawn.

The lower antenna height at DJ7SH had the result that, with the first construction attempt, the precalculated length of the 30-m elements was exactly right. The antenna delivered on all eight bands at the first go.

As can be seen from the SWR charts (Figure B), at DL1BBC the match on 40 m turned out somewhat less favorable. However, this was immediately fixed by changing the antenna height slightly. At DJ7SH, no resonance curve ran above 1.5:1, which was the goal since neither station uses an antenna tuner.

Performance

First contacts were made with both antennas. These showed that the antennas had a good degree of performance for a long wire. Especially the downwards sloping extension elements have a clear advantage over the horizontal basic antenna for DX.

With the first try on 30 m, many contacts were made with the US (East and West coasts), with signal reports between S6 and S7 while running 100 W.

At present, we cannot make a concrete statement about contacts within Europe.

This article makes no scientific claims, but intends to stimulate the long-wire enthusiast, and especially the friends of CW.

Part 2: Adding Another Band

Because the response was unexpectedly great to the publication of the above in *cq-DL*, we went to work again on an extension, as it was worthwhile to add 160 m.

With a half wavelength at 1.835 MHz, we calculated the basic length of the antenna to be 77.65 m (254.75 ft). We tapped the antenna at 25.88 m (84.9 ft) from one end and fed it with 50-ohm coax through a 6:1 balun. The basic antenna of this length was installed horizontally as a reclining L at DL1BBC. The additional elements, with lengths of 4.69 and 9.38 m (15.39 and 30.77 ft), were attached at the balun. This additional Windom for 10 and 21 MHz was again stretched downwards as an inverted V with an angle of about 100°. Here the additional Windom was mounted so that its elements were not extended in the same direction as those of the reclining L, which gave sufficient decoupling (Figures C and D).

For the feed, the Fritzel company made available for testing a new 6:1 balun, series 83, which can also handle high power. The SWR charts (Figure E) were obtained with the wire lengths given in the preceding paragraph. In case builders experience slight resonance shifts, these can be balanced out by lengthening or shortening the additional Windom.

Performance

First contacts were made with the antenna installed at DL1BBC. Here it was once again shown that the antenna has a good degree of performance for a long wire, especially for 1.8 and 3.6 MHz within Europe. The additional Windom again had the degree of performance described in the first part of this article.

The authors welcome questions and exchange of information. (When writing, please include return postage.)

Improved Feed for the Off-Center-Fed Dipole

By Richard A. Formato, K1POO

Theoretical data suggest that the commonly used feedpoint for the off-center-fed dipole (OCFD) may not be the best. The OCFD is an attractive multiband antenna because it's simple, inexpensive and requires no antenna tuner. Improving its performance simply by moving the feedpoint makes the antenna even more attractive. This note illustrates how the feedpoint influences antenna performance by analyzing computer-modeled SWR data for three different feedpoint locations.

The OCFD (shown schematically in Figure 1), consists of a single wire radiator of length L, fed off center a distance D from one end. The usual implementation uses a "1/3 feed," that is, the RF source is located one-third of the way from the end, so that D ≈ L/3. Why the feedpoint should be located there is not exactly clear. The ninth edition of *The ARRL Antenna Book,*[1] for example, observes that there is not much theoretical justification for this choice. Nevertheless, the 1/3-feed is accepted practice for building an OCFD.

Design details for a 1/3-feed three-band OCFD (80, 40 and 20 meters) appear in the 17th edition of *The ARRL Antenna Book.*[2] A 4:1 current balun at the feedpoint matches this antenna to any length of 50-Ω coax. More recently, Bill Wright, GØFAH, described a four-band, 1/3-feed OCFD (40, 20, 15 and 10 meters) fed with 300-Ω ladder line.[3] Matching 50-Ω coax requires a 4:1 balun on 40, 20 and 10 meters, and a 1:1 balun on 15 meters. Four-band operation, therefore, requires switching baluns. Another minor limitation is that the ladder-line length can be only an odd multiple of the wavelength at 21 MHz because the line is used as an impedance transformer. A simpler approach to achieving four-band operation is to feed the OCFD at a different point along its length.

I computer-modeled a 21.03-meter (69-foot) long, 0.2053-cm-diameter (#12 AWG) OCFD in free space. The dimensions are the same as those in the GØFAH design. Free-space results are a good approximation for antennas high enough above the ground (typically a significant fraction of a wavelength). The band-center SWR was computed on 40, 20, 15 and 10 meters at the antenna-input terminals for a feed system impedance of 200 Ω. The theoretical values of input resistance and reactance were used to calculate SWR (the antenna was not assumed to be tuned). Because the feedpoint impedance is 200 Ω, a 4:1 balun is required to feed the antenna with 50-Ω coaxial cable. The results for three different feedpoints appear in Table 1.

For the conventional 1/3-feed (D = 6.98 m), the 40 and 10-meter SWR values are slightly over 2, while the 20-meter SWR is about 1.75. In marked contrast, the 15-meter SWR is off the scale (the actual value > 20). It is this behavior that makes a special feed system necessary on 15 meters, a complication that can be avoided by moving the feedpoint.

When the OCFD feed is located 8.65 meters from one end, the 40, 20, and 10-meter SWRs are somewhat higher than they are with the 1/3 feed, but the 15-meter SWR is very low (≈ 1.2). Moving the feedpoint 1.67 meters closer to the antenna's center of the antenna results in a much better average SWR. And, more importantly, special matching is not required to achieve SWR ≤ 2.5 at the antenna terminals on all bands. Balun and coaxial cable losses, which are inevitable, reduce the SWR at the coax input to even lower levels. For most installations, it is probably reasonable to expect SWR at the transmitter to be less than 2 on all bands.

With the feedpoint located 3.65 meters from one end, the SWR on 40, 20 and 15 meters is excellent. The 40-meter SWR is only slightly above 2, and the 20 and 15-meter SWRs are below 2. The highest SWR occurs on 10 meters, where it's approximately 2.4. Because the SWR is reduced by feed-system losses, it will be less than 2.4 at the coax input. And, because balun and cable losses increase with frequency, the SWR reduction will be greatest on 10 meters where it is needed most. Feeding the antenna 3.65 meters from one end may well provide the best overall four-band performance.

In a specific implementation, the OCFD, like any antenna, must be tweaked for optimum SWR. This is accomplished by adjusting the feedpoint location. Other antennas, nearby metallic objects, and the earth are typical factors that influence antenna performance. Since these factors are not included in the computer model, they must be dealt with empirically by adjusting the antenna on-site. The data presented here provide a starting point for experimenting with different feedpoints. Depending on the total antenna length L, height above ground, earth electrical parameters, and feed system Z_0, it should be possible to operate a single OCFD on four or more bands without an antenna tuner or special feed arrange-

[1]*The ARRL Antenna Book*, 9th Edition (Newington: ARRL, 1960), pp 191 to 192.

[2]R. Dean Straw, N6BV, Editor, *The ARRL Antenna Book*, 17th Edition (Newington: ARRL, 1994), pp 7-20 to 7-21.

[3]Bill Wright, GØFAH, "Four Bands, Off Center," *QST*, Feb 1996, p 65.

Table 1

SWR Versus Feedpoint Placement (see Figure 1)

Feedpoint Distance D (m)	*SWR*
40 Meters	
8.65	2.5
6.98	2.1
3.65	2.0
20 Meters	
8.65	2.0
6.98	1.7
3.65	1.7
15 Meters	
8.65	1.2
6.98	21.3
3.65	1.9
10 Meters	
8.65	2.2
6.98	2.2
3.65	2.4

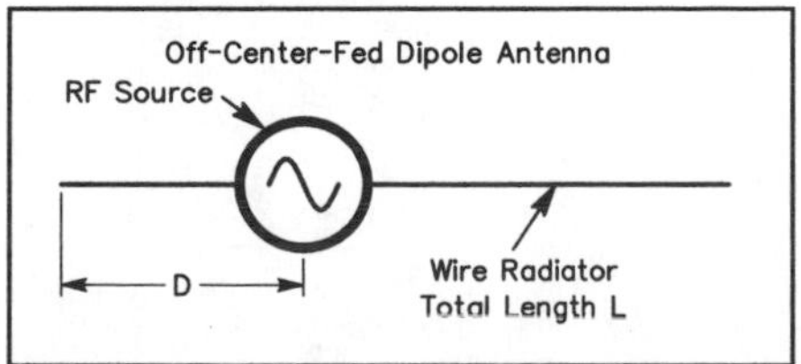

Figure 1—The basic off-center-fed dipole (OCFD).

ment. It's apparent that the OCFD's SWR varies dramatically as the feedpoint is moved, and that the commonly used $^1/_3$-feed is not necessarily the best. Other feedpoints may therefore produce a better antenna.

Dean Straw, N6BV, Senior Assistant Technical Editor, comments:

I've modeled what Richard describes and find that he's basically correct in his assertions. However, I would add a couple of caveats:

- The SWR across an individual band will vary. For example, the 40-meter SWR will rise above the level he describes at the ends of the band, because it is a rather large band, percentage-wise. This antenna is no different from an ordinary center-fed 40-meter dipole in that sense.
- The azimuth patterns for this OCFD will not be symmetrical, as is the case for any such off-center-fed antenna. This is nothing new, but should still be mentioned, particularly for those hopeful folks who want a "one-thing-to-all-people" antenna.
- The ham who tries this approach on 30, 17 or 24 meters may burn out the 200:50-Ω balun transformer. The SWR is very high indeed on those bands.

Off-Center-Fed Dipole Comments, Part 2

By Richard A. Formato, K1POO

In an earlier correspondence,[1] I suggested that the conventional "1/3-feed" used for the off-center-fed dipole (OCFD) is not the best choice. This letter provides additional information, and responds to comments made by Dean Straw, N6BV, at the end of that correspondence. Before addressing the comments, it is important to examine in more detail the SWR data which are the basis of my earlier letter. Figure 1 plots computer-modeled free-space SWR for the prototype 21.03-meter long, 0.0253-cm-diameter OCFD. SWR for a 200-Ω feed system impedance was calculated at the antenna-input terminals every 50 kHz for source frequencies from 5 to 30 MHz. Three feed-point locations were modeled, 3.65, 6.98 (1/3 feed), and 8.65 meters from the end of the antenna; the curves are labeled accordingly.

The most important features of the SWR curves are the locations of minima and maxima, and the corresponding SWR. Table 1 lists minimum SWRs and the frequencies of the minima for the three feed points. For reference, the right-hand columns list the corresponding amateur band and its approximate center frequency.

Table 1
SWR Minima

3.65 m		*6.98 m (1/3-feed)*		*8.65 m*			
Freq (MHz)	*SWR*	*Freq (MHz)*	*SWR*	*Freq (MHz)*	*SWR*	*Band (Meters)*	*Fc*
6.95	1.27	7.05	2.03	7.05	2.49	40	7.15
14.15	1.70	14.05	1.62	14.00	1.61	20	14.20
21.20	1.91	———	———	21.15	1.14	15	21.20
28.25	1.18	28.30	1.30	28.25	1.41	10	28.85

The main point of my previous correspondence was that the 3.65 and 8.65-meter feeds *might* permit four-band operation without a matching network, because these feed points each result in *four* SWR minima between 5 and 30 MHz (see Figure 1). By contrast, the 6.98-meter SWR curve has only *three* minima in that range. The 1/3-feed antenna cannot possibly operate on more than three bands without some sort of matching, which is why the GØFAH design[2] requires two baluns and a specific transmission line length for 15 meters.

How feasible four-band operation is depends only on where the minimum SWRs occur relative to the ham bands, and how low the SWR is across the bands. Table 1 shows that the prototype OCFD has SWR minima very close to the 40, 20, 15 and 10-meter bands. With some on-site tweaking,

[1]Richard A. Formato, K1POO, "Improved Feed for the Off-Center-Fed Dipole," *QST*, May 1996, page 76.

[2]Bill Wright, GØFAH, "Four Bands, Off Center," *QST*, Feb 1996, page 65.

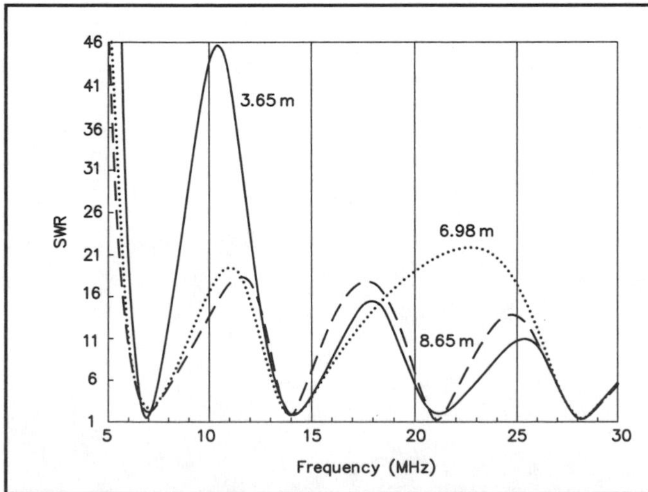

Figure 1—A plot of computer-modeled free-space SWR for the prototype 21.03-meter long, 0.0253-cm-diameter OCFD. SWR for a 200-Ω feed system impedance was calculated at the antenna-input terminals every 50 kHz for source frequencies from 5 to 30 MHz.

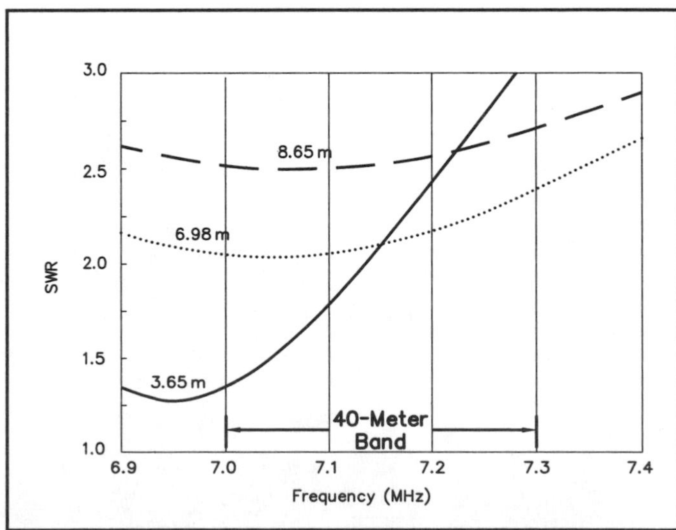

Figure 2—A plot of the prototype OCFD's SWR from 100 kHz below to 100 kHz above the 40-meter band in 5-kHz steps.

it is reasonable to expect that this antenna, fed either 3.65 or 8.65 meters from the end, could provide good SWR across all four amateur bands without a matching network.

The prototype OCFD dimensions were chosen because they correspond to the GØFAH design, thus permitting direct comparison with that antenna, not because they are optimized in any way. In fact, the prototype dimensions are *not* optimum for a four-band OCFD. In free space, the frequencies at which SWR minima occur, and the depth of the minima, are determined by three parameters: radiating-element length and diameter, and feed-point location. Changing any one of these changes both the frequencies of the SWR minima and the minimum SWR values.

Optimizing a four-band, free-space OCFD comes down to determining a set of antenna parameters that produces an acceptably low SWR (typically less than 2) across the 40, 20, 15, and 10-meter bands. Although I have not determined optimum parameter values, the very good predicted performance of the prototype antenna suggests that still better performance is almost certainly achievable. The purpose of my first letter was to encourage experimentation with OCFD designs, which would hopefully advance the state of the art by producing near-optimum designs.

Turning next to the comments at the end of my first correspondence, they are addressed as follows:

SWR Variation Across a Band

In terms of SWR behavior, the OCFD is much different from an ordinary center-fed dipole (CFD). The CFD is intentionally cut (tuned) to place minimum SWR within the band, which is why SWR increases toward the ends of the band (moving away from the minimum). An OCFD may or may not exhibit this behavior, depending on where its SWR minima occur. To illustrate, Figure 2 plots the prototype OCFD's SWR from 100 kHz below to 100 kHz above the 40-meter band in 5-kHz steps. The SWR does *not* increase toward each end of the band. If the SWR minimum is either outside or at one end of the band, as it is in Figure 2, the SWR will increase in one direction (up or down band), but decrease in the other, quite unlike an ordinary CFD. The OCFD's SWR will increase at both band edges *only* when its SWR minimum is inside the band, which generally is not the case.

Azimuth Pattern

The OCFD does indeed have an asymmetrical azimuth pattern, because the radiating element is not symmetrical about the feed point. But feeding the OCFD 8.65 meters from the end provides a *higher* degree of symmetry than the conventional 1/3-feed. Locating the feed 8.65 meters from one end should result in a more symmetrical azimuth pattern than the conventional feed, not less. Of course, feeding the OCFD 3.65 meters from the end increases its asymmetry compared to the 1/3-feed, so that the pattern for this implementation would be expected to be less symmetrical. Even so, pattern asymmetry is not necessarily undesirable. Many operators may want to take advantage of the OCFD's pattern by orienting the antenna to radiate in a preferred direction. This consideration applies to any antenna, even to the single-band CFD, which is an extremely poor radiator in the direction of the antenna axis at low to moderate take-off angles.

Operating the OCFD Out-of-Band

Regardless of where the feed is placed, or how it is implemented, certainly no attempt should be made to operate a 40, 20, 15, 10-meter OCFD on any other band. High SWR conditions may very well result in balun damage.

The data presented here provide additional insight into the advantages of feeding the OCFD at points other than 1/3 of its length from the end. With computer-models for wire antennas widely available, it should be possible to optimize the OCFD in free space and over typical ground so that multiband operation is achievable without a matching network.

Off-Center-Fed Dipole Comments, Part 3

By Roy O. Hill, Jr, W4PID

QST's May 1996 Technical Correspondence column article "Improved Feed for The Off-Center-Fed Dipole" contained a statement that surprised me: "Why the feed point should be located [1/3 of the way from the end] is not exactly clear. The ninth edition of *The ARRL Antenna Book,* for example, observed that there is not much theoretical justification for this choice." It seems to me that there is plenty of theoretical justification. I am not an electrical engineer, and I am going to keep this simple. I won't say anything about computer simulations, ground effects, reactance, unbalanced feed-line currents, or adding 15-meter coverage; and I will use the *lowest* frequencies on the 80, 40, and 20-meter bands with full awareness that the frequencies actually used will be higher than that.

Thirty or 40 years ago, it was more or less generally accepted that the impedance of a half-wave antenna was about 4 kΩ at the ends, about 72 Ω at the center, and that the impedance along the antenna could be accurately represented by a straight line on semi-log graph paper. Using these premises, I plotted the impedance along a half-wave 3.5-MHz antenna on the graph of Figure 3. Then I added 7.0 MHz and 14.0 MHz plots for the same antenna.

All three lines cross at two points at about the 280-Ω impedance mark. These two points are 1/3 of the way (within 0.1 percentage point as I measure it) from the ends of the antenna. The antenna can be fed with 300-Ω line at either of the triple-crossing points and it will work on all three frequencies. You could miss the exact crossing point (or change the frequency) somewhat and still have an impedance between 150 and 600 Ω, which would present an SWR less than 2[4]—an allowable value for most transmitters. All this was pretty widely known in the 1950s. What happened to it? Was all the old low-tech knowledge thrown out when the modern high-tech stuff came along?

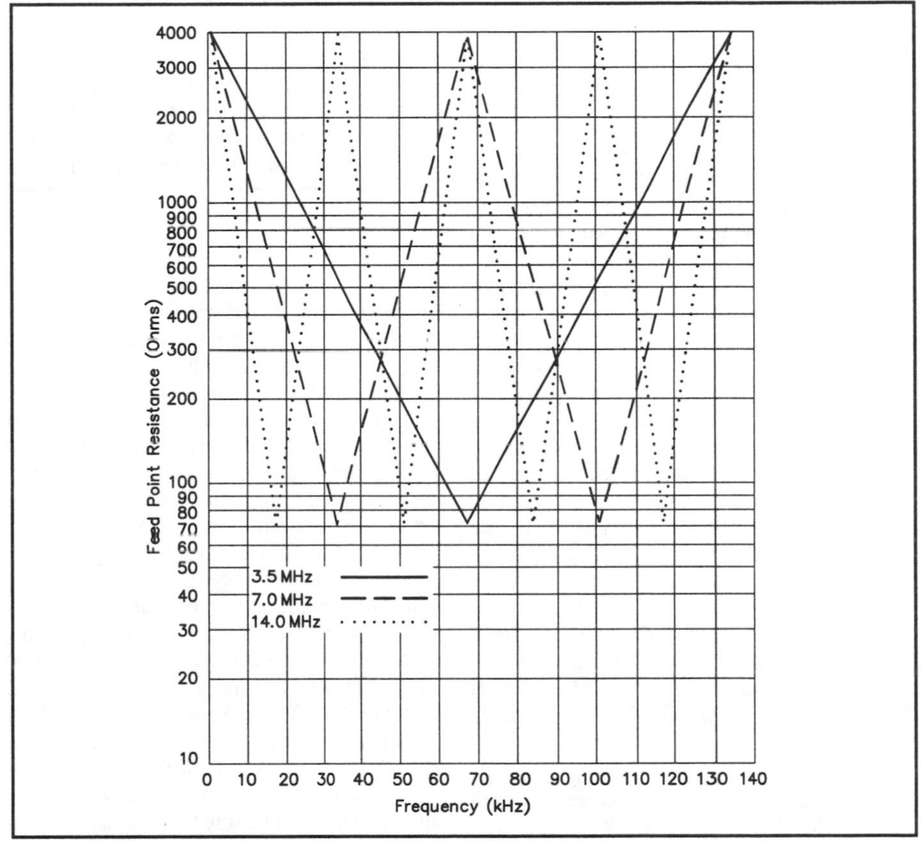

Figure 3—Impedance plot of an OCFD for 3.5, 7 and 14 MHz as a function of distance (in feet) from either end.

Also, the article mentioned that the "1/3-feed" antenna requires no antenna tuner. That is true, of course, for loading the antenna, but back in the days I've been talking about, the ARRL was stressing that you should *not* use a multiband antenna without an antenna tuner because of the danger of harmonic radiation. Do we not worry about harmonic radiation any more?

A little over 30 years ago, using as a guide a graph like the one shown in Figure 3, but using frequencies more like those on which I would be operating, I put up an antenna fed with 300-Ω ribbon 80 feet from one end and 44 feet from the other. I have used this antenna ever since. I do use an antenna tuner with it. The system works fine and causes no RFI or TVI. I used similar antennas for about 12 years before that, at other locations.

I just couldn't pass up that "not much theoretical justification" quoted from an ARRL publication without comment. The 1/3 feed has, not only plenty of theoretical justification, but plenty of practical justification—from use.

[4]This is only accurate if the impedance is purely resistive. With reactive impedances, it is entirely possible to have an SWR higher than 2.—*Zack Lau, KH6CP*

Choke the OCFD

By Dale Gaudier MØAOP/K4DG

After reading with interest Richard (K1POO) Formato's technical correspondence,[3] I would like to share some observations made in the course of modeling, then constructing, an off-center-fed dipole (OCFD) using K1POO's design suggestions.

I first modeled K1POO's designs (overall length of 68 feet 5.5 inches or 21.03 meters) with feed points at 3.65 meters (11 feet 10.5 inches, or 17.4%) and 8.65 meters (28 feet 2 inches, or 41%) from one end using Roy (W7EL) Lewallen's *EZNEC* V1.0. Using K1POO's design criteria (a nominal feed point impedance of 200 Ω) the free-space SWR curves were a good match for those set forth in his correspondence.

Based on the model data, it appeared that the design with a feed point at 3.65 meters (17.4%) from one end gave the best results overall for all four bands (40, 20, 15 and 10 meters).

I then built an antenna based on the 3.65-meter-feed-point model. I constructed the dipole conventionally of #14 AWG stranded copper wire and mounted it approximately 33 feet (10 meters) above ground. A commercial 4:1 balun at the feed point allows matching the 200 Ω antenna feed point to a 50 Ω coax feed line. I used an SWR analyzer attached to the feed line to measure the antenna's resonance and SWR. The results are shown in Table 1.

Table 1
Results without the RF Choke

Band (meters)	*Minimum (MHz)*	*2:1 SWR Range (MHz)*
40	~8.5	(SWR >>3)
20	14.17	13.250 - 15.500
15	21.58	20.420 - 22.350
10	29.69	28.550 - 30.480

The SWR minimum and 2:1 SWR range for the 10 meter band were higher than desirable for an antenna whose purpose is to minimize the need for a tuner. However, the real puzzle was the lack of resonance anywhere near the 40 meter band! The SWR minimum around 8.5 MHz was neither pronounced nor deep, and certainly not what was predicted by computer modeling.

I tried pruning the two dipole legs. This only caused the 8.5-MHz SWR minimum to shift somewhat. Pruning did not produce an acceptable (< 2:1 SWR) minimum anywhere in the 40 meter band; it also decreased the 20, 15 and 10-meter band SWR minima.

It occurred to me that the effects I was observing might be due to unequal currents flowing at the feed point (possibly due to the asymmetrical design of the OCFD). I made a simple RF choke by coiling several turns of the RG-8X feed line close to the feed point and balun. After some initial adjustments of the choke diameter and number of turns, I had my 40-meter resonance! (See Table 2.) The coil is 5½ turns of RG-8X with an inside diameter of 9½ inches (24 cm). Note that the 40-meter band minimum and 2:1 SWR range can be raised by removing part of a turn from the RF choke.

Table 2
Results with the RF Choke

Band (meters)	*Minimum (MHz)*	*SWR Minima*	*2:1 SWR Range (MHz)*
40	7.00	1.1	6.850 - 7.220
20	14.10	1.2	13.520 - 14.420
15	21.22	1.1	20.420 - 22.340
10	28.34	1.4	27.460 - 29.050

As is apparent, the OCFD design of K1POO is surprisingly broadband, and with a little adjustment, will cover virtually all of the 40, 20, 15 and 10-meter bands with an SWR of 2:1 or less. However, an RF choke appears to be necessary for this design to work on its fundamental frequency.

[3]Richard Formato, K1POO, "Off-Center-Fed Dipole Comments, Part 2," Technical Correspondence, *QST*, Oct 1996, pp 72-73.

CHAPTER THREE

MULTIBAND ANTENNAS

By Wesley M. Bell, W7QB From *QST*, August 1963

A Trap Collinear Antenna

Simple 3-Band Radiator with In-Phase Elements

This antenna covers the 15, 20, and 80-meter bands. On the two higher frequency bands the antenna operates with two extended half waves in phase thereby realizing some gain over the dipole operation of a conventional trap antenna.

As your *ARRL Handbook* tells you, broadside gain over a dipole approximately equivalent to doubling transmitting power may be obtained by using a center-fed antenna about $1^1/_4$ wavelengths long (extended double Zepp). Advantage of this is taken in the three-band trap antenna shown in Figure 1. The basic antenna is a dipole for 80 meters. The traps isolate sections of approximately $1^1/_4$ wavelengths for 20 and 15 meters. Since the center of a $1^1/_4$ wavelength wire is not at a current loop, wire is added in the form of a short open-wire feeder to make the total length about $1^1/_2$ wavelengths, thereby bringing a current loop at the point where the system is fed by coax line. A balun is used to couple the unbalanced line to the balanced antenna system.

Trap Construction

The coil and capacitor specifications given under Figure 1 should be adequate for transmitters running at 100 watts input or less. For higher power, the inductance and capacitance values should be the same, but coils should be wound with heavier conductor, and capacitors should be of the transmitting type, such as the Centralab 850SL type. I made my own coils by wrapping a $2^1/_4$-inch form with waxed paper and winding the turns with double strands of No. 18 wire, unwinding one strand and cementing the remaining turns with strips of model-airplane glue. When the glue was dry, the completed coil was slipped off the form. At least one full extra turn should be wound to allow for pruning.

As shown in Figure 2, the capacitor is placed inside the coil, and the terminals of both capacitor and coil soldered to the heads of brass machine screws. These screws serve to hold the assembly central in a plastic waterproof container as shown in Figure 3. The container is the 1-pint size commonly found filled with chip dip, ice cream, potato salad, oysters and whatnot in grocery stores. The screws are fastened in the top cover and bottom of the container with nuts. The container is suspended from an insulator at the appropriate point in the antenna by short lengths of wire dropped from the insulator to the mounting screws where they are secured by a second set of nuts.

After mounting the traps in the containers, they should be resonated to the designated frequencies by carefully pruning the coils while checking with a grid-dip meter. The strain insulator with its wire wraps must be included, as shown in Figure 3, since the insulator capacitance is in parallel with the trap capacitor. (See *ARRL Handbook*.) As resonance is approached, final adjustment can be made by forming what is left of the last turn into a hairpin, and bend-

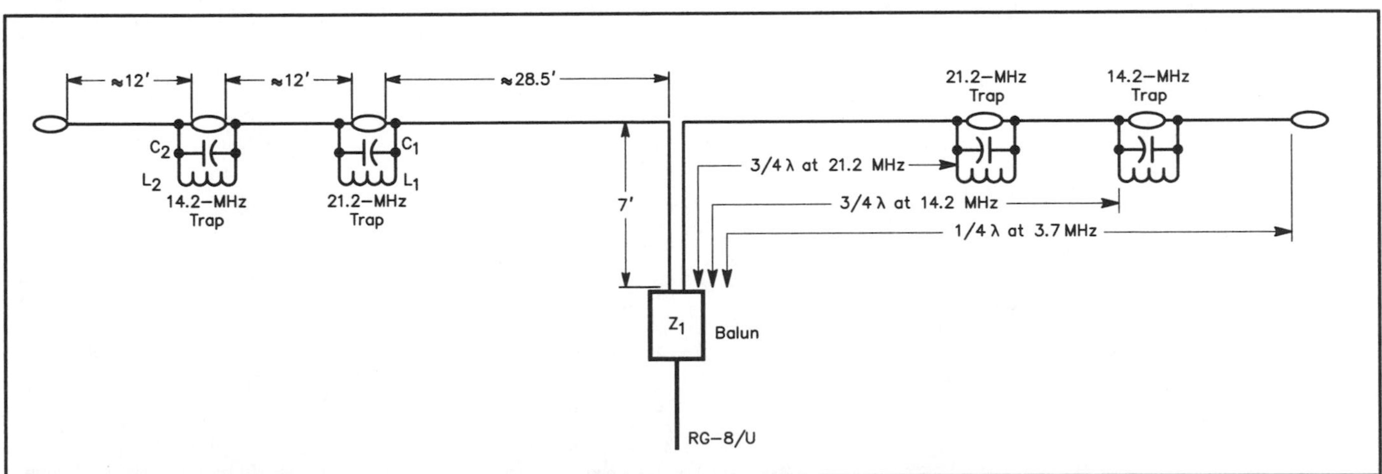

Figure 1—Sketch showing the approximate dimensions of the trap collinear. Wire lengths shown on left side are in terms of feet, while the approximate wavelength equivalents (electrical length) are shown on the right side. Frequencies and approximate wire lengths are for the centers of the three bands as a compromise for full-band coverage. It may be desirable to increase frequency for phone-only operation or lower frequency for c.w.-only operation.

C1, C2—25-pf. 6000-volt disk ceramic. See text.
L_1—Approx. 2 µH-$4^3/_4$ turns No. 18, $2^1/_4$-inches diam., $^3/_8$ inch long, or 5 turns No. 18, 2-inch diam., 16 tpi. See text.
L_2—Approx. 5 µH.—8 turns No. 18, $2^1/_4$-inch dia., $^5/_8$ inch long, or 9 turns No. 18, 2-inch diam. 16 tpi. See text.
Z_1—1 to 1 balun.

Figure 2—Trap components ready to be mounted in weather-proof container.

Figure 3—The completed 21-MHz trap and its supporting insulator ready for checking with a grid-dip oscillator.

ing or twisting the hairpin to alter its inductive relationship to the main part of the coil.

Antenna Adjustment

Antenna resonance can be checked by shorting the ends of the 7-ft. open-wire line and coupling to a grid-dip meter. Initially, the wire lengths should be made a foot or so longer than the lengths shown in Figure 1. Start out with the 21-MHz sections only, anchoring the outer end of each wire section to one side of a 21-MHz trap insulator and connecting it to one side of the trap. An additional insulator should be attached temporarily between the other side of the trap insulator and the antenna-supporting rope. Then gradually shorten the wire until the grid-dip meter shows the desired resonant frequency.

Then add the second sections and traps and adjust similarly for the desired frequency in the 14-MHz band. The end sections of wire are then added and adjusted to show resonance at he desired frequency in the 3.5-MHz band.

In making the antenna adjustments, do not adjust the traps after they have once been set with the g.d.o.; change only the lengths of the wire sections. Devote plenty of time and patience to the adjustments. The job just can't be done correctly in a few minutes.

I use a center supporting pole, and the balun is enclosed in a weather-proof box mounted at the top of the pole.

Results with this antenna have been good. Using it in " inverted-vee" fashion, with one pole at the center and the ends attached to bushes, fences, clotheslines, or whatever else might be handy, I changed directions by simply walking the ends around to different positions. Without half trying, I worked 44 countries with a DX-60. I'm moving to the country soon where I plan to put up several of these antennas. 1 hope to add reflectors too. (Well, a guy can dream, can't he?)

By William J. Lattin, W4JRW From *QST*, November 1972

Antenna Traps of Spiral Delay Line

Most "traps" used in amateur radio multiband antennas are made of lumped inductance and capacitance in parallel. These consist of inductors made of coil stock of No. 12 or No. 14 wire and ceramic capacitors having voltage ratings up to 15,000 volts dc, which are relatively expensive. Vacuum capacitors would be the best, of course, but are also rather expensive.

Another type of trap has a capacitor made of two pieces of aluminum tubing arranged with a small-diameter tube inside a larger tube. Some have polystyrene dielectric, others air dielectric. The *ARRL Handbook* has a very complete description of these types.

Quarter-wave stubs of transmission line can be used for isolating sections of an antenna.[1] Loading coils can be used to modify the harmonic responses of a doublet to the second, third, fifth, seventh, and so on, to obtain a multiband antenna.[2] However, the use of loading coils is quite complicated if more than two bands of frequencies are desired. Traps tuned to the desired resonant frequencies make it much easier to adjust the lengths of the antenna sections, and also to obtain closer spacing between bands than can be obtained with loading coils.

Since quarter-wave sections of transmission line can be used as decoupling stubs for isolation of sections of an antenna, the idea occurred that perhaps quarter-wave sections of spiral delay line (SDL) might be used to make a very simple trap, without lumped capacitance. Spiral delay line is coaxial line with a helical inner conductor.

Construction of SDL Traps

Figure 1 shows the coil which is the helical inner conductor of the spiral delay line, along with the polystyrene tubing and end pieces for the coil, and the aluminum tube. The completed assembly is shown in the title photograph and Figure 2.

For a 28.5-MHz trap, a coil of No. 12 magnet wire was wound on a 1/2-inch rod, 37 turns, close wound. The coil was removed from the 1/2-inch rod and it sprang out to about 3/4-inch OD. The coil was 3 1/4 inches long. Enough wire was used to allow end wires straightened out to be 2 inches long on each end of the coil. Two pieces of 3/4-inch OD polystyrene rod cut 1/2-inch thick were drilled in the center to fit over the No. 12 wire ends and one was slipped over each end of the coil. This coil was placed inside a piece of 1-inch OD × 3/4-inch ID polystyrene tube 4 1/4-inches long, and a piece of 1 1/8-inch OD × 1-inch × 4 1/4-inch long aluminum tube slipped over this. The assembly was held together with No. 6-32 × 5/16-inch screws in holes drilled and tapped just far enough into the polystyrene end pieces to hold the screws in place. The short for this quarter-wave section of spiral delay line was made with a solder lug under one screw with a wire soldered between it and one end of the coil, as shown in the photographs.

The curve in the graph, Figure 3, shows the number of turns of No. 12 wire required for quarter-wave sections of the above construction. A close-wound coil of No. 12 magnet wire has approximately 12 turns per inch. The length of the assembly for a particular frequency can be determined approximately by dividing the number of turns on the coil by 12 to get the length in inches.

The traps were adjusted to frequency through the use of a grid-dip meter (checked on a receiver for accuracy of each frequency). The coil can be changed quite easily to the desired frequency by trimming turns if an extra turn or two is put on for this purpose. The coil can also be wound with spacing between turns and compressed or expanded to get the trap exactly on frequency.

After the assembly is completed and tuned to frequency, the coil can be sealed in the polystyrene tube with polystyrene ce-

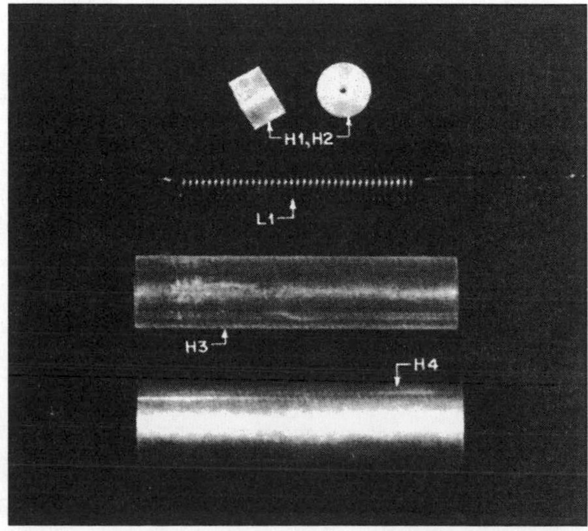

Figure 1 — The parts used in the make-up of a W4JRW SDL trap.
H1, H2 — End pieces of 1/2-inch length of 3/4-inch OD Polystyrene rod with center hole for No. 12 wire.
H3 —1-inch OD × 3/4-inch ID polystyrene tube, length one inch greater than that of coil turns of L1
H4 —1 1/8-inch OD × 1-inch ID aluminum tube, length equal to that of H3.
L1— Close-wound coil of No. 12 enam. or magnet wire, 3/4-inch OD to fit inside H3. See Figure 3 and text for turns information.

[1]Lattin, "Multiband Antennas Using Decoupling Stubs," *QST*, December, 1960, p. 23.
[2]Lattin, "Multiband Antennas Using Loading Coils," *QST*, April, 1961, p. 43.

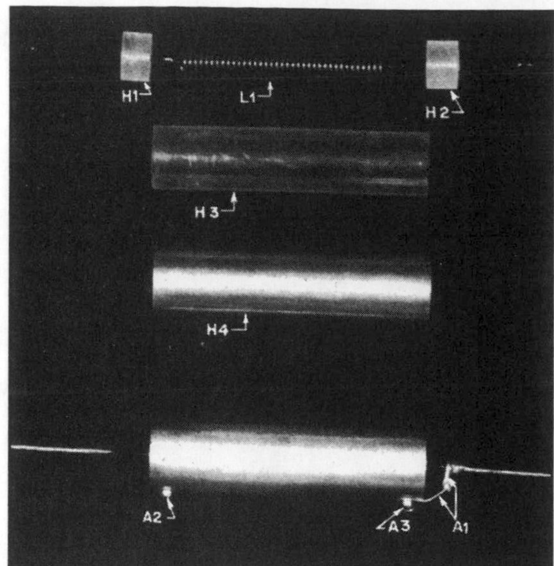

Figure 2—Assembly of the SDL traps. See Figure 1 and text for identification of parts not listed below.
A1—Solder lug and short length of wire assembled and soldered to short one end of H4 to one end of L1.
A2, A3—See text. The screw at A2 must not contact the wire of L1.

ment or coil dope. An inert gas could be sealed inside quite easily, but there seems to be no particular advantage to this.

Trap Ratings and Performance

The thickness of the polystyrene tube used was 1/8 or 0.125 inch. The average voltage rating for polystyrene is given in various handbooks as 500 volts per mil (.001 inch). This would be 62,500 volts for the thickness of 1/8 inch. The maximum power rating of these traps has not been determined. They have been used with a 2-kW PEP ssb transmitter without any failures from either voltage breakdown or heating. Larger wire, polystyrene tubing, and aluminum tubing can be used, of course, but the curve of Figure 3 will be different. Formulas for characteristics of spiral delay lines can be found in radio handbooks and textbooks in which this type of line is described.

The resonant impedance of the traps was measured and found to be approximately 100,000 ohms. For comparison, a trap made of No. 12 wire and a ceramic capacitor gave about the same resonant impedance. Several lumped-constant 15-meter traps borrowed from triband beams were measured and values from 16,000 to 28,000 ohms were found!

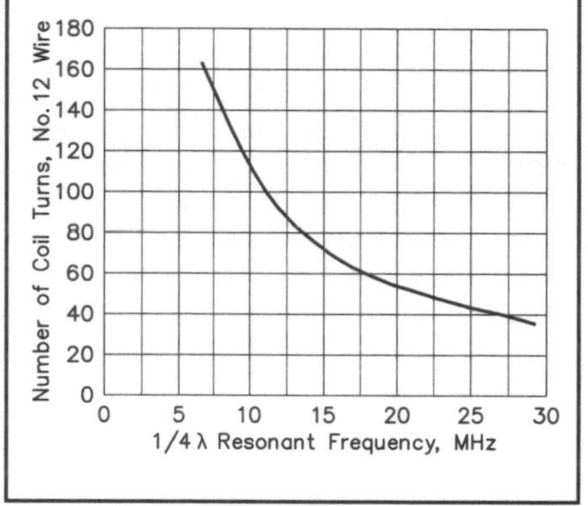

Figure 3—Resonant frequency of spiral-delay-line trap versus number of turns in coil.

The SDL traps can be used in beam antennas, of course, with suitable mechanical modifications to fit the aluminum tube used in the beam. Spiral delay line can be used for other purposes, such as matching transformers, phasing, and any place where coax line is used but short dimensions are needed. The Zo (characteristic impedance) is a function of wire size, diameter, and spacing of the helical coil, dimensions of the insulator and external aluminum tube. Measurement with an rf bridge indicated a Zo of about 250 ohms for the construction used in these traps.

Figures 4 and 5 give the dimensions of two doublet antennas experimented with here. The antenna of Figure 4 has two 7.2-MHz traps and resonates at 3.9 and 7.2 MHz. The antenna of Figure 5 has eight traps, two each for 10 meters, 15 meters, 20 meters, and 40 meters. Resonances are at 3.9, 7.2, 14.3, 21.3, and 28.6 MHz. The flat-top portions were made of No. 12 solid copper wire. The feeder used was RG-8/U. No balun was used in our experiments. It was found that if a trap was not tuned exactly to frequency, it could still be used by changing the wire lengths in the antenna adjacent to the trap to get the desired antenna resonance. An antenna shortened and using traps is sharper in resonance than a full-length doublet. This is generally very well known, but perhaps bears repetition.

SWR curves are also shown in Figures 4 and 5 for these two antennas. During measurements, the antennas were supported in the center about 30 feet high and were 20 feet high at the ends. Measurements were made at the transmitter with 100 feet of RG-8/U coax between the transmitter and the antenna.

It is advisable to support doublet antennas at the center as well as at the ends, with strain relief at the ends – a simple arrangement of a screw eye, plastic rope, and a sash weight or a brick will do.[3] Since RG-8/U coax is fairly heavy, the center support is helpful to reduce the strain on the antenna. The breaking load of No. 12 copper wire is given in handbooks as 197.5 pounds for soft or annealed wire and 261.6 pounds for medium hard-drawn wire. In a high wind any type of support such as trees, towers, push-up masts, and so on, may move a few inches, putting thousands of pounds of tension on a wire stretched between them. Is there any ham who hasn't broken a wire antenna stretched between two trees when no strain reliefs were used?

One spiral-delay-line trap was tested with a hoist and concrete blocks for weight and didn't break at 200 pounds. As the No. 12 wire used in the antenna was softer than

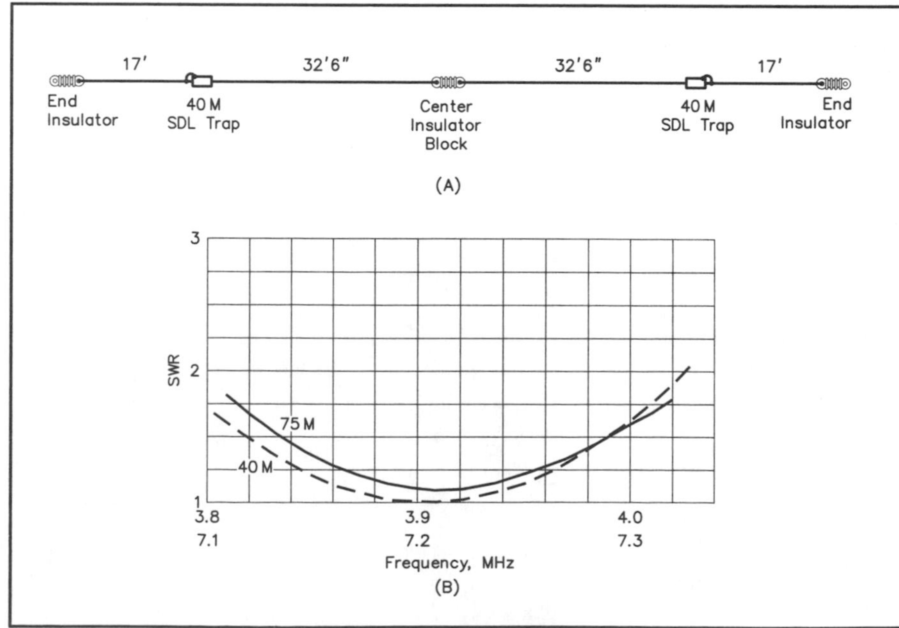

Figure 4—At A, dimensions for a 2-band spiral-delay-line antenna, resonant at 3.9 and 7.2 MHz (not drawn to scale). At B, the measured SWR values with this antenna.

[3] [EDITOR'S NOTE: To reduce wear and eventual breaking of the plastic-rope halyard, a pulley should also be used; large-diameter types sold in hardware stores as clothesline pulleys are economical and quite satisfactory.]

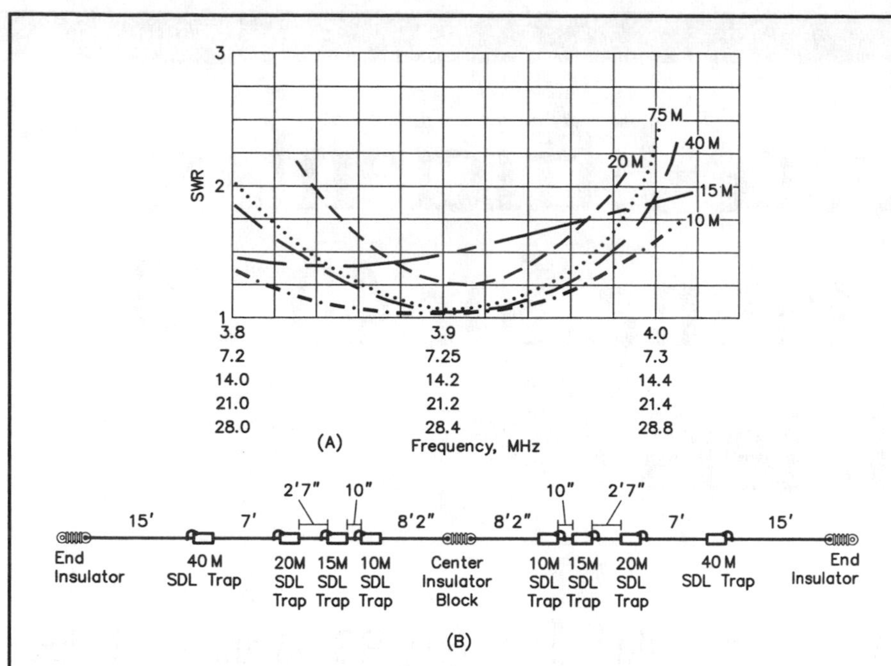

Figure 5—At A, the measured SWR values for a 5-band SDL antenna and at B the dimensions for this antenna.

the No. 12 magnet wires in the SDL traps, it appeared that the antenna wire would probably stretch before the wire in the traps. We have had the antenna up for almost two years, supported by three large trees, with strain reliefs at the ends. The antenna was not damaged at all by an 85-mph wind during a storm which bent the top section of our guyed crank-up tower into an inverted U shape with the triband beam hanging down. We did have one ice storm, but it wasn't severe enough to lift the strain-relief weights at the ends of the antenna.

If one desires, he can make the traps stronger by using two screws 180 degrees apart at each end, or even three screws at 120-degree spacing. Also the ends could be made of copper-weld wire soldered to the inner coil. The materials are not difficult to obtain. Most cities now have plastic supply distributors, and also aluminum tubing suppliers. The magnet wire can be obtained at a motor repair shop or from an electrical supply distributor.

By A. C. Buxton, W8NX From *QST*, July 1992

Build a Space-Efficient Dipole Antenna for 40, 80 and 160 Meters

A new trap design, using only RG-58 and PVC pipe, yields better space efficiency than conventional coaxial traps.

These days more than ever before, many hams who want to work the low bands need an effective antenna that fits on a small lot. I'll show you how to build a shortened dipole for 160, 80 and 40 meters using improved coaxial-cable traps that I call Super Traps. The antenna, which covers the three ham bands below 7.3 MHz, is about the same length as a fullsize 80-meter dipole. If you install the antenna as an inverted V with a 90° included angle, the baseline length is 88 feet. The antenna uses traps that are easily constructed, rugged and weatherproof. They use no exposed capacitors or inductors.

You can feed the antenna directly with balanced 75-Ω line or via a 1:1 balun with either 50- or 75-Ω coaxial cable. Feed-line length is not critical. The antenna resonates at 1.865, 3.825 and 7.225 MHz. I installed such an antenna on my lot as an inverted V, with the apex 38 feet high and the ends at about 15 feet.

As part of this project, I developed a BASIC-language computer program[1] for trap design; a listing is available from the ARRL.[2] You can use this program to design these traps for frequencies of your choice, but you don't need a computer to make the antenna described here.

Figure 1 shows the antenna layout. The antenna is made of #14 stranded wire and two pairs of coaxial traps. Construction is conventional in most respects, except for the high inductance-to-capacitance (L/C) ratio that results from the unique trap construction. Two recent *QST* articles give tips on dipole construction and feeding.[3,4]

The traps use two-layer windings of the core (dielectric and center conductor) of RG-58 coaxial cable. Coaxial cable with flexible, rugged stranded-wire center conductors is preferable to that with a more brittle solid-wire center conductor. Figure 2 shows the traps. The 3.8-MHz trap is shown with the weatherproofing cover of electrical tape removed to show the construction details.

Precautions and Trap Specifications

With this trap-winding configuration, there are two thicknesses of core dielectric material between adjacent turns, which doubles the breakdown voltage of the traps. The transformer action of the two windings gives a second doubling of the trap-voltage rating. Thus, the trap voltage rating is 5.6 kV (four times RG-58's 1.4-kV rating). Conventional coaxial-cable traps made of RG-58 have a rating of 2.8 kV.

The 7-MHz traps have 33 µH of inductance and 15 pF of capacitance, and the 3.8-MHz traps have 74 µH of inductance and 24 pF of capacitance. The trap Qs are over 170 at their design frequencies, as measured on a Boonton Q meter.

These traps are suitable for operation at the 1-kW power level. When making the traps, *do not* use RG-8X or any other foam-dielectric cable. Winding such cables on small-diameter forms causes the center conductor to migrate through the dielectric toward the inside, decreasing the breakdown rating and compromising trap performance. The core diameter also differs from that of RG-58.

Construction

Although these traps are similar in many ways to other coaxial-cable traps, the shield winding of the common coax-cable trap has been replaced by an outer winding that fits snugly into the grooves formed by the inner layer. Capacitance is reduced to 7.1 pF per

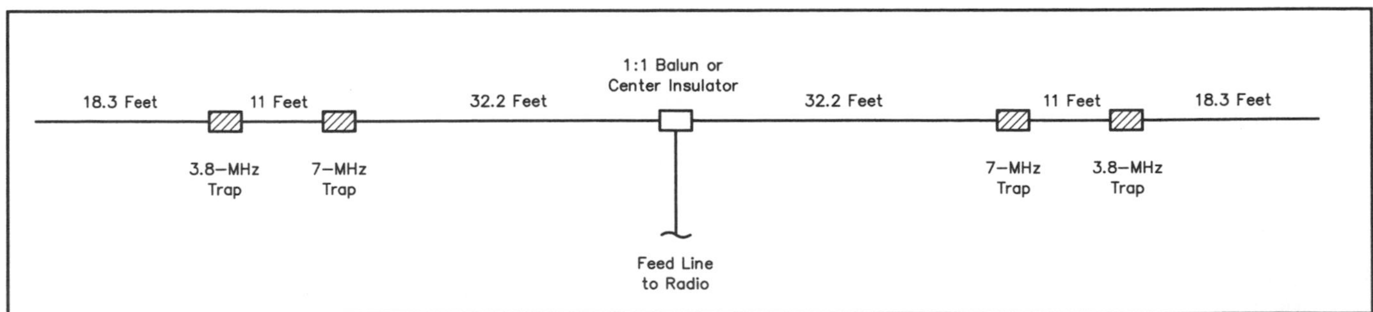

Figure 1—The shortened dipole resonates in the SSB portions of the 40, 80 and 160-meter bands. The antenna is 124 feet long.

Figure 2—The improved coaxial-cable traps use two layered windings to provide an unusually high inductance-to-capacitance ratio, higher Q, and twice the breakdown voltage of single-layer traps. The 3.8-MHz trap is shown without its protective electrical-tape wrap to show the details of trap construction. This construction method makes for simple, lightweight, rugged and weatherproof traps.

foot, compared to 28.5 pF per foot with conventional coax traps made from RG-58. Trap reactance can be up to four times greater than that provided by conventional coax-cable traps.

The coil forms are cut from PVC pipe available at plumbing-supply stores. The 7-MHz trap form is made from 2-inch-ID pipe with an outer diameter of 2.375 inches. The 3.8-MHz trap form is made from 3-inch pipe with an outer diameter of 3.5 inches. The 7-MHz trap uses a 12.3-turn inner winding and an 11.4-turn outer winding. The 3.8-MHz trap uses a 14.3-turn inner winding and a 13.4-turn outer winding. All turns are closewound. The inner-trap frequency is 7.17 MHz and the outer-trap frequency is 3.85 MHz.

If you are unable to get PVC forms of exactly the same diameters as those called for here, compensate for the effect of form-size differences by taking advantage of the fact that the number of turns varies inversely with the form diameter. Thus, if the form diameter you use is, say, 5% larger than mine, reduce the number of turns by 5%. If necessary, add or remove fractions of a turn at the end of the outer winding. If you have a computer, you can use the BASIC programs[5] to calculate the exact number of turns for other form diameters. Stay as close as you can to the prescribed diameters because too much deviation changes the loading effect of the traps. A small change in trap loading may require a change in the lengths of the tip segments beyond the traps.

Use a #30 (0.128-inch diameter) drill for the feed-through holes in the PVC coil forms. The start and end holes of the 7-MHz traps are spaced 1.44 inches center to center, measured parallel to the trap center line. The holes in the 3.8-MHz traps are 1.66 inches apart. Wind the traps with a single length of coax core. The unspliced lengths are 17.55 feet for the 7-MHz traps and 28.45 feet for the 3.8-MHz traps. These lengths include the trap pigtails and a few inches for fine tuning.

Strip the jacket from the coax. This is easily done using a wood vise with wide jaws to hold the cable while cutting the jacket longitudinally with a sharp knife or razor. The coax outer conductor (braid) is best removed by pushing (not pulling) it off.

Use electrical tape to keep the turns of the inner-layer winding closely spaced during the winding process. This counteracts the tendency of the tension in the outer-layer winding to spread the inner–layer turns. Stick the tape strips directly to the coil form before winding and then tightly loop them over and around the inner layer before winding the outer layer. Use six or more tape strips for each trap.

If possible, check the resonant frequencies of your traps with a dip meter. Try to maintain an accuracy of 50 kHz or better.

For low-noise reception, erect the antenna as close to horizontal as possible. If you let the ends of the antenna droop toward the ground, as I have done with my inverted-V installation, you may have to accept a somewhat higher noise level in the interest of structural simplicity and reduced baseline length. Some feel that the inverted V configuration is better for DXing than a horizontal dipole at the same height. For an inverted V with a 90° included angle (legs that slope downward at 45°), you'll need a minimum apex height of about 55 feet and a baseline length of 88 feet. Get the apex as high as you can and keep the ends at least 10 feet above the ground for safety.

Configuration and Performance Trade-Offs

You seldom get something for nothing. This antenna proves no exception to that rule. As with all trap dipoles, this one has less-than-ideal bandwidth due to the loading effect of the traps. This is the price paid for multiband coverage and physical shortening. This antenna covers 65 kHz of 160 meters, 75 kHz of 80 meters and the entire 40-meter band with SWRs under 2:1. The bandwidth limitations on 160 and 80 meters can be largely offset with an antenna tuner.

It is also important to recognize that the traps are used in low-current portions of the antenna, minimizing I^2R trap losses. A relatively high radiation resistance is therefore also retained.

Good luck with your low-band antennas!

Notes

[1] I wrote the program in GWBASIC 3.2. It uses generic BASIC commands and can be easily converted for use with computers other than IBM PCs and compatibles.

[2] For a copy of the BASIC program, send a business-size SASE to the ARRL Technical Department Secretary, 225 Main St, Newington, CT 06111-1494. Request the July 1992 *QST* BUXTON BASIC PROGRAM listing.

[3] J. Healy, "Antenna Here is a Dipole," *QST*, Jun 1991, pp 23-26.

[4] J. Healy, "Feeding Dipole Antennas," *QST*, Jul 1991, pp 22-24.

[5] See note 2.

Trap Construction Information for Al (W8NX) Buxton's July 1992 Dipole

Several readers have written asking for clearer instructions on how to build the antenna traps described by A. C. Buxton, W8NX, in his July 1992 *QST* article, "Build a Space-Efficient Dipole Antenna for 40, 80 and 160 Meters." For those of you who didn't write, but aren't sure how to build the traps, here's what you need to know.

The trap-winding technique is deceptively simple. Each trap is simply comprised of one winding atop another, wound in the same direction. First, drill four holes in the form for wire entry and exit, as described in July *QST*. Starting from inside the form, pass the wire through the hole labeled "1" in Figures 1 and 2. Wind a layer of the inner conductor on the form. Then, run the wire end down through a second hole into the form (just below the one labeled "EXIT" in Figure 2) and, inside the form, bring the wire back to a point adjacent to where the first winding started. Bring the wire up through the hole marked "2" and wind another layer on top of the first, in the same direction as the first winding. Pass the wire down through the hole labeled "EXIT" and you're done. The tape you see in the photos, described in the article, helps hold the bottom winding's turns together and the top winding in place as you're assembling the trap.

The wires passing through holes in the form are strain-relieved, but use the two mounting holes you drilled in the form to support the trap, just to be on the safe side.
–Rus Healy, NJ2L

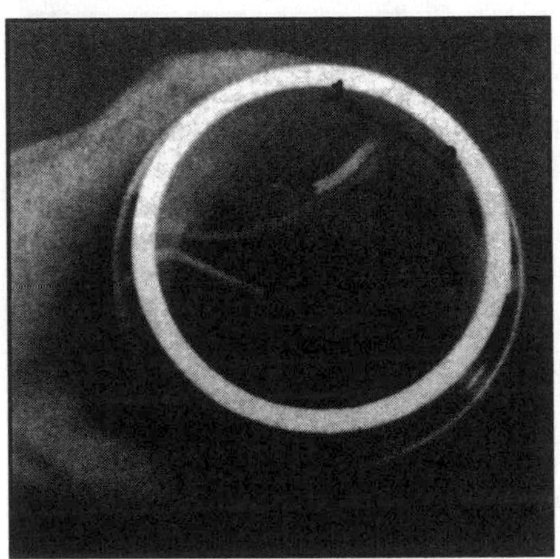

Figure 1—An inside view of a W8NX two-layer trap shows how the windings enter and exit the form. Two holes at each end of the form pass the windings into and out of the form.

Figure 2—An outside view of a partially assembled W8NX trap. The bottom winding starts at hole "1" and reenters the form just below the "EXIT" hole. The wire then comes back through the inside of the form to hole "T," and is used to make a second winding atop the first. It reenters the form at the "EXIT" hole.

By Al Buxton, W8NX From *QST*, August 1994

Two New Multiband Trap Dipoles

W8NX details a new coax trap design used in two multiband antennas; one covering 80, 40, 20, 15 and 10 meters, and the other covering 80, 40, 17 and 12 meters.

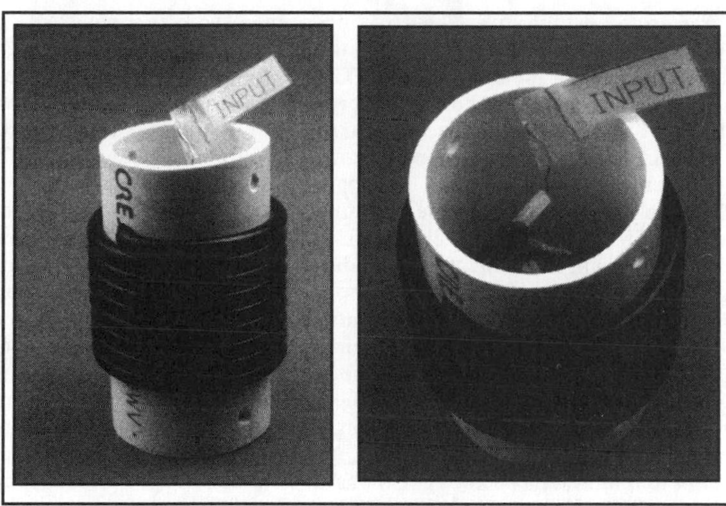

Over the last 60 or 70 years, amateurs have used many kinds of multiband antennas to cover the traditional HF bands. The availability of the 30, 17 and 12-meter bands has expanded our need for multiband antenna coverage. A fortunate few have the space and resources for multiband antennas like rhombics or long Vs, but many hams have employed inverted-L long wires or parallel dipoles. Old-timers will recall the offcenter-fed *Windom* of the '30s–the first version using a single-wire transmission line, and the later design using two-wire feed line. Over the years, random-length dipoles with open-wire feeders and associated tuners have been used successfully as multiband antennas. The G5RV multiband antenna is a specialized example of this approach.[1]

The *log periodic array* represents a kind of brute-force approach to the goal of achieving coverage of multiple HF ham bands. It seems inefficient because of the large gaps between our relatively narrow amateur HF bands.

Over the last few decades, two factors have affected the development of multiband antennas—the popularity of low-impedance (usually 50-Ω) coaxial feed lines, and the appearance of untuned, 50-Ω solid-state amplifiers. The impedance of an antenna is relatively low only at its fundamental frequency and at odd-order harmonics. Although antenna tuners are often necessary to resonate an antenna system, the quest for expanded multiband coverage with simple antennas continues.

At the end of the 1930s, a different technological approach appeared in the form of resonant traps in antennas. The *Mims Signal Squirter* is the grandfather of modern day tribanders.[2] This article discusses in detail an innovative trap design employed in two multiband dipoles.

One W8NX Trap Design—Two Multiband Dipoles

Two different antennas are described here. The first covers 80, 40, 20, 15 and 10 meters, and the second covers 80, 40, 17 and 12 meters. Each uses the same type of W8NX trap—connected for different modes of operation—and a pair of short capacitive stubs to enhance coverage. The new W8NX coaxial-cable traps have two different modes: a high- and a low-impedance mode. The inner-conductor windings and shield windings of the traps are connected in series in the conventional manner for both modes. However, either the low- or high-impedance point can be used as the trap's output terminal. For low-impedance trap operation, only the center conductor turns of the trap windings are used. For high-impedance operation, all turns are used, in the conventional manner for a trap. The short stubs on each antenna are strategically sized and located to permit more flexibility in adjusting the resonant frequencies of the antenna.

Figure 1 shows the configuration of the 80, 40, 20, 15 and 10-meter antenna. The radiating elements are made of #14 stranded copper wire. The element lengths are the wire span lengths in feet. These lengths do not include the lengths of the pigtails at the balun, traps and insulators. The 32.3-foot-long inner 40-meter segments are measured from the eyelet of the input balun to the tension relief hole in the trap coil form. The 4.9-foot segment length is measured from the tension relief hole in

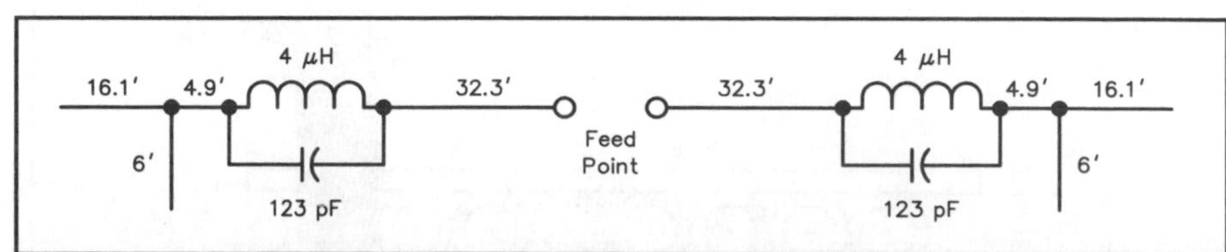

Figure 1—A W8NX multiband dipole for 80, 40, 20, 15 and 10 meters. The values shown (123 pF and 4 μH) for the coaxial-cable traps are for parallel resonance at 7.15 MHz. The low-impedance output of each trap is used for this antenna.

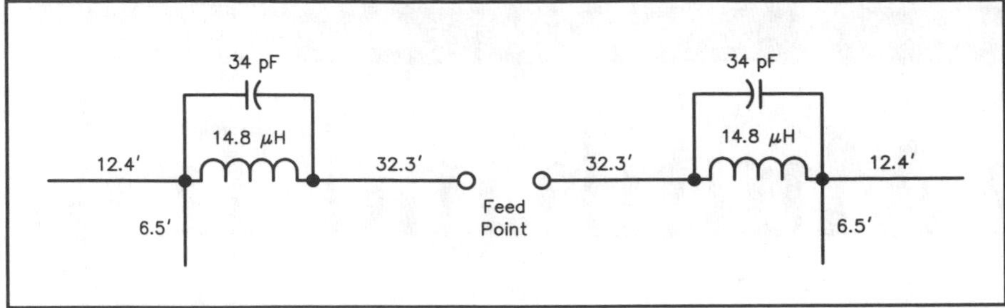

Figure 2—A W8NX multiband dipole for 80, 40, 17 and 12 meters. For this antenna, the high-impedance output is used on each trap. The resonant frequency of the traps is 7.15 MHz.

the trap to the 6-foot stub. The 16.1-foot outer-segment span is measured from the stub to the eyelet of the end insulator. The coaxial-cable traps are wound on PVC pipe coil forms and use the low-impedance output connection. The stubs are 6-foot lengths of 1/8-inch stiffened aluminum or copper rod hanging perpendicular to the radiating elements. The first inch of their length is bent 90° to permit attachment to the radiating elements by large-diameter copper crimp connectors. Ordinary #14 wire may be used for the stubs, but it has a tendency to curl up and may tangle unless weighed down at the end. I recommend that you feed the antenna with 75-Ω coax cable using a good 1:1 balun.

This antenna may be thought of as a modified W3DZZ antenna[3] (shown for many years in various ARRL publications) with the addition of capacitive stubs. The length and location of the stub give the antenna designer two extra degrees of freedom to place the resonant frequencies within the amateur bands. This additional flexibility is particularly helpful to bring the 15 and 10-meter resonant frequencies to more desirable locations in these bands. The actual 10-meter resonant frequency of the W3DZZ antenna is somewhat above 30 MHz, pretty remote from the more desirable low frequency end of 10 meters.

Figure 3—Schematic for the W8NX coaxial-cable trap. RG-59 is wound on a $2^3/_8$-inch OD PVC pipe.

Figure 2 shows the configuration of the 80, 40, 17 and 12-meter antenna. Notice that the capacitive stubs are attached immediately outboard after the traps and are 6.5 feet long, 0.5 foot longer than those used in the other antenna. The traps are the same as those of the other antenna, but are connected for the high-impedance output mode. Since only four bands are covered by this antenna, it is easier to fine tune it to precisely the desired frequency on all bands. The 12.4-foot tips can be pruned to a particular 17-meter frequency with little effect on the 12-meter frequency. The stub lengths can be pruned to a particular 12-meter frequency with little effect on the 17-meter frequency. Both such pruning adjustments slightly alter the 80-meter resonant frequency. However, the bandwidths of the antennas are so broad on 17 and 12 meters that little need for such pruning exists. The 40-meter frequency is nearly independent of adjustments to the capacitive stubs and outer radiating tip elements. Like the first antennas, this dipole is fed with a 75-Ω balun and feed line.

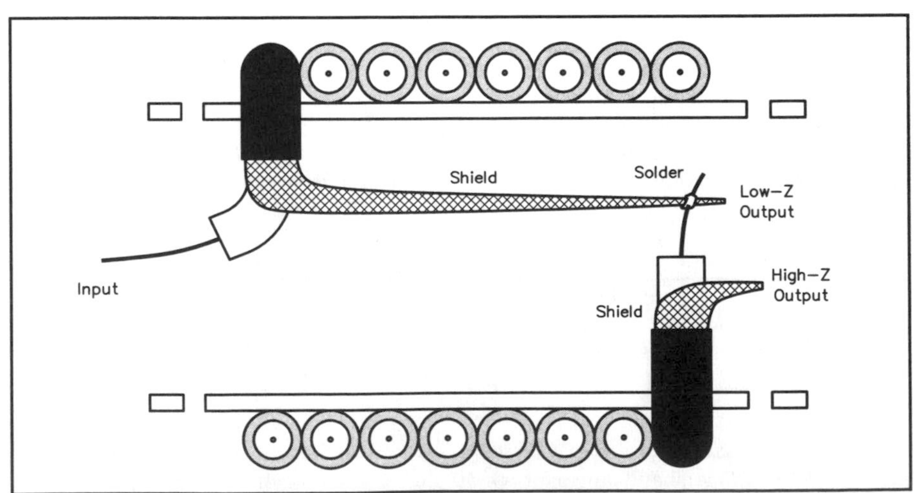

Figure 4—Construction details of the W8NX coaxial-cable trap.

Figure 3 shows the schematic diagram of the traps. It explains the difference between the low and high-impedance modes of the traps. Notice that the high-impedance terminal is the output configuration used in most conventional trap applications. The low-impedance connection is made across only the inner conductor turns, corresponding to one-half of the total turns of the trap. This mode steps the trap's impedance down to approximately one-fourth of that of the high-impedance level. This is what allows a single trap design to be used for two different multiband antennas.

Figure 4 is a drawing of a cross-section of the coax trap shown through the long axis of the trap. Notice that the traps are conventional coaxial-cable traps, except for the added low-impedance output terminal. The traps are $8^3/_4$ close-spaced turns of RG-59 (Belden 8241) on a $2^3/_8$-inch-OD PVC pipe (schedule 40 pipe with a 2-inch ID) coil form. The forms are $4^1/_8$ inches long. Trap resonant frequency is very sensitive to the outer diameter of the coil form, so check it carefully. Unfortunately, not all PVC pipe is made with the same wall thickness. The trap frequencies should be checked with a dip meter and general coverage receiver and adjusted to within 50 kHz of the 7150 kHz resonant frequency before installation. One inch is left over at each end of the coil forms to allow for the coax feed-through holes and holes for tension-relief attachment of the antenna radiating elements to the traps. Be sure to seal the ends of the trap coax cable with RTV sealant to prevent moisture from entering the coaxial cable.

Also, be sure that you connect the 32.3-foot wire element at the start of the inner conductor winding of the trap. This avoids detuning the antenna by the stray capacitance of the coaxial-cable shield. The trap output terminal (which has the shield stray capacitance) should be at the outboard side of the trap. Reversing the input and output terminals of the trap will lower the 40-meter frequency by approximately 50 kHz, but there will be negligible effect on the other bands.

The title-page photos show a coaxial-cable trap. Details of the trap installation are shown in Figure 5. This drawing applies specifically to the 80, 40, 20, 15 and 10-meter antenna, which uses the low-impedance trap connections. Notice the lengths of the trap pigtails: 3 to 4 inches at each terminal of the trap. If you use a different arrangement, you must modify the span lengths accordingly. All connections can be made using crimp connectors rather than by soldering. Access to the trap's interior is attained more easily with a crimping tool than with a soldering iron.

Antenna Patterns

The performance of both antennas has

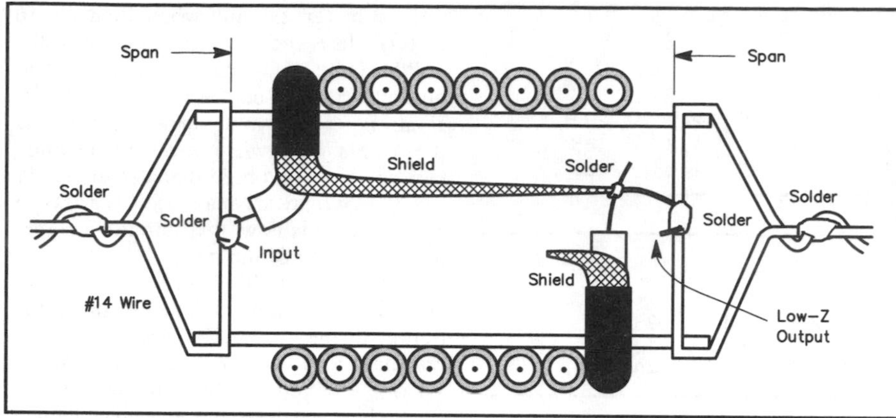

Figure 5—Additional construction details for the W8NX coaxial-cable trap.

been very satisfactory. I am currently using the 80,40, 17 and 12-meter version because it covers 17 and 12 meters. (I have a tribander for 20, 15 and 10 meters.) The radiation pattern on 17 meters is that of 3/2-wave dipole. On 12 meters, the pattern is that of a 5/2-wave dipole. At my location in Akron, Ohio, the antenna runs essentially east and west. It is installed as an inverted V, 40 feet high at the center, with a 120° included angle between the legs. Since the stubs are very short, they radiatc little power and make only minor contributions to the radiation patterns. The pattern has four major lobes on 17 meters, with maxima to the northeast, southeast, southwest, and northwest. These provide low-angle radiation into Europe, Africa, South Pacific, Japan and Alaska. A narrow pair of minor broadside lobes provides north and south coverage into Central America, South America and the polar regions.

There are four major lobes on 12 meters, giving nearly end-fire radiation and good low-angle east and west coverage. There are also three pairs of very narrow, nearly broadside, minor lobes on 12 meters, down about 6 dB from the major end-fire lobes. On 80 and 40 meters, the antenna has the usual figure-8 patterns of a half-wavelength dipole. I have some pattern distortion and input impedance effects from aluminum siding on my house. Nevertheless, DX is easily workable on either of these antennas using a 100-W transceiver, when the high-frequency bands are open.

Both antennas function as electrical half-wave dipoles on 80 and 40 meters with a low SWR. They both function as odd harmonic current-fed dipoles on their other operating frequencies, with higher, but still acceptable, SWR. The presence of the stubs can either raise or lower the input impedance of the antenna from those of the usual third and fifth harmonic dipoles. Again, I recommend that 75-Ω, rather than 50-Ω, feed line be used because of the generally higher input impedances at the harmonic operating frequencies of the antennas.

The SWR curves of both antennas were carefully measured. A 75 to 50-Ω transformer from Palomar Engineers was inserted at the junction of the 75-Ω coax feed line and my 50-Ω SWR bridge. The transformer prevents an impedance discontinuity, with attendant additional undesired line reflections appearing at the 75 to 50-Ω junction. The transformer is required for accurate SWR measurement if a 50-Ω SWR bridge is used with a 75-Ω line. No harm is done to any equipment, however, if the transformer is omitted. Most 50-Ω rigs operate satisfactorily with a 75-Ω line, although this requires different tuning and load settings in the final output stage of the rig or antenna tuner. I use the 75 to 50-Ω transformer only when making SWR measurements and at low power levels. The transformer is rated for 100 W, and when I run my l-kW PEP linear amplifier the transformer is taken out of the line. (I hope my absent-mindedness doesn't catch up with me some day!)

Figure 6 gives the SWR curves of the 80, 40, 20, 15 and 10-meter antenna. Minimum SWR is nearly 1: 1 on 80 meters, 1.5:1 on 40 meters, 1.6:1 on 20 meters, and 1.5:1 on 10 meters. The minimum SWR is slightly below 3:1 on 15 meters. On 15 meters, the stub capacitive reactance combines with the inductive reactance of the outer segment of the antenna to produce a resonant rise that raises the antenna input resistance to about 220 Ω, higher than that of the usual 3/2-wavelength dipole. An antenna tuner may be required on this band to keep a solid-state final output stage happy under these load conditions.

Figure 7 shows the SWR curves of the 80, 40, 17 and 12-meter antenna. Notice the excellent 80-meter performance with a

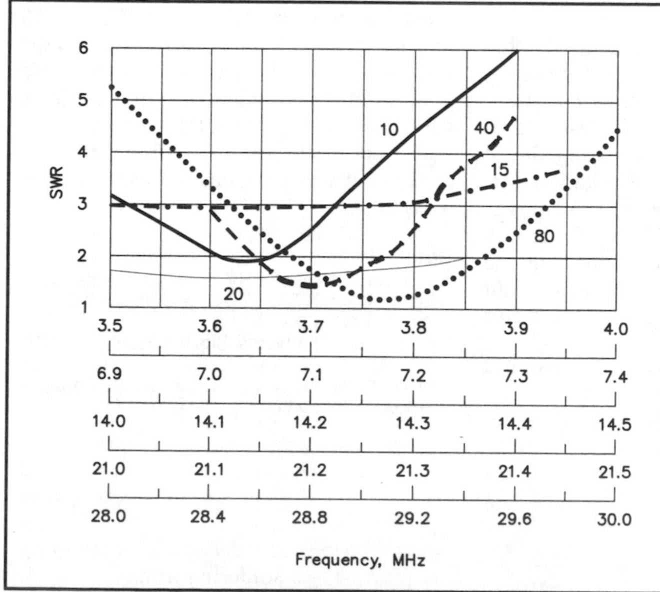

Figure 6—Measured SWR curves for an 80, 40, 20, 15 and 10-meter antenna, installed as an inverted-v with 40-ft apex and 120° included angle between legs.

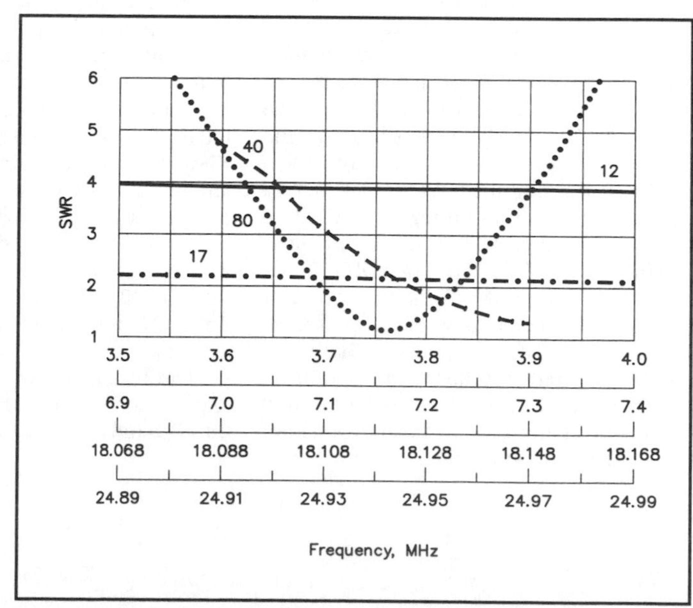

Figure 7—Measured SWR curves for an 80, 40, 17 and 12-meter antenna, installed as an inverted-v with 40-ft apex and 120° included angle between legs.

Table 1
Trap Q

Frequency (MHz)	3.8	7.15	14.18	18.1	21.3	24.9	28.6
High Z out (Ω)	101	124	139	165	73	179	186
Low Z out (Ω)	83	103	125	137	44	149	155

Table 2A
Trap Loss Analysis: 80, 40, 20,15, 10-Meter Antenna

Frequency (MHz)	3.8	7.15	14.18	21.3	28.6
Radiation Efficiency (%)	96.4	70.8	99.4	99.9	100.0
Trap losses (dB)	–0.16	–1.5	–0.02	–0.01	–0.003

Table 2B
Trap Loss Analysis: 80, 40, 17, 12-Meter Antenna

Frequency (MHz)	3.8	7.15	18.1	24.9
Radiation Efficiency (%)	89.5	90.5	99.3	99.8
Trap losses (dB)	–0.5	–0.4	–0.03	–0.006

nearly unity minimum SWR in the middle of the band. The performance approaches that of a full-size 80-meter wire dipole. The short stubs and the very low inductance traps shorten the antenna somewhat on 80 meters. Also, observe the good 17-meter performance, with the SWR being only a little above 2:1 across the band.

But notice the 12-meter SWR curve of this antenna, which shows 4:1 SWR across the band. The antenna input resistance approaches 300 Ω on this band because the capacitive reactance of the stubs combines with the inductive reactance of the outer antenna segments to give resonant rises in impedance. These are reflected back to the input terminals. These stub-induced resonant impedance rises are similar to those on the other antenna on 15 meters, but are even more pronounced.

Too much concern must not be given to SWR on the feed line. Even if the SWR is as high as 9:1, *no destructively high voltages will exist on the transmission line.* Recall that transmission-line voltages increase as the square root of the SWR in the line. Thus, 1 kW of RF power in 75-Ω line corresponds to 274 V line voltage for a 1:1 SWR. Raising the SWR to 9:1 merely triples the maximum voltage that the line must withstand to 822 V. This voltage is well below the 3700-V rating of RG- 11, or the 1700-V rating of RG-59, the two most popular 75-Ω coax lines. Voltage breakdown in the traps is also very unlikely. As will be pointed out later, the operating power levels of these antennas are limited by RF power dissipation in the traps, not trap voltage breakdown or feed-line SWR.

Trap Losses and Power Rating

Table 1 presents the results of trap Q measurements and extrapolation by a two-frequency method to higher frequencies above resonance. I employed an old, but recently calibrated, Boonton Q meter for the measurements. Extrapolation to higher frequency bands assumes that trap resistance losses rise with skin effect according to the square root of frequency, and that trap dielectric losses rise directly with frequency. Systematic measurement errors are not increased by frequency extrapolation. However, random measurement errors increase in magnitude with upward frequency extrapolation. Results are believed to be accurate within 4% on 80 and 40 meters, but only within 10 to 15% at 10 meters. Trap Q is shown at both the high- and low-impedance trap terminals. The Q at the low-impedance output terminals is 15 to 20% lower than the Q at the high-impedance output terminals.

I computer-analyzed trap losses for both antennas in free space. Antenna-input resistances at resonance were first calculated, assuming lossless, infinite-Q traps. They were again calculated using the Q values shown in Table 1. The radiation efficiencies were also converted into equivalent trap losses in decibels. Table 2A summarizes the trap loss analysis for the 80, 40, 20, 15 and 10-meter antenna and Table 2B for the 80, 40, 17 and 12-meter antenna.

The loss analysis shows radiation efficiencies of 90% or more for both antennas on all bands except for the 80, 40, 20, 15 and 10-meter antenna when used on 40 meters. Here, the radiation efficiency falls to 70.8%. A 1-kW power level at 90% radiation efficiency corresponds to 50-W dissipation per trap. In my experience, this is the trap's survival limit for extended key-down operation. SSB power levels of 1 kW PEP would dissipate 25 W or less in each trap. This is well within the dissipation capability of the traps.

When the 80, 40, 20, 15 and 10-meter antenna is operated on 40 meters, the radiation efficiency of 70.8% corresponds to a dissipation of 146 W in each trap when 1 kW is delivered to the antenna. This is sure to burn out the traps—even if sustained for only a short time. Thus, the power should be limited to less than 300 W when this antenna is operated on 40 meters under prolonged key-down conditions. A 50% CW duty cycle would correspond to a 600-W power limit for normal 40-meter CW operation. Likewise, a 50% duty cycle for 40-meter SSB corresponds to a 600-W PEP power limit for the antenna.

I know of no analysis where the burnout wattage rating of traps has been rigorously determined. Operating experience seems to be the best way to determine trap burn-out ratings. In my own experience with these antennas, I've had no traps burn out, even though I operated the 80, 40, 20, 15 and 10-meter antenna on the critical 40-meter band using my AL-80A linear amplifier at the 600-W PEP output level. I have, however, made no continuous, keydown, CW operating tests at full power purposely trying to destroy the traps!

Summary

Some hams may suggest using a different type of coaxial cable for the traps. The dc resistance of 40.7 Ω per 1000 feet of RG-59 coax seems rather high. However, I've found no coax other than RG-59 that has the necessary inductance-to-capacitance ratio to create the trap characteristic reactance required for the 80, 40, 20, 15 and 10-meter antenna. Conventional traps with wide-spaced, open-air inductors and appropriate fixed-value capacitors could be substituted for the coax traps, but the convenience, weatherproof configuration and ease of fabrication of coaxial-cable traps is hard to beat.

Notes

[1] L. Varney, "The G5RV Multiband Antenna... Up-to-Date," *The ARRL Antenna Compendium, Vol. 1*, p 86.

[2] M. Mims, "The Mims Signal Squirter," *QST*, Dec 1939, p 12.

[3] "Five Band Antenna," *The ARRL Antenna Book*, 16th Edition, pp 7-10 to 7-11.

By William C. Gann, W4NML From *QST*, May 1966

A Center-Fed "Zepp" for 80 and 40

Fast QSY for the Phone-C.W. Operator

The center-fed "Zepp" antenna is reviewed by W4NML, showing how complete coverage of a single band is made easy by using old concepts. Although the author shows how to use the Zepp on 80 and 40, only, the system can be used from 80 through 10 meters by employing an all-band transmatch.

Multiband antennas fed with resonant feeders were very popular in the pre-coax cable days. This article is presented to review a good, but seemingly forgotten system. This antenna should be of interest to traffic and contest operators, and to the casual operator who likes to use both the c.w. and phone portions of the 80- and 40-meter bands.

Our section s.s.b. net meets on 3965 kHz and the c.w. net meets on 3575 kHz. Many schemes were tried to make one antenna usable on both ends of the band so that a low s.w.r. could be maintained while securing efficient operation at the different frequencies. None of the antennas tried would permit an excursion of more than 300 kHz without a serious s.w.r. problem between the transmitter and the line. Getting from 80 to 40 meters with such an antenna was even more perplexing. The writer's dilemma was finally solved by the installation of the old reliable center-fed Zepp antenna.

Choosing the Dimensions

In order to use the antenna on 40, 75, and 80 meters, tuned feeders are required[1]. So that the feeders can be matched to the transmitter, a transmatch is used at the "shack" end of the line. Parallel tuning is used to minimize the complexity of the transmatch. This requires that the transmission line presents a high impedance to the transmatch on both bands.

The charts in the handbooks did not give a set of Zepp antenna dimensions that were suitable for the author's installation. Because of the existing tower, which would permit the antenna to be supported at the 50-foot level, and because the ham shack was adjacent to the tower, the prescribed feeder lengths were not practical. A graph was plotted to show the frequency extremes to which the antenna would be tuned, showing the minimum and maximum impedance points across the bands. It was determined that the combined length of one leg of the feed line and one section of the dipole would be 114 feet.[2] A length of 53 feet was used for the feed line and each leg of the driven element was cut to 61 feet.

To broaden the antenna's response, the driven element's effective area was made larger by paralleling two lengths of No. 12 copper wire as shown in Figures 1 and 2. With this arrangement, the Q of the antenna is lower, permitting the operator to QSY approximately 200 kHz without readjusting the transmatch.

Construction Notes

The driven element and the feed line are made from No. 12 copper wire. The 4-inch wide ceramic spreaders used to hold the feeder wires apart are made by the E. F. Johnson Co. Light-weight poly spacers are used to spread the driven-element wires and are sold as TV "clothespins" by the Telco Co. (Figure 2). All of the spreaders are attached to the No. 12 wire by short pieces of No. 18 copper wire. The distance between the spreaders is 4 feet for both the driven element and the feed line.

Sections of 1 × 4-inch lumber are used to hold the feeders away from the steel tower (Figure 3). Each piece is 24 inches long, notched at one end, and is fastened to the tower with U-bolts. Porcelain telephone-type insulators are attached to the feed-line end of each board, offering low-loss anchor

[1]Center-fed Antennas, *A.R.R.L. Antenna Book*, Chapter 6.

[2] This length is generally 145 feet for operation in the 3.5- to 30-MHz range when the antenna is mounted horizontally (no droop), away from steel towers, and with a singlewire driven element.—*Editor*

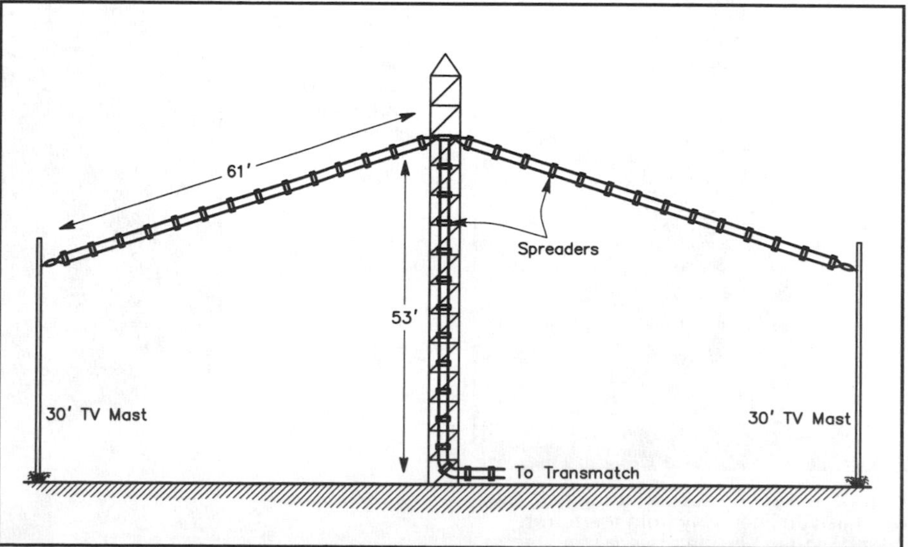

Figure 1—Layout of the 2-band Zepp antenna. Dimensions for each part of the antenna are shown. Feed point is anchored to one of the wooden support arms (See Figure 3).

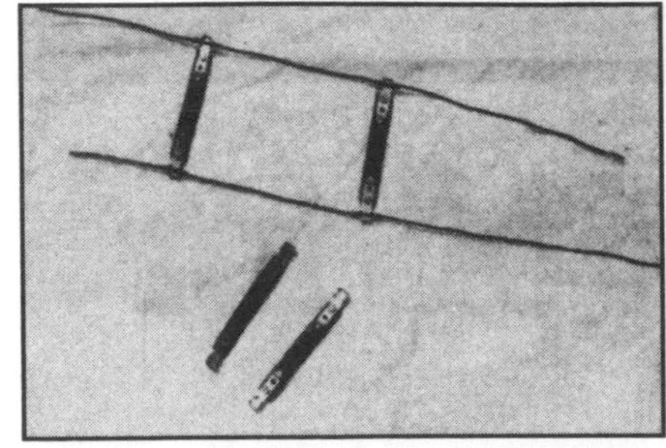

Figure 2—Details showing how the driven element spreaders are attached to the No. 12 wire. No. 18 copper wire is wrapped above and below each spreader to hold it in place.

points for the transmission line. The uppermost support arm, at the 50-foot level, is used as a mount for the center of the driven element.

The far ends of the antenna are supported by 30-foot TV masts. A pulley and halyard arrangement is used for raising and lowering the ends of the antenna. Because the end supports are not as high as the feed point of the antenna, the dipole has a slight droop, but this does not seem to impair the performance.

The transmission line is brought into the operating position by means of feed-through insulators, mounted on a plywood strip which fits under a partially raised window. Insulated No. 12 house wire is used between the feed-through insulators and the transmatch.

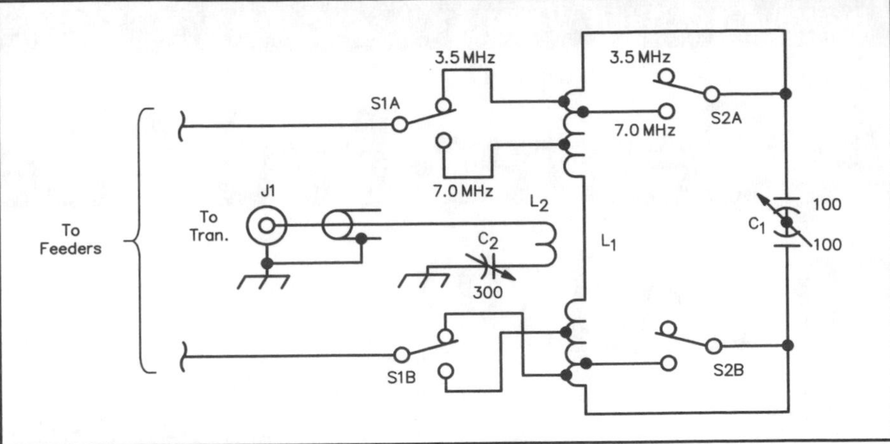

Figure 4—Schematic diagram of the W4NML transmatch.

C_1—100 pf. per section transmitting variable. (Split stator type with 0.175-inch spacing between plates.)
C_2—300-pf. variable capacitor(0.078 spacing or greater).
J_1—SO-239 coax connector.
L_1—56 turns, 3-inch diam., 8 turns per inch coil. S_1 taps are 9 turns from ends of coil for 80 meters, and are 22 turns from ends of coil for 40 meters. S_2 taps are 5 turns from ends of coil for 80 meters and 17 turns from ends of coil for 40 meters. (Air Dux 2408T or Polycoils 1779 usable.)
L_2—8 turns of Air Dux 2408T (center portion of L_1).
S_1, S_2—Ceramic rotary, 2 poles, 2 positions, 2 sections.

Transmatch

Ideas for the author's tuner (Figure 4) were taken from the excellent transmatch article by McCoy.[3] Band changing is made possible by a large ceramic switch (origin unknown) which was obtained at a hamfest. An identical switch is used for selecting the taps for the feed line (Figure 5). Coil L_1 contains 56 turns of No. 14 wire, is 3 inches in diameter, and has 8 turns-per-inch (Air Dux 2408T). A stationary link, L_2, at the center of L_1, contains 8 turns of No. 14 wire and is a part of the Air Dux coil from which L_1 is made. The link is tuned with a 300-pf. variable capacitor. The author did not have a unit of the correct type, so two 150-pf. capacitors were parallel-connected (mounted under the chassis). Capacitor C_1 is a 100-pf-.per-section variable

[3] *QST*, July 1965.

Figure 3—Wooden support arms hold the transmission line away from the tower. Telephone-type insulators are mounted at the end of each board to make the feed line secure.

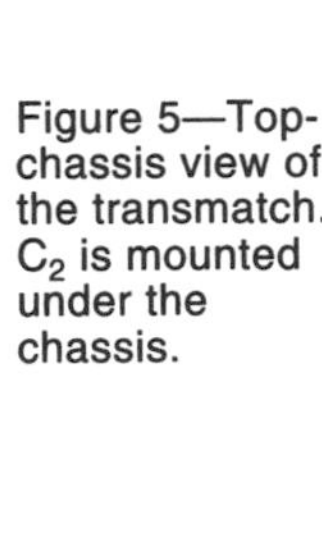

Figure 5—Top-chassis view of the transmatch. C_2 is mounted under the chassis.

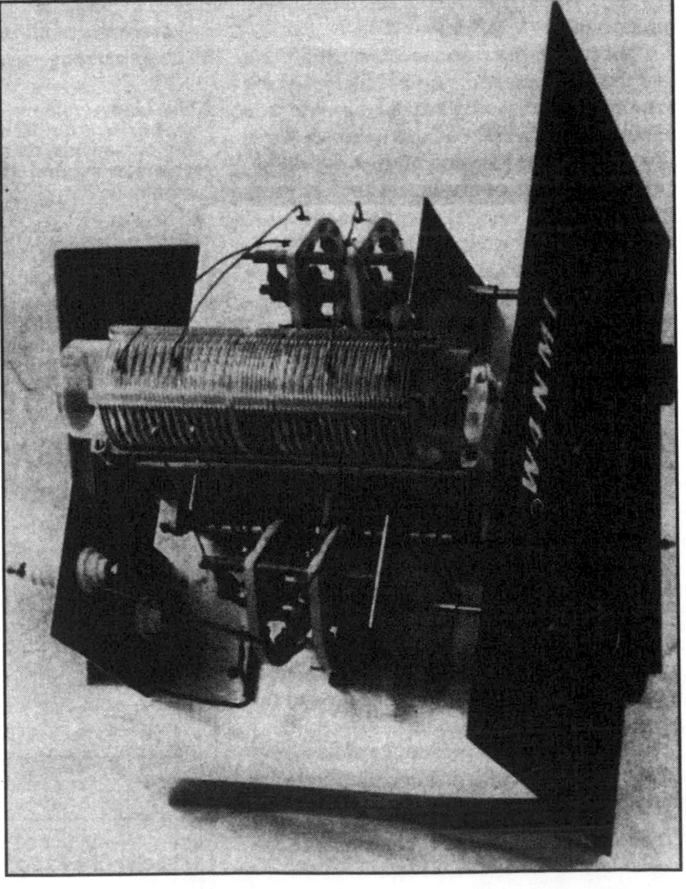

with wide spacing. To give L_1 some rigidity, it is mounted on a plexiglass tube which is supported by the frame of C_1 with standoff insulators.

Results

While using clip leads, the correct tap points for the feeders were found by operating the transmitter through a Collins wattmeter and tuning C_1 and C_2 for zero reflected power. The transmatch permitted the transmitter to "see" 50 ohms in any part of either band. After establishing the correct tap points for the feed line, permanent connections were made between L_1 and the switches.

Next, the tuner was used with the 30L-1 amplifier at an output level of 700 watts. After a 30-minute QSO, no evidence of coil heating could be detected.

When compared to other antenna systems used by the author, the new skywire showed improved performance. It was believed that some sacrifice in efficiency would result from changing to the new antenna. Happily, it was found that we could have our cake and eat it too! Extended use indicated that the performance was, indeed, better than with previous antennas used.

I wish to thank three friends for their help in making this article possible: K4WWN for his tower climbing and photography work, K4ADK for building the transmatch cabinet, and Roy LeCrone for additional darkroom and photographic assistance.

Although this antenna system is an old standard, it may be the answer to your QSY problems. The cost is nominal and the results are most rewarding.

All-Band Antenna

With reference to the article on a Center-fed Zepp for 80 and 40 in May 1966 *QST*:

I set out to accomplish several things with an antenna to be installed on a California lot which runs east-west:

1) One pole.
2) No guys.
3) Good for short skip up and down the West Coast on 80, 75 and 40 meters.
4) Throw lobes across populated DX areas on 20 and 15 meters with a fairly low radiation angle.
5) Allow some onmidirectional DX on 40 meters.
6) Keep away from anything with critical antenna length or critical tuning.
7) Minimum cost.

Figure 1 shows the arrangement I ended up with. Results have been exceptional for a simple system of low height.

I use No. 14 wire for the antenna. The feeders are also No. 14. The center is mounted on an unguyed wooden pole about 34 feet high. Each section of the antenna is 65 feet long and the ends are only 14 feet high. Thirty-foot feeders are used with series tuning on 80 and 40 and parallel tuning on 20, 15 and 10 meters. Loading from 3.5 to 30 MHz is excellent and not at all critical in tuning. The fact that the feeders are less than 1/8 wave on 80 allows reactance to be tuned out in the feeder-tuning arrangement on that band. The antenna is a bit long for the high end of 75 meters, but tuning there is good (this length was picked because of the slightly longer physical length required on the upper bands for end effect).

I have the antenna itself running east-west, giving some directivity north-south for QSOs with short skip up and down the west coast on 80 and 40 meters (it was found in an earlier antenna . . . vertical . . . that a vertical was not satisfactory for high-radiation-angle short-skip operation). On 20 and 15 the lobes tend to cut across major population DX areas. The tilt of the wire, which lowers the vertical radiation angle plus apparently some lobe addition, seems to give better results in the desired DX directions on 20 and 15 meters than 33-feet-high half-wave horizontal antennas oriented in the correct directions. Quite a bit of omnidirectional DX has been worked on 40 meters, undoubtedly because of the antenna tilt. DX operation on 15-meter c.w. has been really exceptional. Quite often I hook a DX station through the pileup when local beam stations miss (power output is about 150 watts). Since 10 has opened up I have used the antenna quite a bit on that band with very good results for both North American and DX contacts.

As a result of playing around on 160 meters with the bottom of the feeders connected together and working the antenna as a "T" against ground, I decided to see what happened when it was operated as a top-loaded vertical on 80 and 40 meters. The feeders were tied together and the antenna worked against a ground consisting of two 8-foot rods in water-soaked earth. On 80 this places the maximum-current point directly at the top of the vertical section (1/8 wave long) and on 40 gives the effect of a 1/4-wave vertical with maximum current at the bottom. This arrangement gave much better results than the Zepp where low-angle radiation was required, and less effective results than the Zepp where medium- and high-angle radiation was required. An exception is directly off the ends on 80 meters, where the vertical and Zepp seem to give the same results. The Zepp arrangement is therefore now used for short and medium skip on 80 and 40 and the vertical arrangement for long skip or DX. In receiving, the signal-to-noise ratio decreases greatly with the vertical arrangement (vs horizontal), thus somewhat offsetting the overall advantage of the vertical for DX operation. If this condition is extreme, I use the Zepp for receiving and the vertical for transmitting, for DX operation. It appears that the vertical transmitter and horizontal receiver is by far the best DX arrangement for metropolitan areas, but most likely the vertical for both would be best for rural areas where the QRN is lower. – *Dave Hardacker, W6PIZ*

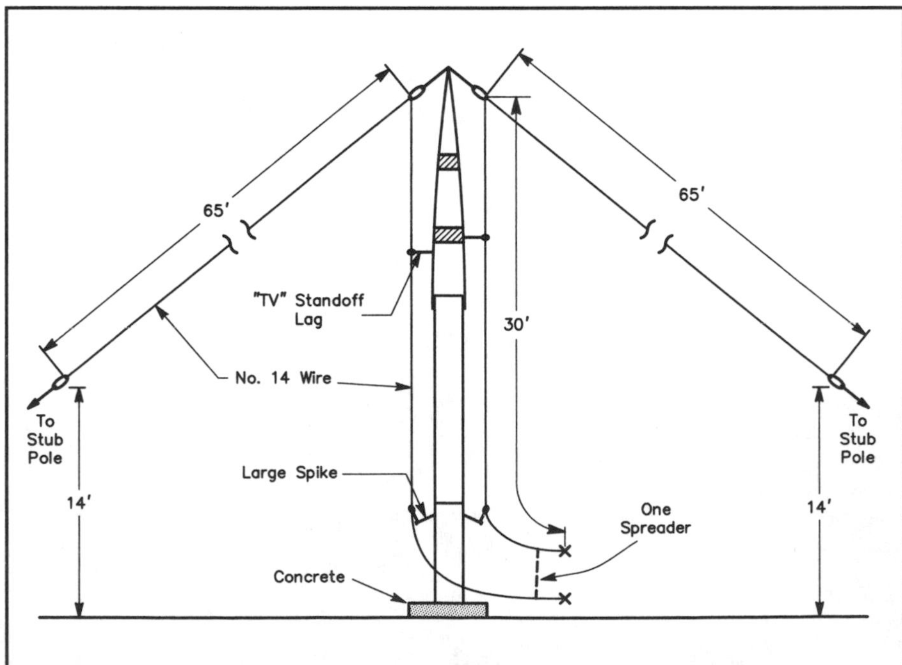

Figure 1–The W6PIZ all-band antenna. Power is applied through a series- or parallel-tuned link-coupled matching circuit at X-X for operation on 3.5 through 30 MHz (see text). For 2 MHz, and also for certain types of work on 3.5 and 7 MHz, points X-X can be connected together and the antenna worked against ground.

By William J. Lattin, W4JRW From *QST*, December 1960

Multiband Antennas Using Decoupling Stubs

Substituting Transmission Line Sections for Lumped-Constant Traps

Since W4JRW obtained a patent on this multi-frequency antenna system nearly ten years ago we can't call it "new," but at least it should be welcome news to those seeking a simple way to get good radiation on several bands, Shorted 1/4-wavelength stubs provide r.f. insulation and also serve as part of the antenna.

Since amateurs usually desire to operate on more than one band, several methods have been devised to use a single antenna on several bands. The earliest arrangements employed various combinations of feeder lengths, antenna lengths, and series or parallel tuning of the coupling circuit. Later on, the use of parallel-tuned "traps" with lumped constants which act as insulators at a particular frequency was invented.[1] A practical arrangement of this system for amateur use was developed[2] and is in rather wide use.

It is well known that the parallel-tuned circuit and quarter-wavelength shorted stub of Figure 1 are very similar electrically. Both configurations show a high impedance across points A and B. However, if a stub is connected to an antenna in this manner it does not act as an insulator but rather as a phase changer. The collinear antenna uses such stubs to operate a series of halfwave sections in phase.

There is a different connection possible for the stub, that is from A to C, which will result in insulator action or decoupling in an antenna.[3] For instance, shorted stubs a quarter wavelength long at 28 MHz can be attached to the ends of a 28 MHz dipole as in Figure 2. The 28 MHz dipole is effectively isolated or decoupled from the balance of the antenna which can be made long enough to resonate at 14, 7 or 3.5 MHz. If another pair of stubs is added for 14 MHz,

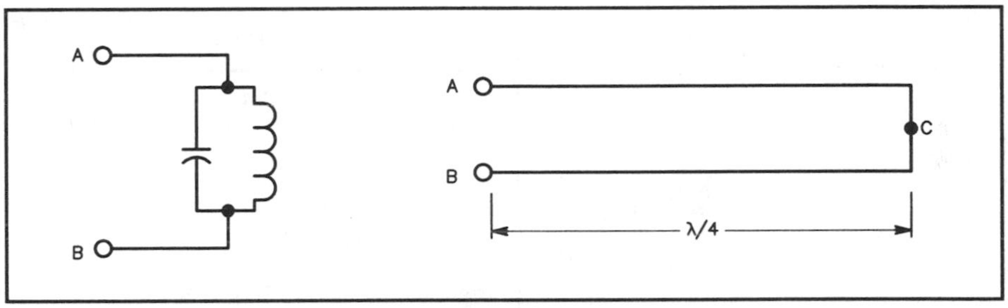

Figure 1—A parallel-tuned circuit has a high impedance at its resonant frequency, and so does a 1/4-wavelength shorted transmission line.

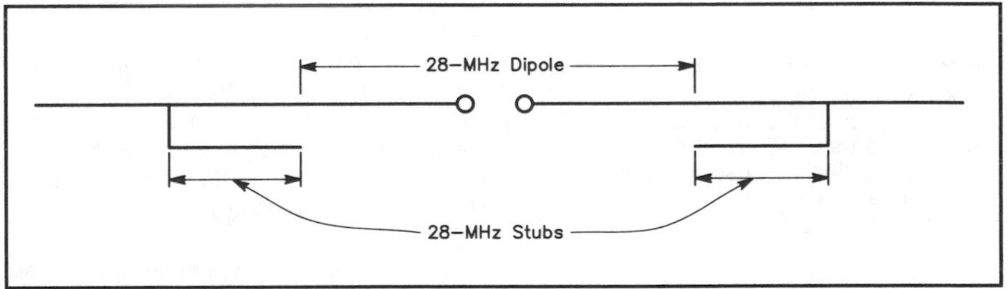

Figure 2—A two-band antenna for 28 MHz and some lower frequency. The center portion is an ordinary 10-meter dipole. The shorted stubs are 1/4 wavelength long at 28 MHz and look like an open circuit at that frequency when connected to the dipole as shown. Extensions on the ends of the stubs can be used to resonate the antenna at any frequency less than half of 28 MHz.

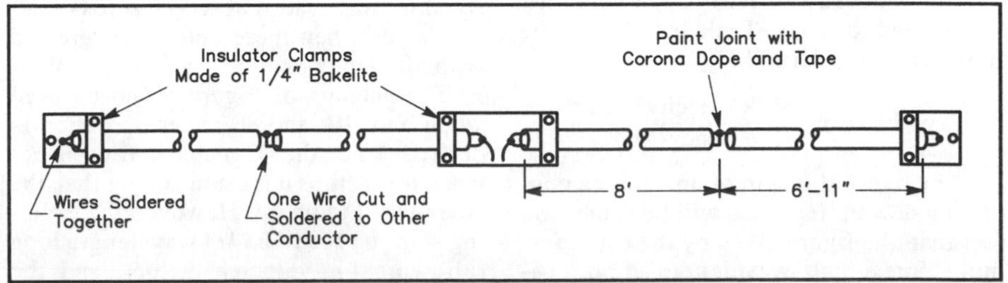

Figure 3—Construction and dimensions of an antenna for 10 and 20 meters using 300-ohm tubular Twin-Lead for both the dipole and stubs. Either a 50- or 75-ohm transmission line can be connected at the center of the dipole.

[1]Morgan, "A Multifrequency Tuned Antenna System," *Electronics*, August, 1940.
[2]Buchanan, "The Multimatch Antenna System," *QST*, March, 1955.
[3] Lattin, Patent No. 2,535,298.

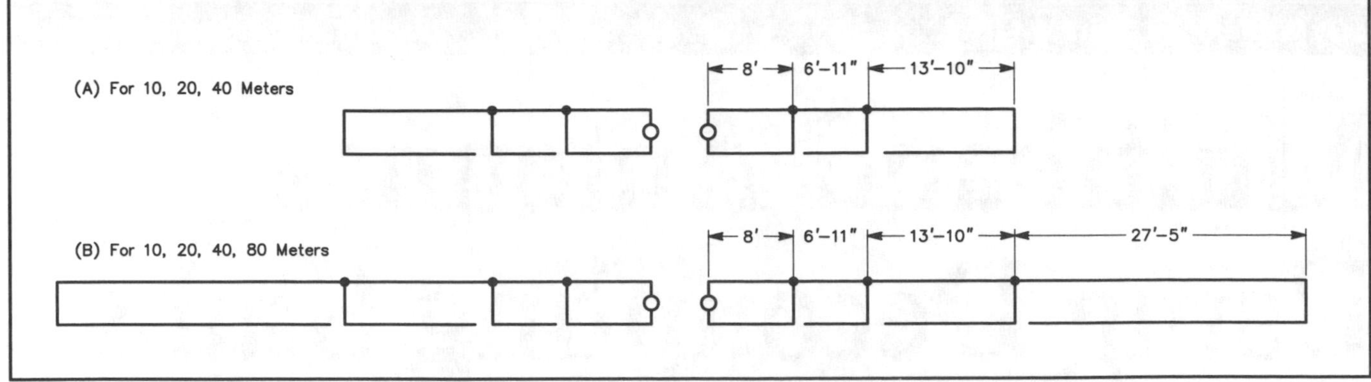

Figure 4—Dimensions of stub-decoupled antennas for 10, 20 and 40 meters and 10, 20, 40 and 80 meters made of tubular Twin-Lead. Either antenna can also be used on 15 meters where the 40-meter section is 3/4 wavelength long.

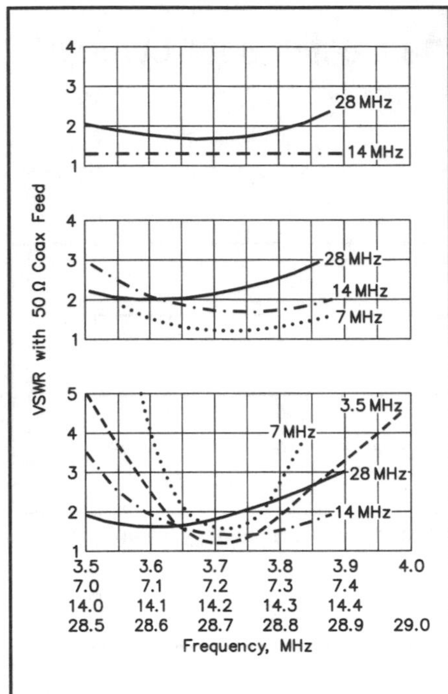

Figure 5—From top to bottom, SWR characteristics of the antennas shown in Figures 3, 4A and 4B. A 50-ohm coaxial transmission line was used, and the measurements were made with a Micromatch.

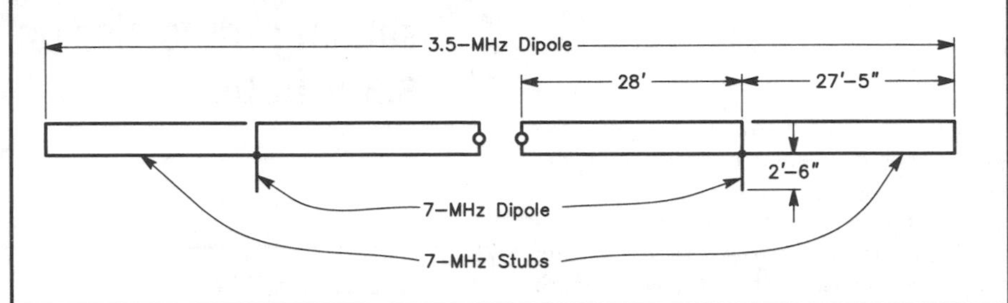

Figure 6—A stub-decoupled antenna for 40 and 80 meters. In this case wires must be hung from the ends of the 40-meter dipole to resonate the antenna in that band.

there will be isolation at both 28 and 14 MHz, and a 10-20-40-meter or 10-20-80-meter antenna can be made.

The stubs can be made of open-wire line, Twin-Lead, or coax. Their lengths can be found from the formula

$$\text{Length (feet)} = \frac{246 \times \text{Velocity Factor}}{\text{Frequency (MHz)}}$$

The over-all length of an antenna containing decoupling stubs will be somewhat less than the figure given by the usual formula for a half-wavelength dipole — *Length* (feet) = 468/*Frequency* (MHz). For instance, an antenna for 10 and 20 meters must be 29 feet, 10 inches long for resonance at the lower frequency, whereas the formula gives a length of 33 feet.

If open line with a velocity factor of nearly unity is used for the stubs, the over-all length of a two-band antenna would be nearly a full free-space wavelength at the higher frequency and the whole antenna would resonate at something *less* than half that frequency. Very fortunately, the velocity factor of 300-ohm tubular Twin-Lead (0.8) gives such lengths for the stubs that, in most cases, adding the stub makes the antenna resonate at just half the original frequency.

Figure 3 shows how tubular Twin-Lead can be used for the antenna itself as well as the stubs and includes dimensions for 10- and 20-meter operation. The foam-filled type of Twin-Lead is recommended to keep out moisture. Lengths for three- and four-band antennas using the same construction are given in Figure 4. Figure 5 indicates the standing-wave ratios observed across various bands when these antennas were fed with 50-ohm coax.

The antenna of Figure 6 can be used when only 40- and 80-meter operation is desired. Since the 40-meter portion is not made up of stubs it must be longer than the antenna of Figure 4A. However, the isolating stubs must still be 1/4 wavelength long (allowing for velocity factor), and the whole antenna would resonate at a frequency below 3.5 MHz if the stubs were simply added to the ends of the 7-MHz dipole. To get around this, the dipole is shortened until the whole antenna tunes to 80 meters. Then resonance at 40 meters is restored by adding extra lengths of wire at the stub junctions. These wires are short and can just hang down from the antenna as shown.

Any of the antennas which will operate on 40 meters can be used on 15 meters as the 40-meter stubs will be approximately 3/4 wavelength long and will provide decoupling . The result is equivalent to operating a 7-MHz dipole at three times its resonant frequency, and we have found the s.w.r. is usually not lower than 3 to 6 when using 40-meter antennas of any type on 15 meters[4].

The power rating of the antenna will depend on the insulation at the stub junctions. These junctions can be painted with corona dope and covered with vinyl tape. It has been our experience over several years that the insulation will not break down with a kilowatt-input transmitter, 100 percent modulated, except when wet or very damp. In this case, the input should be reduced to

[4] Theoretically, a center-fed antenna working on its third harmonic shouldn't be more than about 50 percent higher in resistance than on the fundamental. One would expect an s.w.r. on the order of 2 to 1 rather than such high figures. – Ed.

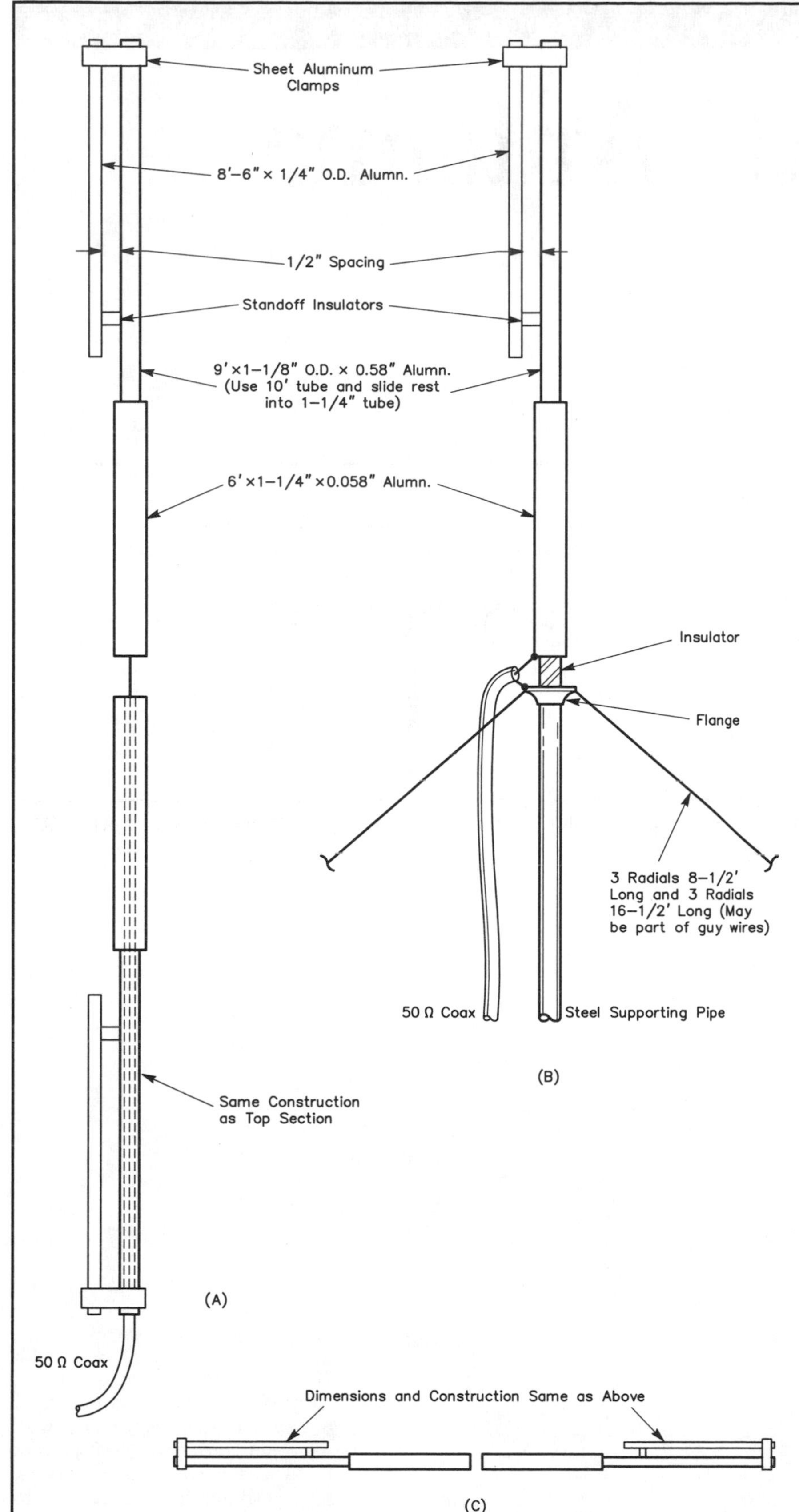

Figure 7—Dimensions and suggested construction for coaxial, ground-plane and tubing dipole antennas for 10 and 20 meters. The arrangement in A might be mounted with standoff insulators attached to the 1¼-inch sections near the center of the antenna. The dipole in C could be closed at the center and fed with a gamma or "T"-matching system. Similarly-constructed parasitic elements could be added to make a multiband beam.

perhaps 500 watts unless special precautions have been taken to seal up the junctions at the open ends of the stub. Of course, on the lowest band for which the antenna is designed the stubs do not have voltage across them and will not be subject to breakdown or flashover. The high voltage across the open end of a stub occurs only at the resonant frequency of that stub.

Figure 7 shows the construction of several 10 and 20-meter antennas which have been built and the dimensions required for resonance in these bands. The spacing between the rods forming the shorted stubs is not critical—the same lengths were obtained with 1-inch instead of 1/2-inch spacing. Insulators should be made of low-loss material. Reflectors and directors for a multiband beam could be made up the same way.

"All-Band" Antenna

Figure 1 is a sketch of an "all-band" antenna system that I have been using with success for some time. The idea is not a new one, having appeared in *QST* at least 10 years ago. However, I feel that there are many newcomers since that time who would be interested in a simple system that can be fed with a single 70-ohm transmission line.

The arrangement consists of dipoles, cut for each band and connected in parallel at the center. Although I have not checked standing-wave ratios, the results seem to indicate that it gets out as well as a bunch of individually-fed doublets. If you haven't tried it, you're in for some surprises.– *R. L. Cope, W8MOK*

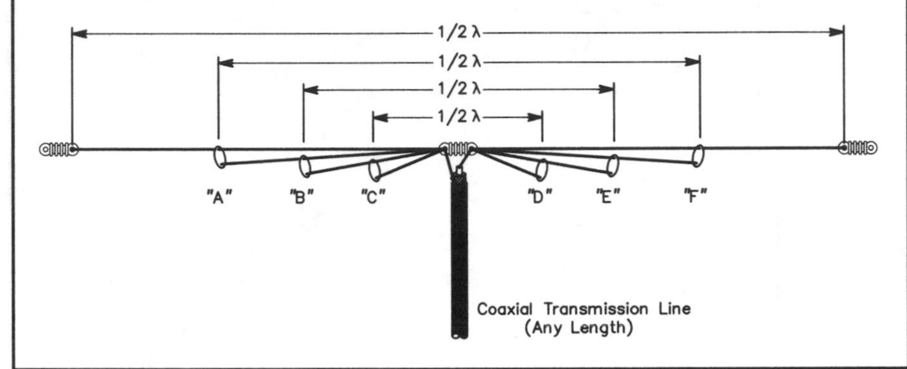

Figure 1–Sketch of W8MOK's "all-band" antenna. Egg insulators are used at points "A" through "F."

Three-Band Matching System for a Forty-Meter Doublet

A common method for energizing a half-wave antenna is to feed it at the center with parallel-conductor TV lead-in, or Twin-Lead as it is usually called, and to use an open stub for matching the 50- to 70-Ω antenna resistance to the 300-Ω impedance of the line. However, this technique, as described in the latest edition of *The ARRL Antenna Book*, generally gives proper matching on only one band.

After a number of trial-and-error calculations on a Smith Chart, along with lots of cut-and-try experimenting, I devised a three-stub matching scheme so that I could operate my 40-meter doublet on 40, 20 and 15 meters. Figure 1 illustrates this method and gives the lengths of the stubs and their positions along the feed line. The dimensions shown are for standard Twin-Lead, with a velocity factor of 0.82. All of the stubs are open at the ends and are made from the same type of line as the feed line. Note that the two lower stubs are connected at the same point on the feed fine.

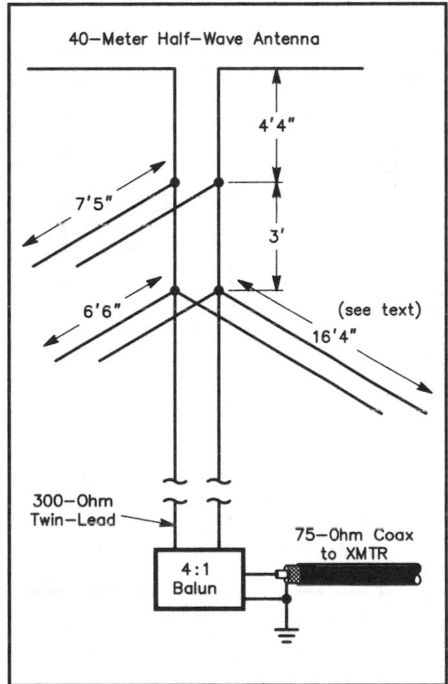

The length of the longest stub is fairly critical. It should first be cut to 17 feet and then trimmed no more than two inches at a time, until the SWR is minimum in the center or the desired portion of the 20-meter band. The feed fine can then be matched with a 4:1 broad-band balun to a 75-Ω coaxial cable from the transmitter. With my antenna, the described matching system gives an SWR of less than 2.5 to 1 over all of the three bands, with minimum values of 1.3 to 1 on 40 and 15 meters, and 1.7 to 1 on 20 meters. As with any multiband antenna, one must guard against harmonic radiation. – *Frank Stuart, K7UUC*

A "Z" Antenna For the 10-160 Meter Bands

One of my interests in ham radio is designing and constructing antennas for both general amateur use and for the Army MARS system. For the amateur who has limited space, I have designed a "Z" antenna that covers the bands from 10 through 160 meters. It is easily constructed from wire. Spreaders for the transmission line are fabricated from Lucite strips or Plexiglas rods. (Refer to the yellow pages of telephone directories for the names of dealers who handle Plexiglas or the equivalent.) For two no. 14 wires, a 2-inch (51-mm) spacing is adequate.

Although a height of 100 feet (30 meters) is indeed desirable for this antenna, hams who settle for elevations between 30 and 50 feet (9 and 15 meters) will still obtain good results. The angles α between the wire segments will depend on individual situations such as the placement of trees or other supports. Generally, the wider the angle, the better the performance.

W1NH, the New Hampshire SCM, who is really into antennas, says my design is "FB." My evaluation of the antenna is that the"aerial" is great.–*John N. MacInnes III, WB1FPD*

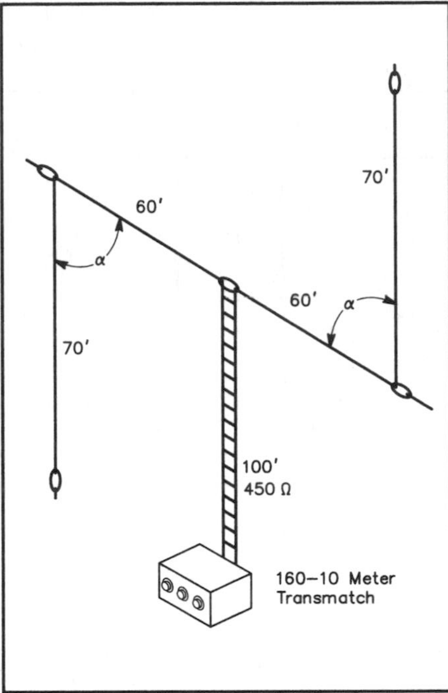

A 10- to 160-meter horizontal Z antenna. Inexpensive no. 14 copper-clad (electric fence) wire may be used. For complete information on open-wire transmission lines, see *The ARRL Antenna Book*, any recent edition.

By Taft Nicholson, W5ANB/AAR6AG From *QST*, November 1981

Compact Multiband Antenna Without Traps

Looking for an inexpensive, easy-to-build, portable antenna to use with your tube-type transmitter? This first cousin of the G5RV may be the aerial of your dreams.

Do you need an antenna for portable operation? I do! My preference is for making things as simple as possible. I don't like traps and matching units. The antenna shown in Figure 1 has no traps, needs no matching unit for the 10-, 20- and 40-meter bands (when used with vacuum-tube PAs), is lightweight and installs easily. What more could you ask for?

Construction

Construction is simple. Cut two lengths of stranded copper wire (such as Radio Shack 278-1292) to 44 feet, 2 inches each (m = ft × 0.3048). Attach 36 feet, 8 inches of 300-ohm twin lead as shown in Figure 1. Coaxial cable attaches to the other end of the twin lead at the points marked A and B in the diagram. I wound 7 feet, 2 inches of RG-58/U coaxial cable into an rf choke to minimize problems with rf flowing on the outside of the coaxial cable. This length of cable in the choke evolved from an attempt to match the antenna to the transmitter for operation on 15 meters.

Alternatively, you could use open-wire feeders in place of the twin lead. If you choose that method, you will need to make the section 42 feet, 6 inches long. Or you could attach the twin lead or openwire conductors to a matching unit. When using a matching unit, there is no particular merit in the lengths given in the diagram.

SWR

The SWR of the antenna is less than 3:1 on 10, 20 and 40 meters. I have no difficulty loading transmitters with tube-type PAs (e.g., Galaxy V, Swan 350 or Drake T4X). This antenna will work with some transmitters on 15 meters, but tuning is quite critical.

However, 80-meter operation is a problem. I have not been able to obtain full output power with these transmitters on frequencies below 3.750 MHz. The SWR measures between 5:1 and 8:1 for the lower part of the band. I constructed a loading coil to go between the choke and the twin lead (Figure 2). The coil consists of 44 inches of twin lead wound on a 7/8-inch diameter form (my left thumb). Once the coil was wound, I removed it from my thumb and used electrical tape to secure it. For 80-meter operation, attach ends C and D to the choke and ends E and F to the twin lead. Remove the coil for operation on the other bands. Banana plugs and sockets can be used to facilitate the insertion and removal of the coil.

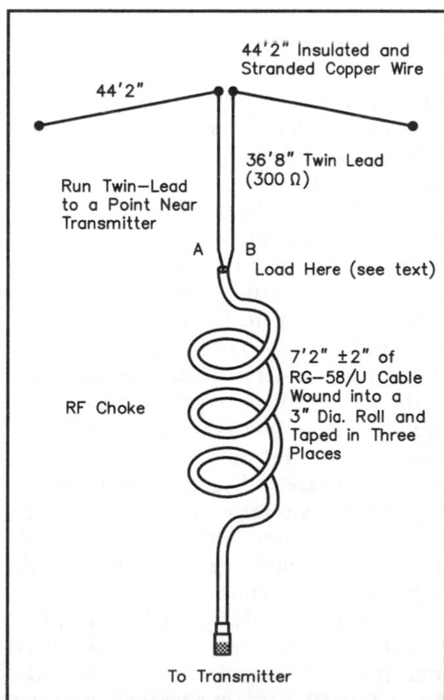

Figure 1—Diagram of the compact multiband antenna. For 80-meter operation the loading coil is inserted at points A and B. Banana plugs and jacks may be added here to facilitate insertion and removal of the loading coil.

Installation

The antenna should be as high as practical. I've had satisfactory results with the center of the antenna only 25 feet above ground, with the ends tied to fences or other convenient supports. Telescoping TV mast sections make a good support if nothing else is available. The legs of the antenna serve as two of the guy wires. One or two additional guy supports should be added (nonconducting material such as nylon rope is best).

This compact multiband antenna works satisfactorily on all bands from 20 through 80, and 10, meters. It has no traps and requires no matching unit when used with tube-type equipment. I have used it for portable operation in and out of the country. It is easy to pack, carry and erect. Perhaps you might want to try one. I think you'll like it! A brief discussion of the theory of operation follows in the appendix.

[Editor's Note: A description of the G5RV appears in the RSGB *Radio Communications Handbook.* Gray described it in the June issue of *Ham Radio Horizons.* A similar design was depicted in the Collins Radio manuals of the 1930s.]

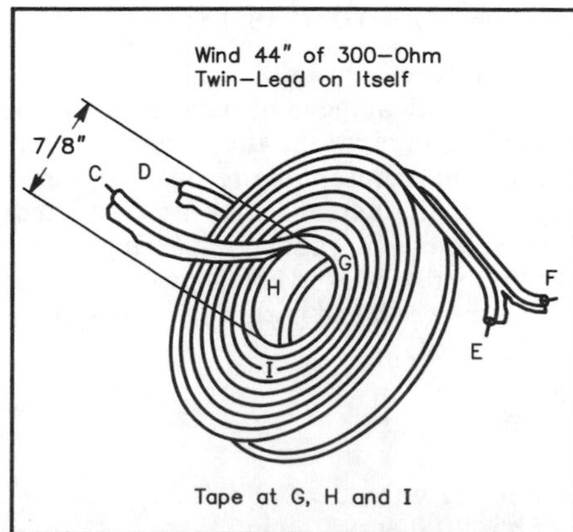

Figure 2—Loading coil for 3500-3750 kHz operation. Points E and F on the coil are connected to the coaxial cable. Points C and D are connected to the twin lead.

Appendix

This multiband antenna evolves from two connected transmission lines with critical length and ratios of surge impedances. The system is self-resonant at a fundamental frequency and at most of the even harmonics and several of the odd harmonics. The first five are 2nd, 4th, 5th, 7th and 8th.

Consider the transmission lines in Figure 3.

The two lines are of equal length, "ℓ" and different surge impedances, Z_o and Z_s.

Looking into the lines:

$$X_o = \frac{Z_o}{\tan(2\pi\ell/\lambda)} \text{ and } X_s = Z_s \tan(2\pi\ell/\lambda)$$

where

λ = wavelength

X_o = reactance looking into open line

X_s = reactance looking into shorted line

From the theory of resonant circuits, we know if we connect the lines the system will be resonant at all frequencies where $\overline{X}_o = \overline{X}_s$ provided the two reactances are of equal value and opposite signs. The open and shorted line provide this condition except at some harmonics.

Joining the lines as depicted in Figure 4 we find that

$$\frac{Z_o}{\tan(2\pi\ell/\lambda)} = Z_s \tan(2\pi\ell/\lambda)$$

$$\frac{Z_o}{Z_s} = \tan^2(2\pi\ell/\lambda) = \tan^2(360°\ell/\lambda)$$

If the angle $2\pi\ell/\lambda$ is made 60°, then the amplitude of the tangent at 120° or second harmonic will be the same. This will be true for 240° (4th harmonic) and 300° (5th harmonic). Similarly, these harmonic responses will continue at discrete angles above 360°, e.g., 7th, 8th, 10th, 11th, 13th, 14th and so on. The signs of the tangents wash out when squared.

The angle $2\pi\ell/\lambda$ becomes 60° by making $\ell = \lambda/6$ (1/6 of a wavelength) at the fundamental frequency.

$$\frac{Z_o}{Z_s} = \tan^2(360° \times 1/6)$$

$$= \tan^2 60° = (1.73)^2 = 3$$

Therefore $Z_o = 3Z_s$. This equation makes practical the multiband antenna because Z_o can represent the antenna proper and Z_s can represent the resonant feeder.

Z_o for the antenna may be computed from formulas in radio engineering handbooks or textbooks. For a piece of wire above the earth and parallel to it as shown in Figure 5C.

$$\frac{Z_o}{2} = 138 \log \frac{4h}{d}$$

For no. 12 wire, d = 0.08081 in.

Let h = 20 feet or 240 in.

$$\frac{Z_o}{2} = 138 \times \log\left(\frac{960}{0.08081}\right)$$

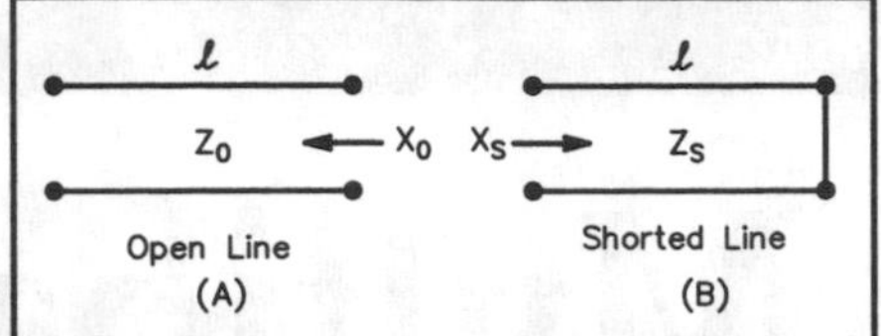

Figure 3—Two identical lengths of transmission line. Looking into the lines, the opposite ends are open and shorted at A and B, respectively.

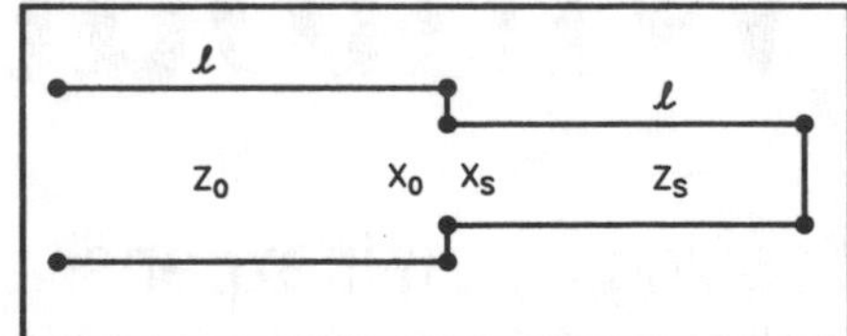

Figure 4—The two transmission lines from Figure 3 are joined to form one line.

138 log (11879) = 562 ohms

This value is not critical. One can use 300-ohm twin lead or 400-ohm open line with

If the wire size had been no. 18,

$$\frac{Z_o}{2} = 605 \text{ ohms}$$

If we use no. 12 wire, then we would have 562 ohms on one side and + 562 ohms on the other side.

$$Z_o = \frac{Z}{2} + \frac{Z}{2} = 1124 \text{ ohms}$$

From the formula $\frac{Z_o}{Z_s} = 3$

$$Z_s = \frac{Z_o}{3} = \frac{1124}{3} = 374.6 \text{ inches}$$

good results. The harmonics will be displaced somewhat, but with variable tuning of the transmitter the system can be brought on frequency.

The above indicates that $Z_o/2$ varies with antenna height, wire size and configuration. The function is logarithmic and a lot can be done to the antenna before Z_o changes very much. The inverted V works well; just use the $Z_o/2$ formula for a horizontal wire and let h be the average height of the inverted V. A formula for surge impedance can be worked out for most any configuration, including a vertical. If the reader is interested in feeding a vertical antenna, he is referred to LaPort,[1] which has the fundamental information for finding surge impedance or characteristic impedance of antennas.

The system could be used for a single-ended antenna fed with a balanced transmission line with a balun at each end. Another possibility for a vertical is the use of a two-wire, grounded, open transmission line, as discussed in LaPort's book.[2] The ground system would be critical.

When experimenting with these multiband lines, it is convenient to have some

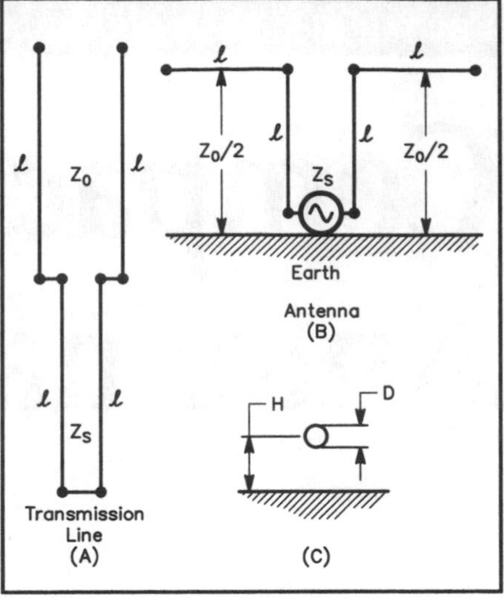

Figure 5—At A and B, the open portion of the transmission line evolves into the flat-top portion of the antenna. At C, diagram illustrating the formula for calculating Z_o for the antenna.

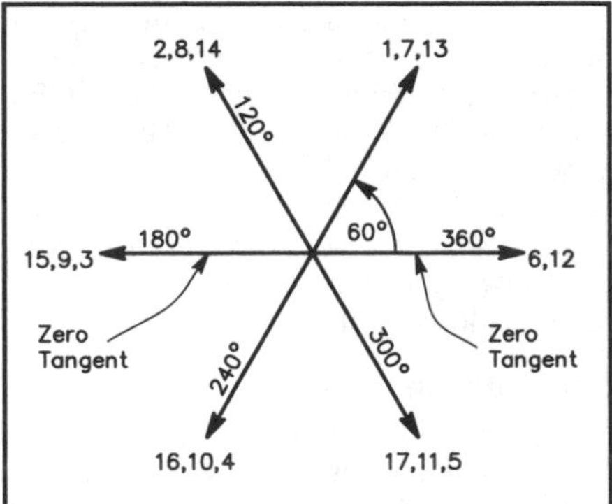

Figure 6—Angular position chart useful for determining "stock numbers" to apply to the chart.

"stock" numbers to apply to the lines (see Figure 6). One-sixth of a wavelength is one-third of a half wavelength. A convenient length for a half wave on 80 meters is 135 feet. One-third of that is 45 feet or $\lambda/6$ for 80 meters. One-sixth of a wavelength on 40 meters is 22-1/2 feet. When you are designing an antenna, these lengths need to be multiplied by the propagation constant of the line. After construction and testing, the dimensions can be pruned for end effect, etc.

When operated as a transmission line, the system as described may have application in end-feeding half-wave antennas, especially two half waves in phase. The system transforms a high impedance to a low impedance as a quarter-wave line will; however, it will do this at several even harmonics, in contrast to the quarter-wave line that is only responsive to odd quarter wavelengths.

The author wishes to thank Walt Maxwell, W2DU, for his detailed analysis of the theory section of this article.

Notes

[1]E. LaPort, *Radio Antenna Engineering* (New York: McGraw-Hill Book Co., 1952).

[2]See note 1.

By Bill Wright, G0FAH From *QST*, June 1995

Five Bands, No Tuner

Enjoy some of the advantages of a multiband, ladder-line-fed antenna without an antenna tuner.

Reading "The Doctor is IN" (*QST*, January 1995) reminded me that the search continues for a simple backyard antenna. A wire *dipole* antenna fed at the center with 450-Ω ladder line is a good choice. The ladder line keeps your losses low—even at moderately high SWRs. All you need is an antenna tuner and you're in business. No coils or traps necessary.

But can you do away with the tuner and still keep the ladder line? That would certainly make life simpler. To achieve this, your transceiver needs to "see" an impedance that looks reasonably close to 50 Ω on as many bands as possible. Without an antenna tuner acting as the middleman between the 450-Ω ladder line and your 50-Ω radio, this could be a problem.

Table 1
Calculated SWRs for a 94-foot Dipole Fed with 41 Feet of 450-Ω Ladder Line

Note: Although the antenna is cut for the CW portions of the band, expect similar results at other frequencies.

Frequency	*SWR*
3.56	7.6:1
7.1	2.4:1
14.2	1.5:1
18.1	2:1
24.9	1.5:1
29	2.4:1

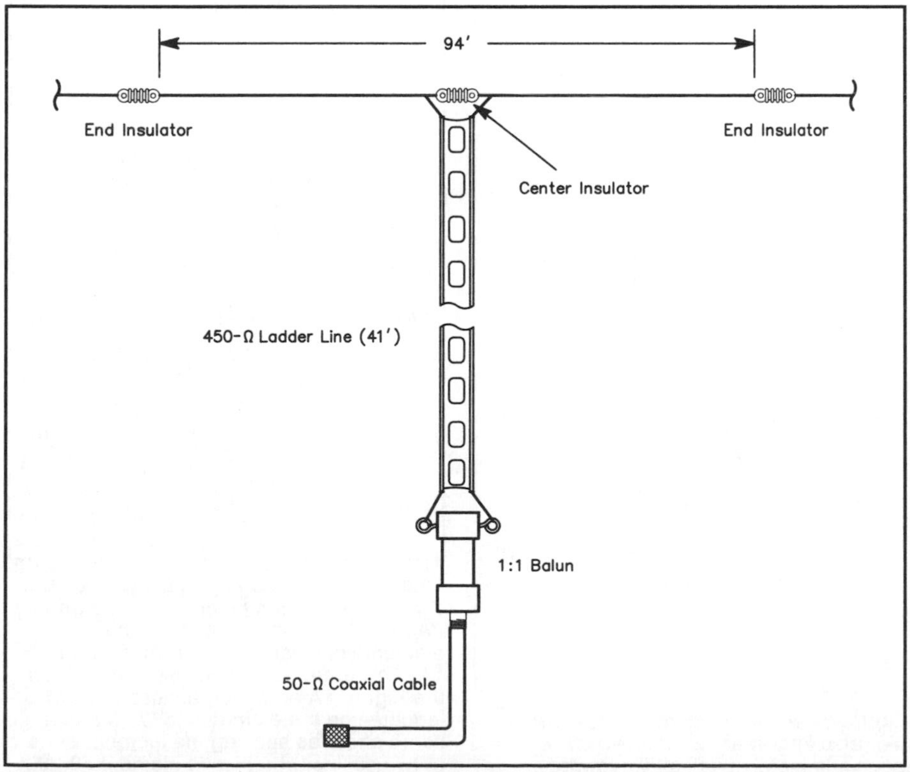

Can It Be Done?

A few years ago I attempted to design an antenna that would work on several HF bands from 80 to 10 meters. Full details were published in the spring 1992 edition of *SPRAT*, the journal of the G-QRP club.

My inspiration was the venerable G5RV. I took a 94-foot-long dipole and fed it with ladder line (see Figure 1). By cutting the ladder line to a specific length and using a 1:1 balun to make the transition to coaxial cable, I found that I could get close to 50-Ω (and thus achieve reasonably low SWRs) on at least five bands: 40, 20, 17, 12 and 10 meters (see Table 1).

The on-air results were better than I expected. My radio was happy and I didn't need to meddle constantly with an antenna tuner.

Of course, you'll need an antenna tuner to work the bands where the SWR exceeds 3:1. A simple tuner will do the job, though. Because you're using *unbalanced* coax ahead of the balun, you won't need one of the more expensive tuners designed for *balanced* feed lines.

For best results, put your antenna as high as possible. If the ends must bend downward to accommodate the size of your lot, don't worry. Run the ladder line to your balun and take your coax from there to your radio. Keep the coax portion as short as possible.

Conclusion

By eliminating the antenna tuner completely, you lose the flexibility of loading your ladder-line-fed antenna on virtually any band. In return, however, you gain the convenience of operating on several bands without making tuner adjustments each time you change frequency. Your losses are held to a minimum, which means that most of the power your radio generates is radiated by your antenna. Not a bad compromise! Of course, the results I achieved will vary when used at other locations. Still, it's a simple, fun project in the experimental spirit of Amateur Radio.

Figure 1—The wire dipole antenna is 94 feet in length. If you don't have 94 feet of open space, don't hesitate to droop the ends of the dipole to make it fit. Feed the antenna with 41 feet of 450-Ω ladder line that is connected to a 1:1 balun. From the balun to your radio, use 50-Ω coax.

By Rick Olsen, N6NR From *QST*, March 1993

The NRY: A Simple, Effective Wire Antenna for 80 through 10 Meters

Known as a *broadside collinear curtain array*, this antenna is simple to build, rakes in DX signals and has gain over a dipole on all the bands it covers!

Hams' thoughts ultimately turn to antennas. So it was with me a while ago. I had just moved to the emerald beauty of the Pacific Northwest and began pondering the 160-foot-tall Douglas firs in my backyard. I had considered putting up a crank-up tower, but decided that neighbors, architectural committees and crank-ups don't mix. Then I thought about topping a tree. Nope–I moved here because I love these trees.

I started sniffing around for alternatives. I obtained a copy of fellow ARRL Technical Advisor Roy (W7EL) Lewallen's antenna modeling program, *ELNEC*, along with a copy of *The ARRL Antenna Book*, and began to incubate the seeds of what would become my most enjoyable antenna project since I built a four-element quad back in 1974.

Hanging ropes from big trees is not a new idea.[1,2] I began by pacing off the distances between the monsters in my yard so I would know how much space I had to work with and how I could optimally direct my signal to places like Europe, Africa and the Caribbean.

Next came the selection of a configuration. Because the trees are approximately 120 feet apart, my first choice was a multielement collinear array. But a good friend, Terry Conboy, N6RY (in *NRY*, I'm the *NR* and he's the *RY*), suggested stacking a pair of *extended double Zepps*. I filed his suggestion away in my memory for the time being. I started my analysis.

The ARRL Antenna Book, on page 8-32, describes the basic function of the collinear array. In sum, it concentrates energy perpendicular to the wire, hence the term *broadside array*. This antenna's physical layout and pattern are shown in Figure 1. Such an antenna's gain can be increased by varying the end-to-end spacing of the elements.[3]

Chapter 8 goes on to explain three- and four-element arrays. Disappointment set in when I sized a four-element in-line array for 20 meters: It wouldn't fit between the trees! What *would* fit, however, was a "two-over-two." In fact, depending on the vertical separation, this array might even have some gain over the four-element collinear arrangement.[4]

My choice was made. I wanted to maximize performance on 20 meters and take what I could get on the other bands. I chose two 1-λ, center-fed wires vertically spaced 5/8 λ apart. According to *ELNEC*, with the array 100 feet high at the top and modeled

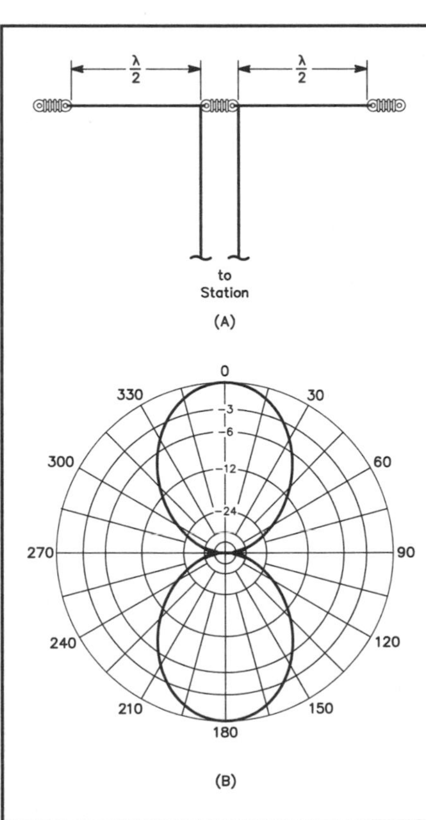

Figure 1—A two-element collinear array (A), also known as *two half-waves in phase*, is merely two half-wave wire elements placed end to end. This antenna has about 1.5 dB gain over a dipole in the same surroundings, and B shows its free-space pattern.

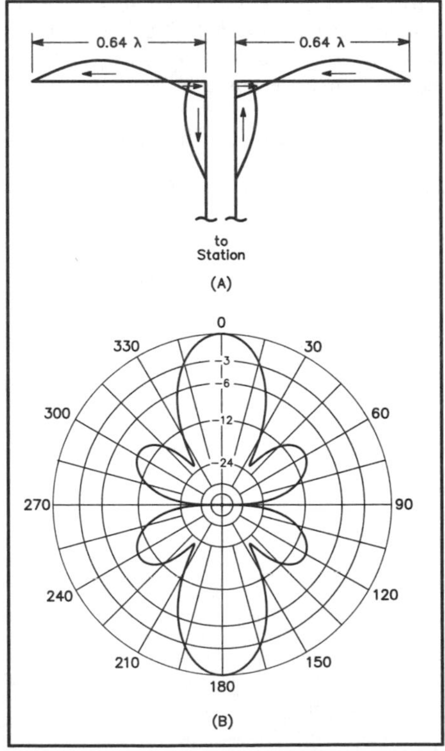

Figure 2—At A, the physical layout of and current distribution on an *extended double Zepp* (EDZ), a special case of the two-element collinear array shown in Figure 1A. This antenna exhibits more gain than the Figure 1A antenna because lengthening the elements to 5/8 λ has the same effect as spacing the element ends of two collinear half-wave elements 0.28 λ apart. (See Note 3.) At B, the antenna's azimuth-plane pattern in free space. This antenna has about 3 dB gain over a dipole in the same surroundings.

over lossy ground, this antenna's gain is approximately 12 dBi.[5] Not bad! The only problem is that this antenna's 20-meter feedpoint impedance is very high.

Reenter N6RY and the extended double Zepp. Take a look at Figure 2A. The extended double Zepp is simply a 1.25-λ, center-fed wire (two end-to-end 5/8-λ elements). There's a hidden message here! Have you found it?

Well, at first, I didn't either. Terry explained to me that the extended double Zepp (EDZ) is essentially two half-wave collinear elements spaced just over 1/4 λ apart. The extended double Zepp's added gain over two half waves in phase comes *not* from the extra 0.28 λ of wire between them, but from the *increased end-to-end spacing of the outermost pair of halfwave elements.* (Radiation from the inner 0.28-λ actually works *against* that from the outer halfwaves, but the gain increase from the new element spacing more than overcomes this disadvantage.) Figures 1B and 2B are similar. The EDZ has broader lobes and useful smaller lobes, both of which are attractive for a fixed array.

Sure enough, it was right in front of me all the time in the *Antenna Book* (page 8-34, fifth paragraph). In fact, the *whole darned array* was right in front of me in Chapter 8! I just needed some help piecing it all together. I modeled this array every way I could imagine, looking for the best possible performance. My wife didn't see me in the evenings for two weeks! Figure 3 shows the final configuration.

Figures 4, 5 and 6 show some of the results of my modeling. I used a design frequency of 14.2 MHz, a height of 100 feet above ground (for the top wire), elements made of #14 copper wire, and ground coefficients typical of the Pacific Northwest. All the elements in this array are fed in phase, as described in a bit.

As the figures show, the antenna's gain and patterns are impressive. On the 80- and 40-meter bands, the antenna has remarkably dipole-like patterns and gains and seems to work at least as well as a dipole in the same height range. On the higher bands, it works *much* better than a dipole. Its predicted gains in my installation are 9, 11 and 14 dBi at 7, 10.1 and 14 MHz, respectively. At 18.1 MHz (Figure 6) and above, the array begins to take on long-wire properties. It has several major lobes in directions closer to the wire axis, which the lower-frequency patterns don't provide.

To help you evaluate the NRY's patterns, Figures 4-6 superimpose the patterns of halfwave dipoles cut for 10.1, 14 and 18.1 MHz, respectively, on the NRY's patterns. These comparisons assume that the dipole is the same height as the NRY's feed point (halfway between the elements, or about 22 feet lower than the top wire), and over the same ground.

You may wonder about this antenna's effectiveness as a function of height. According to the *Antenna Book*, arrays like this have the best gain when the lower element is at least 1/2 λ above the ground (33 feet at 14 MHz). And, of course, as with all horizontally polarized antennas, this array's radiation angle decreases as you raise the antenna, which makes for better DX performance. The bottom line is that for DX work, you don't have to get the array up 100 feet for it to work well—you just need to put it up as high as you can. For closer-in coverage, it will work well at lower heights.

Another advantage of this array is that its feed-point impedances are manageable. To maintain my goal of broadband use, I decided to feed the antenna as shown in Figure 3. This is the "lazy-H" configuration described starting on page 8-37 of *The ARRL Antenna Book*. The attractiveness of this configuration is that it doesn't rely on specific phasing-line lengths to work properly. To the extent made possible by the different mutual impedances between each element and the ground, the phase relationship between elements remains constant regardless of frequency.

Final Dimensions

The antenna's physical dimensions (86 feet, 10 inches long and 43 feet, 5 inches high, not counting insulators) are easy to remember because the height is half the length. Because the antenna is basically 1.25 (5/4) λ long, center-fed and vertically spaced 5/8 λ apart, you can easily scale it to whatever band you like. For instance, to scale it to 10 meters, divide all the dimensions by two (28 MHz/14 MHz = 2).

Construction

Now for the really fun stuff: putting this thing up in the air! Building the antenna is a breeze. It took me all of about an hour to cut and solder the wires and attach the tethers. The harder part was preparing the trees.

If you have a friend like Bernie Olshausen, N6RUX, you have it made. He's a crack shot with a slingshot, similar to that described by Wade Calvert, WA9EZY, in *QST* a couple of years back.[6]

Once Bernie landed the fishing lines over the right tree limbs, the next job was to pull up some lightweight twine (we used seiner twine—the stuff fish nets are made of) with the fishing line. This is important; rope sometimes snags as it goes over limbs, and you don't want to try to pull it over with 6-pound-test fishing line! With the twine, we pulled up some 800-pound-test nylon rope with a pulley on one end, using another piece of nylon rope through the pulley to act as a catenary for the array.

It's important to use a strong, weather-resistant catenary rope to hang the array, because the antenna loses its desirable properties quickly when it takes on the shape of a V. You can expect a 2-dB gain

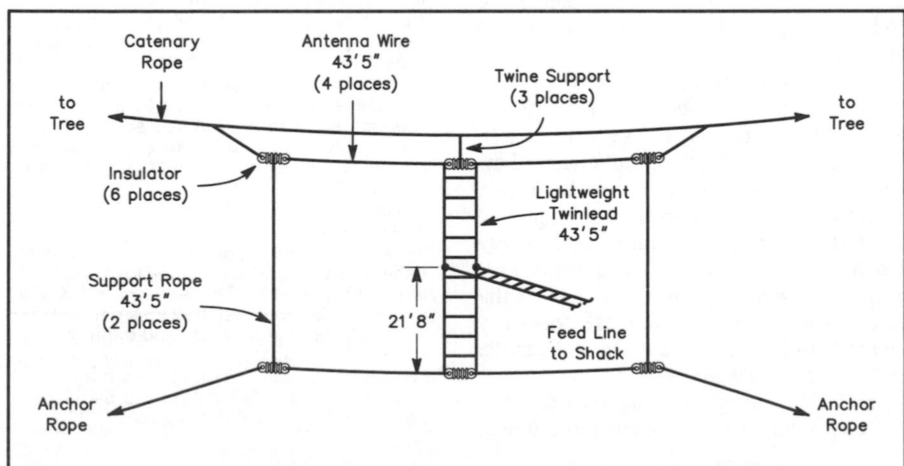

Figure 3—The broadside collinear curtain array—the NRY—in its final configuration. This antenna is cut for 20 meters, but works well on all the HF ham bands. The array hangs from a catenary rope via three short lengths of twine. The lower end insulators are supported by ropes to those of the top wire, and the center insulators are joined by the phasing line, which is 43 feet, 5 inches of parallel-wire feed line. The feeder to the radio is made of the same type of line connected at the halfway point on the phasing line.

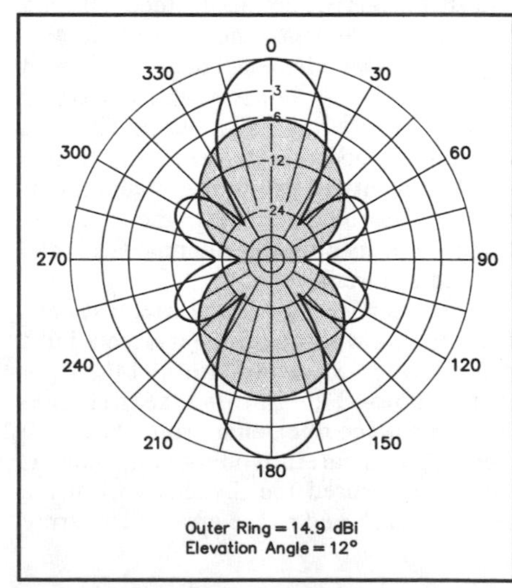

Figure 4—The NRY's 14-MHz azimuth-plane pattern with the bottom wire at 56.5 feet and top wire at 100 feet. For comparison, the shaded part shows the pattern of a half-wavelength, 14-MHz dipole at 78 feet (the same height as the NRY's feed point).

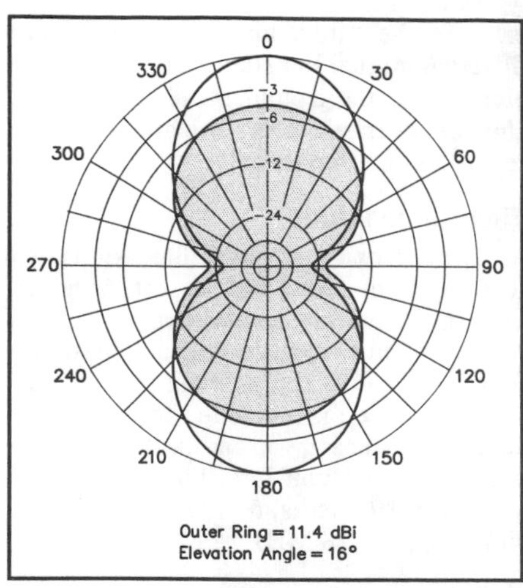

Figure 5—The NRY's 10.1-MHz azimuth-plane pattern with the bottom wire at 56.5 feet and top wire at 100 feet. For comparison, the shaded part shows the pattern of a half-wavelength, 10.1-MHz dipole at 78 feet (the same height as the NRY's feed point).

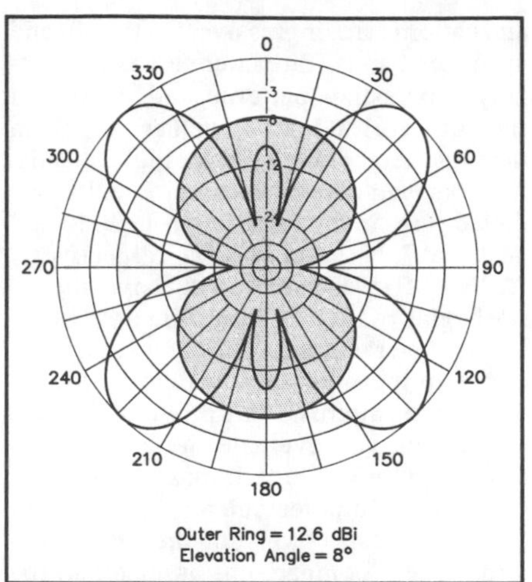

Figure 6—The NRY's 18.1-MHz azimuth-plane pattern with the bottom wire at 56.5 feet and top wire at 100 feet. For comparison, the shaded part shows the pattern of a half-wavelength, 18.1-MHz dipole at 78 feet (the same height as the NRY's feed point).

decrease at 20 meters with a 20° variance from the flat-top configuration. The rope, therefore, must be strong enough to minimize droop and maintain integrity in the sun and wind. We'll get back to the rope in a minute.

For the phasing harness I used low-loss, 300-Ω twinlead. Any parallel-wire feeder should work fine. Don't use coaxial cable in the feed system. It's too heavy, unbalances the array and has very high loss when operated at a high mismatch.[7]

To assemble the feed system, I used porcelain insulators and equal lengths of 300-Ω line. I then cut the #14 antenna wire to the proper length and soldered it to the feed system in a standard dipole configuration.

Next came the tethers. For an antenna almost 87 feet wide and 43 feet tall, you need to provide some means of maintaining the proper element spacings. I used equal lengths of seiner twine (nylon or dacron rope would also work fine) to vertically space the element ends. I also attached plenty of line to the lower end insulators so I could tie them to the appropriate fence posts once the antenna was in the air. See Figure 3.

The catenary rope is the trickiest part. The antenna won't hang right if you don't attach the top three insulators in the right spot. I know—I did it wrong the first time!

I cut a piece of catenary rope 95 to 100 feet long and carefully marked the center. Next, I measured the distances from the center to each end of the array. (The array itself is a handy measuring line.) Then I added about 10 more inches and marked the line. I'll tell you why in a minute.

Then I cut three pieces of twine, each about 24 inches long. At the center of the catenary rope, I carefully opened the rope weave and passed about 6 inches of twine through it, then let the catenary rope retake its natural shape. Wrapping the twine around the rope several times on either side of the joint relieved the tension placed on the rope introduced by passing the twine through it.[8] I then tied the other end of the twine to the top center insulator so that the spacing between the rope and the insulator is 12 inches.

After doing that, I attached the twine at the end marks on the catenary rope in the same manner. The distance to the end insulators is 18 to 20 inches. This causes the ends to hang off the rope at an angle and provides strain relief.

If you use a braided rope (with a weave you can't open to pass twine through it), one good alternative is to use insulators in the catenary rope (at the points where I attached the twine) as tie points for the array.

Now you're ready to haul the antenna up on the lanyards. Consult your US Navy *Marlinspike Seamanship* manual (just kidding, of course) and join the ends of the catenary to the lanyards that are hopefully by now through the pulleys, up in the trees. Use a strong knot that won't slip.

Before hoisting the array into the sky, *make sure* that the connections to the feed harness at the top and bottom are the same (ie, that the left elements both attach to the same wire in the twinlead). This thing does not work right if the top and bottom are fed 180° out of phase!

Now you're ready to haul it up and connect the feed line to the antenna tuner. Check to make sure that the symmetry and top-to-bottom wire spacings are within reasonable limits, then tie the bottom tethers off so that the bottom wire is as flat as you can make it.

You're Done!

This is quite an effective antenna. I've installed two of them: One is boresighted on the Middle East and works like gangbusters into Europe as well; the other favors the Caribbean and most of the US. In the first three months after putting them up, I worked 105 countries on CW, including some rare ones, with a 100-watt transceiver. I'm surprised at how easily I can bust pileups—especially on 30 and 20 meters.

I hope you have an opportunity to try this array on for size. Don't limit yourself to just the backyard, either: It's a great Field Day antenna, to say the least. It also comes in handy for emergency work where you need a gain antenna.

You don't have to be a rocket scientist to design effective antennas! Your old friend *The ARRL Antenna Book*, and other ARRL publications, like Walt (W2DU) Maxwell's *Reflections*, the *Handbook*, the *Antenna Compendium* series and Wilfred Caron's *Antenna Impedance Matching*, are serious reference guides and sources of numerous ideas. It doesn't hurt to have your favorite antenna-modeling software loaded on your PC, either!

Notes

[1]D. Brede, "The Care and Feeding of an Amateur's Favorite Antenna Support—the Tree," *QST*, Sep 1989, pp 26-28, 40.

[2]J. Hall, ed, *The ARRL Antenna Book*, Chapter 22 and pp 4-3 and 15-3.

[3]*The ARRL Antenna Book*, Fig 8-38, p 8-32, illustrates how gain varies with the spacing of the two adjacent element ends.

[4]The *Antenna Book*'s Fig 8-45, p 8-35, suggests that a spacing of 5/8 to 2/3 λ yields good broadside gain. For the four-element array, upwards of 8.5 dBi gain is obtainable.

[5]dBi means decibels relative to an *isotropic radiator* (point source) in free space. It *does not* denote gain relative to a point source above real ground, or a dipole in any surroundings.

[6]W. Calvert, "The EZY-Launcher," *QST*, Jun 1991, pp 34-35.

[7]*The ARRL Antenna Book* shows how coaxial-cable loss varies with matched loss and SWR in Fig 18, p 24-18. (The *Antenna Book* also contains a table of matched loss per 100 feet as functions of frequency and cable type in Fig 26, p 24-18).

[8]If you use braided rope for the catenary, you'll need to use another method to secure the seiner twine to the catenary rope.

CHAPTER FOUR

LOOP ANTENNAS

By John Griggs, W6KW From *QST*, March 1982

Dual Full-Wave Loop Antenna

Achieving gain from an antenna at 7 and 3.5 MHz normally requires a rather large piece of real estate, a high tower or both. To obtain significantly improved performance over a dipole at these frequencies, and to do it inside an average city lot, is a goal worth pursuing! With this in mind I decided to replace my inverted-V dipoles on 7 and 3.5 MHz with a dual full-wave loop antenna, one inside the other.

I had noticed the strong signals on 7 MHz from Pat Kearins, W7UI (now a silent key), and from Carl Winter, W6OAW, on 3.9 MHz, and was interested to find that both were using full-wave loops. Both signals would pin my S-meter and my QTH is well over 200 miles from either station. The antenna at W7UI was a loop suspended by a very high supporting structure. This loop resembled a square, with one corner facing up, one facing the ground, and the other two corners pulled outward by guy wires.

My attention was attracted to a horizontal loop antenna used by another ham on 3.9 MHz that produced extremely strong signals. It was a square loop, 65 feet on a side, with each leg parallel to the ground but only 14 feet high. Fed by a tuned line, it functioned well, but it was most effective for relatively short ranges, up to 200 miles or so.

Of more interest to me, however, was the 75-meter rectangular loop used at W6OAW. I learned that this was a dual antenna, i.e., a full-wave rectangular loop for 3.9 MHz suspended about 40 feet above ground, with another full-wave loop for 7 MHz inside the first, in the same plane. The 3.9-MHz loop is a closed circuit, whereas the 7-MHz loop is an open circuit. This permits operation on 10, 15 and 20 meters as well as 40 meters when used with open-wire feed line and a matching network. The outer loop is fed with coaxial cable.

Design Planning

In considering loop antennas for my lot, which is 75 feet wide by 125 feet deep, I found that I could use two 40-foot high pipe masts, 130 feet apart along a diagonal across my lot. I duplicated the W6OAW arrangement, except that I hung the antenna vertically and made both loops closed-circuit designs. I could use them only on the bands for which they were cut, 7.2 MHz and 3.8 MHz. Each is fed with a quarter-wave matching section of RG-59/U, which provides a 50-ohm match for the RG-8/U cables leading in to my operating position.

Construction

See Fig. 1 for construction details and dimensions. Matching the 104-ohm antenna impedance to the 50-ohm line impedance requires a quarter-wave coaxial line having a characteristic impedance of 72 ohms. The formula for a ¹/₄-wave section is:

$$\frac{246}{f\,(\text{MHz})} = \ell\,(\text{ft}) \qquad \text{(Eq. 1)}$$

This result must be multiplied by the velocity-factor of the coaxial cable, 0.66 for RG-59/U 73-ohm cable. The quarterwave section was determined to be 42.7 feet for 3.8 MHz and 22.6 feet for 7.2 MHz. These connect to RG-8/U cables for a 50-ohm match to the transceiver.

The top section of the 3.8-MHz loop is fed at the center because of the length of the quarter-wave line, but the shorter matching section for the 7.2-MHz loop allows me to feed the bottom portion of that antenna.

The inner loop is supported from the corners of the outer loop by means of nylon rope and suitable insulators. It is best to use lightweight, high-quality insulators and wire no larger than no. 14 to reduce the weight of the array. Pulleys (and rope) will be required at the tops of the 40-foot poles and also at the 18-foot level. The fact that the bottom horizontal section of the antenna narrowly misses the top of my house roof seems to have no deleterious effect on its operation.

Measure the antenna sections carefully, and cut them to length. Pay particular attention to locate the feed point at the exact center of the horizontal wire sections. This provides horizontal polarization. Feeding the loop at the center of either vertical section will provide vertical polarization.

Tuning

If trimming is needed to resonate the antenna to your favorite operating frequency, cut or add equal length pieces at each vertical section. Do not shorten the sides to the point at which the top and bottom sections are less than 0.1 wavelength apart. Careful pruning of the horizontal sections would be required if such a problem develops.

Information given to me by those using this antenna seems to confirm my results. A typical comment is "Boy are you ever loud!" A gain of 1.8 dB over a dipole is claimed for this type of antenna. The principal difference appears to be a lower angle of radiation. While it is possible to work stations off the ends of the antenna, maximum radiation occurs broadside to the element. What I like about the antenna is the ability to hear weak stations. It is outstanding in this regard.

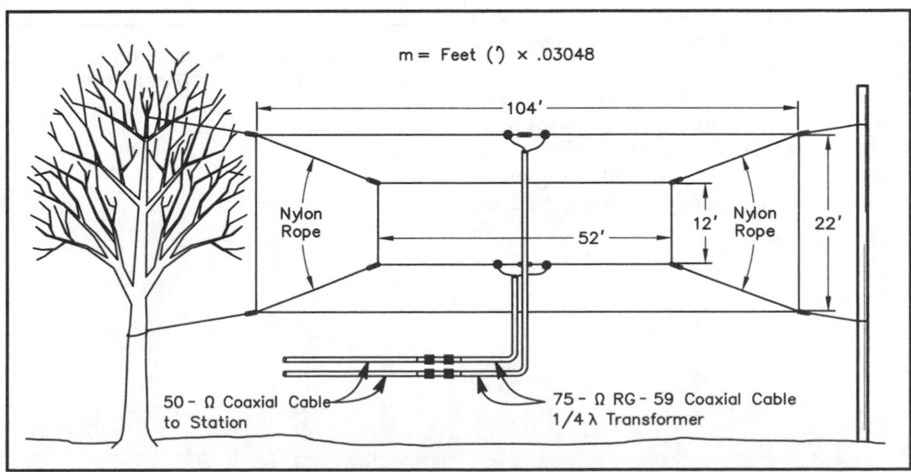

Fig 1—Dimensions for the dual full-wave loop antenna. Input impedance is on the order of 104 ohms. A quarter-wave matching section of RG-59/U is used to provide a match to 50 ohms.

A Gain Antenna for 28 MHz

Give your 10-meter signal a boost with this simple antenna.

Although in coming years the 10-meter band won't provide the excitement it did at the peak of sunspot cycle 22, DX openings will still occur, especially during spring and fall. And who knows? With the large number of 10-meter operators and wide availability of inexpensive, single-band radios, the band may be the hangout for local ragchews that it was before the advent of 2-meter FM.

I'd like to present a simple antenna for 10 meters that provides gain over a dipole or inverted V. Similar designs have been published before, but because I never hear them on the air, I think they must not be fully appreciated. The antenna is a resonant, rectangular loop with a particular shape. It provides 2.1 dB gain over a dipole at low radiation angles when mounted well above ground. This represents a power increase of 62%. The antenna is simple to feed–no matching network is necessary. When fed with 50-Ω coax, SWR is close to 1:1 at the design frequency. SWR is less than 2:1 from 28.0 to 28.8 MHz for an antenna resonant at 28.4 MHz.

The antenna is made from #12 wire (see Figure 1). For horizontal polarization at 28.4 MHz, the loop is 73 inches wide and 146 inches high (just larger than 6×12 feet). Feed the antenna at the center of the lower wire. Coil the coax into a few turns near the feedpoint to provide a simple balun. A coil diameter of about a foot will work fine. You can support the antenna on a mast with spreaders made of bamboo, fiberglass, wood, PVC, or other nonconducting material. You can use aluminum tubing both for support and conductors, but you'll have to readjust antenna dimensions for resonance.

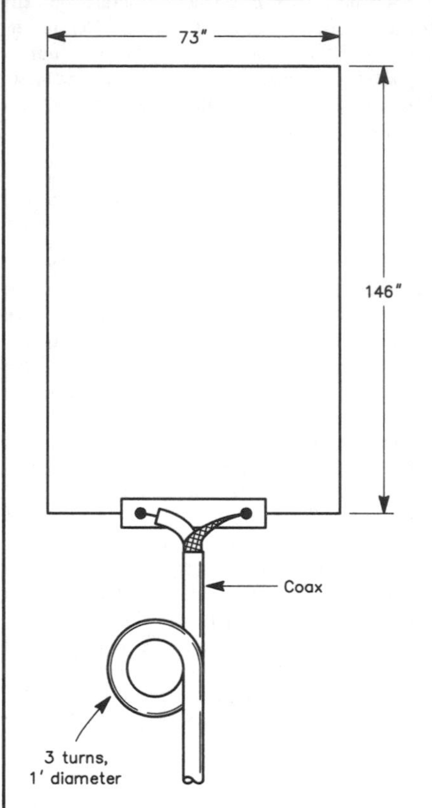

Figure 1—Construction details of the 10-meter rectangular loop antenna.

This rectangular loop has two advantages over a resonant square loop. First, a square loop has just 1.1 dB gain over a dipole. This is a power increase of only 29%. Second, the input impedance of a square loop is about 125 Ω. You must use a matching network to feed a square loop with 50-Ω coax. The rectangular loop achieves gain by compressing its radiation pattern in the elevation plane. The azimuth pattern is slightly wider than that of a dipole (it's about the same as that of an inverted V). A broad pattern is an advantage for a general-purpose, fixed antenna. The rectangular loop provides bidirectional gain over a broad azimuth region.

You should mount the loop as high as possible. To provide 1.7 dB gain at low angles over an inverted V, the top wire must be at least 30 feet high. The loop will work at lower heights, but its gain advantage disappears. For example, at 20 feet the loop provides the same gain at low angles as an inverted V.

A small, 3-element Yagi can provide 6 dB gain over a dipole and great rejection of signals to the rear. If you can install a beam and rotor, you'll find it much more effective than a loop. But for a simple, cheap, gain antenna that can be thrown together quickly, the rectangular loop is hard to beat.

Note: I used the AO 6.0 Antenna Optimizer program to automatically optimize the dimensions of a rectangular loop for maximum forward gain and unity SWR. I used NEC/Wires 1.5 to verify the design with the Numerical Electromagnetics Code.

10-Meter "Lazy Quad"

Among your readers that like to tinker with antennas there may be some that would like to try the antenna shown in Figure 1. The basic idea came from a station using a somewhat similar configuration on 15 meters, and I make no claim of originating the idea. However, I have never heard of anyone using the antenna on 10 meters. Since it is currently nameless, and has features similar to both a Lazy H and a quad, perhaps it should be known as a "Lazy Quad."

The major advantages of the antenna are: (1) extreme simplicity, (2) feasibility of installation on a light unguyed pole, (3) small horizontal space requirements (as compared to a horizontal dipole), (4) low QRN in receiving (as compared to a vertical dipole), (5) ability to withstand high wind loads and (6) broadband operation.

When looking at loop 1 we see a horizontally-polarized full-wave loop radiating broadside, with maximum currents along the top and bottom horizontal wires H_1 and H_2. When looking at loop 2 we see a similar horizontally-polarized loop with maxiinum current on wires H_2 and H_3. When looking at wires H_1 and H_3 we notice two in-phase horizontal wires spaced a half wave, both carrying maximum current. Minimum currents appear on the vertical portions of the loops where the tuning stubs are inserted.

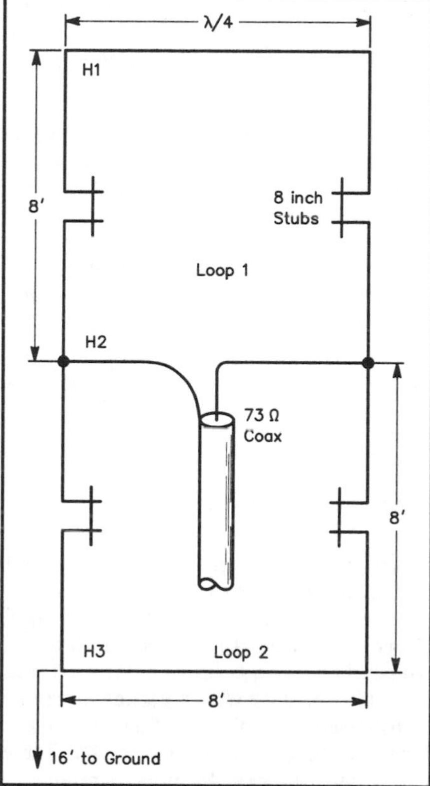

Figure 1—W6PIZ's "Lazy Quad" for 10 meters. The loops are No. 14 wire, with the horizontal sections mounted an 8-foot lengths of 1×2 wood. The adjustable stubs are of the same type wire and are self-supporting since the length is only 6 inches. The antenna at W6PIZ is mounted on an unguyed wooden pole 32 feet high.

I operate mostly on 10-meter c.w. and have the antenna peaked at 28,050 MHz. In my specific case the length of each stub is 6 inches for this frequency. Loading is almost constant from 28 to 30 MHz.

The coax feed line is run straight down the pole (wood) to the ground, and there is little antenna effect on the feeder.

Results have been consistently better than had been obtained with vertical or horizontal dipoles previously installed at the same effective height at the same location. Indications are that it outperforms some beams of the same approximate height, perhaps because of the broad vertical pattern (which allows longer QSOs under critical skip conditions), as compared to the beams.

Perhaps some of your readers might be urged to try this basic idea with a reflector of the same basic configuration as the driven element. It would appear that the Double Lazy Quad would be anything but lazy in operation. — *Dave Hardacker, W6PIZ*

The Full-Wave Delta Loop at Low Height

You'll be surprised at the results you'll get from a full-wave loop at low heights.

Property size and antenna-support height are ever-present concerns of the urban amateur. Many good antennas are untried because the radio amateur is unable to imagine how a large wire antenna could be squeezed onto a small lot. Certainly, this is typical in the case of full-wave loop antennas. But there is no rule that dictates using a symmetrical loop. It can be distorted rather severely without spoiling the performance. The same philosophy is appropriate with regard to height above ground and the plane in which the antenna is erected. In most instances, a less-than-optimum full-wave loop will outperform a dipole or inverted V antenna that is close to the ground in terms of wavelength. It is possible that such a loop will give comparable or better performance than a vertical antenna that is less than 90 degrees (with respect to ground), or one with a substandard ground screen.

We want to discuss the Practical considerations of loops that can be supported from low supports on small pieces of property. The results we have obtained are noteworthy with respect to all-around "solid" communications within and outside the USA. Perhaps you will be inspired to unroll some wire and try a loop at your QTH.

Some Loop History

Loops were used first as receiving antennas. While single- and multiturn small loops worked well for receiving, they were not satisfactory for transmitting: They were inefficient in terms of gain, and the feed impedance was generally a fraction of an ohm, making them difficult to match. The losses were significant. But, it was possible to use a compact loop (less than 0.5 wavelength) for receiving in place of a fullsize version that could require thousands of feet of conductor. One of us owned a portable broadcast-band receiver in the 1930s. The loop antenna was stored in the lid of the cabinet, and needed to be mounted atop the radio during reception periods! The radio was heavy: it weighed 91 pounds, including the various dry batteries.

Receiving loops continued to be useful for many years in the commercial services, especially for LF and VLF applications. Amateurs also used them (and continue to do so) for improved reception on 160 and 80 meters. The signal-to-noise ratio of receiving loops is markedly better than that of vertical antennas, and they are directional.[1] Many successful 160-meter DXers owe their success to the use of receiving loops with low-noise preamplifiers. Practically, these loops are the next best thing to Beverage antennas.[2]

Loop Characteristics

What are some of the advantages of a closed, full-wave loop? Perhaps number 1 on the list is the lack of need for a ground screen. The matter of effective height above ground is still a consideration, but we need not lay a ground-radial system as would be the case with a vertical antenna. Consideration number 2 is that a full-wave loop (depending on the shape) has some gain over a dipole. Number 3 relates to noise factor. A closed loop is a much "quieter" receiving antenna than are most vertical and some horizontal antennas.

To illustrate this point, the 160-meter antenna at W1FB is a 3/8-wavelength inverted-L with twenty 3/8-wave radials. Since this is essentially a vertically polarized antenna, it is noisy (man-made and atmospheric noise). There are times when an S9 signal is unreadable because of the ambient noise being S9 or greater in strength. Upon switching to the 75-meter Delta loop, the same signal will rise above the noise by 1 or 2 S units, while the noise and signal will drop well below S9. For example, the received signal may drop to S6 on the loop, but the noise will decline to S4.

Feed-point selection will permit the choice of vertical or horizontal polarization. Various angles of radiation will result from assorted feed-point selections. The system is rather flexible when we want to maximize close-in or faraway communications (high angle versus low angle). Figure 1 illustrates various configurations that can be used. The arrangement at C is used at W1SE, and the shape at D is being applied

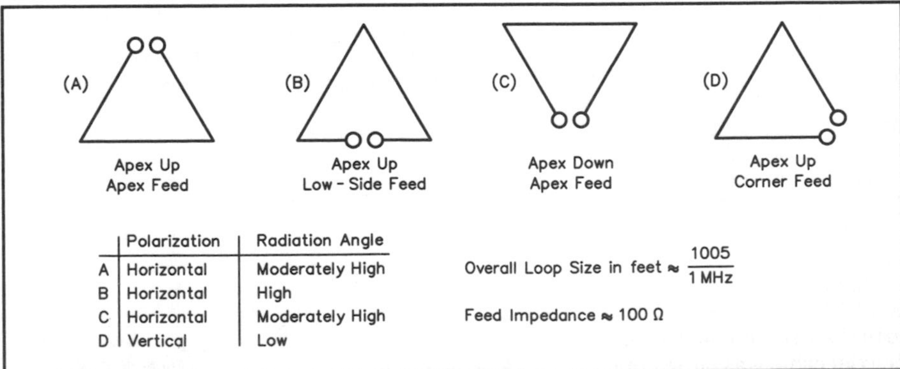

	Polarization	Radiation Angle
A	Horizontal	Moderately High
B	Horizontal	High
C	Horizontal	Moderately High
D	Vertical	Low

Figure 1—Various configurations for a full-wave Delta loop. Radiation angles and polarization are affected by the feed-point placement and location of the apex.

at W1FB. Both antennas are cut for 80-meter operation. The bandwidth at resonance is on par with that of a dipole. A Transmatch is used for matching the system to the transmitter in those parts of the band (75 and 80 meters) where the SWR is too high to deal with.

Our loops are not deployed in a vertical plane, owing to the lack of tower height. A 60-foot tower and 50-foot tree support the W1SE antenna. A single 50-foot tower is used at W1FB. Both loops are tilted away from the supports at roughly 45 degrees (Figure 2). This shows the present W1FB system. The loop is broadside northeast and southwest for maximum radiation in those directions at 80 meters. More on this later.

When these low-to-the-ground experiments began in the summer of 1983, we were joined by Bill Martinek, W8JUY, near Traverse City, Michigan. Bill experimented with various loop configurations so that he and W1FB could make signal comparisons locally and afar. He finally adopted the W1SE format with the apex down (Figure 1C, with the flat top strung between two 50-foot trees). In order to keep the loop completely vertical (not sloping), he chose a triangle that was not equilateral. The upper side of his triangle is substantially longer than the two downward sides. His signal on 75 meters is consistently 10 to 20 dB stronger than with his inverted V. The point of this discussion is that you need not use an equilateral triangle if it will not fit on your property. Erect whatever you can, then give it a try!

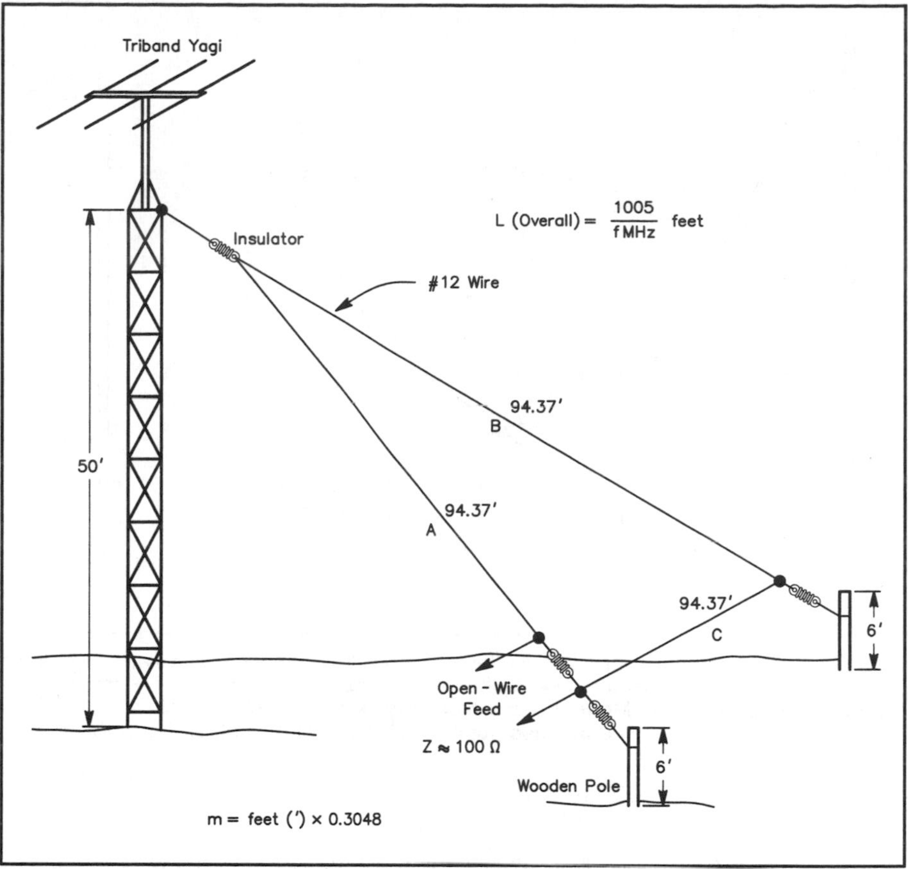

Figure 2—A tilted Delta loop for 80 meters is used at W1FB. The tower height is only 50 feet. Homemade open-wire line is used as the feeder to permit multiband use with vertical polarization and a low radiation angle.

Feed Methods

A Q section is used for feeding the W1SE loop. A Q section is a quarter-wavelength line with an impedance that is somewhere between the antenna feed impedance and that of the feed line. Calculation is a simple matter:

$$Z\ (\text{Q section}) = \sqrt{Z1\ Z2}\ \text{ohms} \qquad \text{(Eq. 1)}$$

where Z1 is the antenna impedance, and Z2 is the feeder impedance in ohms. In this case, assuming approximately 100 ohms for the antenna. feed impedance, we would have $\sqrt{100 \times 50} = 70.7$ ohms for the Q-section impedance. This represents a close match to 52-ohm coaxial cable. The Q-section length (made from RG-59/U) can be determined from L(feet) = 246 V/f(MHz), where V is the velocity factor of the coaxial line for the matching section. (The length should be verified using a dip meter.) For operation at the W1SE-chosen frequency of 3.825 MHz, the calculation calls for a Q section of 42 feet 5 inches (Figure 3).

Open-wire feed is used at W1FB (Fig 6B) to permit multiband operation through 10 meters. Unfortunately, a short run of RG-8/U was needed to bring the feed line to the ham station–under the driveway. The coaxial cable was buried in the ground for this reason. A homemade 4:1 toroidal balun transformer (two stacked T200-2 Amidon

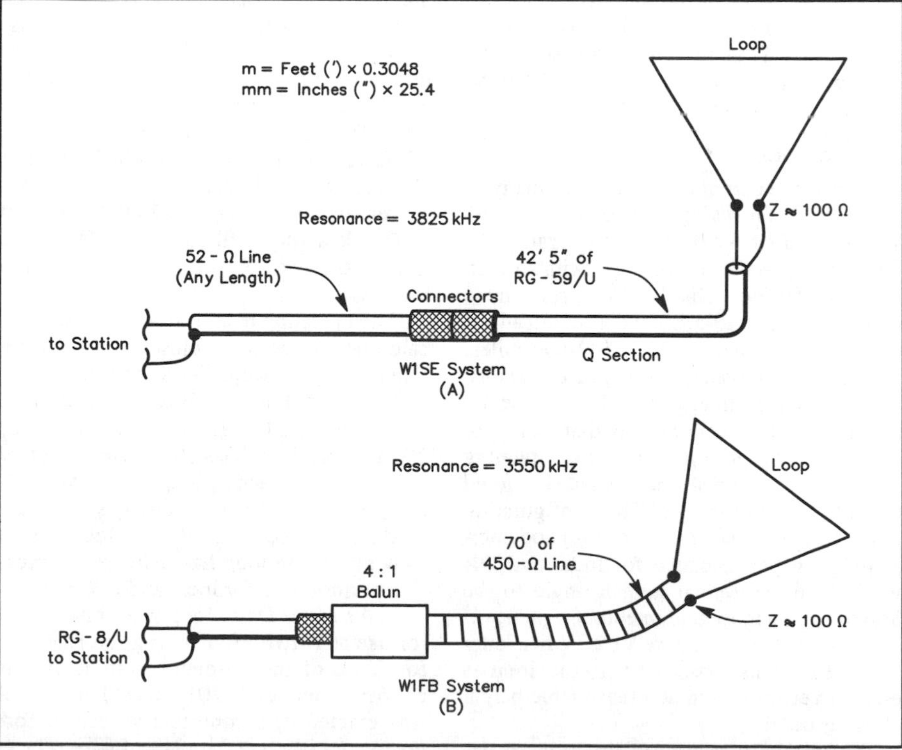

Figure 3—At A is the feed method used at W1SE. A coaxial Q section closely matches the 100-ohm feed impedance to a 52-ohm coaxial line. Illustration B shows the W1FB feed arrangement. Open-wire line, a balun transformer and a short length of RG-8/U cable permit multiband use with a Transmatch. Ideally, the open-wire line would continue all the way to the Transmatch, and the balun transformer would be located at the Transmatch.

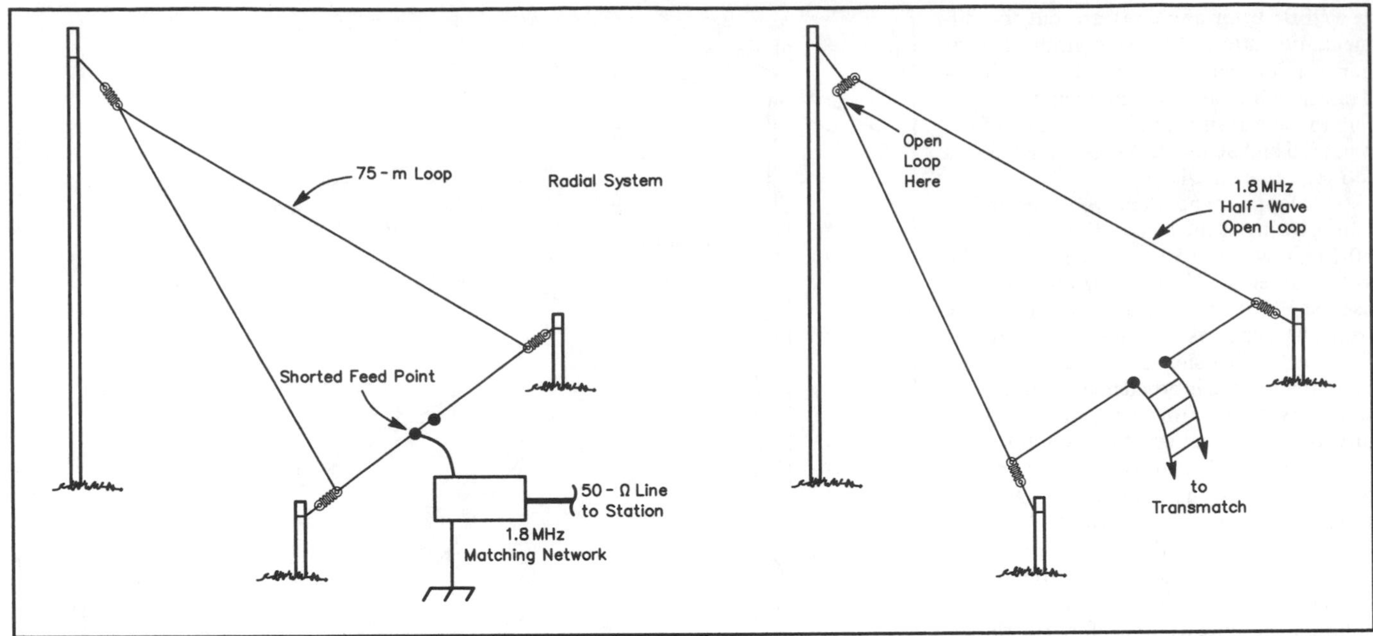

Figure 4 — Two methods for using a full-wave loop at half frequency. A switching arrangement could be applied at the feed point to change from a closed, full-wave loop to the first configuration seen here. The method at A performs as a ¼-wavelength radiator, but a ground screen is required. Method B is satisfactory as a ½-wavelength open loop for half-frequency use. It requires opening the loop at the electrical point opposite the feed point. A relay could be used for this purpose.

cores and Teflon-insulated no. 14 wire) was enclosed in a weatherproof box and mounted on one of the support poles for the 450-ohm open-wire line. The RG-8/U was run underground from that location (about 25 feet). Ideally, the openwire line would have been brought into the house, where it would be matched to the station gear with a Transmatch. Fortunately, the SWR at loop resonance is 1.3:1 without the Transmatch in use.

Performance

This is the part of our article that many of you have been waiting to read. Well, the W1FB results have been entirely gratifying. The loop replaced an inverted V with an apex height of 50 feet. This led to a pronounced improvement in all-around communications on 75 and 80 meters out to 500-600 miles. But, the loop proved to be very effective also for DX communications to Europe on 80 meters. The first version was that of Figure 1B. Although the antenna was outstanding for close-in 75- and 80-meter work, it offered dismal DX performance. The configuration at D of Figure 1 seems to offer a good compromise in performance for local and DX work. The theoretical launch angle to the horizon at the loop fundamental frequency is 10 degrees, as reported by VE2CV in a letter to W1FB. This assumes that the loop is erected vertically and at a reasonable height above ground.

Harmonic operation of the loop, as depicted in Figure 1D, is superb. At times it outperforms the trap tribander atop the tower during DX operation to Europe and Africa. The loop shows an average 6-dB signal increase on 20 and 15 meters in the favored direction, owing to the gain and lower radiation angle of the loop. Radiation at the harmonics is in the plane of the loop rather than broadside to it. This makes it ideal for contacts into Africa. It is perhaps the most effective 40-meter DX antenna that has been used at W1FB from northwest lower Michigan. The Transmatch is required on all harmonic frequencies other than 18.111 MHz, where W1FB has been conducting propagation studies with Bill Orr, W6SAI, Prose Walker W4BW, Bob Haviland, W4MB and Stu Cowan, W2LX, under special experimental/research licenses (KM2XQV). The loop has worked very well on 24.9 MHz as well during these tests. At 18.111 MHz, the SWR is 1.4:1.

The operating results at W1SE also indicate that a tilted loop, close to the ground, functions quite well. With loop resonance at 3825 kHz, the 2:1 SWR points occur at 3734 and 3934 kHz, respectively. This 200-kHz bandwidth spectrum can be shifted up or down the band by lengthening or shortening the loop conductor and Q section accordingly. From the W1SE location in Newington, the loop has delivered impressive performance for local and DX work.

A 40-meter Delta loop was constructed for use at W1SE after noting the fine performance of the 80-meter system. It was cut for resonance at 7016 kHz. This model was erected in a completely vertical format, using 143 feet 3 inches of wire. The Q section is 23 feet 2 inches long. The apex (feed point) is 4 feet above ground. The SWR on 40 meters is less than 2:1 across all of the band. The 80- and 40-meter W1SE loops showed resonance slightly apart from the design frequency, perhaps because of the proximity of the antennas to ground. Resonance on 40 meters was checked as 7050 kHz. Both loops are performing better for local and DX contacts than any of the many antenna types tested at W1SE. We would be even more impressed if we could elevate our Delta loops so the lower portions were a half wavelength or greater above ground.

In Conclusion

There is no rule that dictates the shape of a full-wave loop. The triangular format is convenient for mounting the radiator. If the apex is at the top, only one high support structure is needed. You may have one or more tall trees that can be used as supports. Circular, square or rectangular shapes have been used by many amateurs and the results were good. Certainly, a loop is an impressive receiving antenna, in terms of noise reduction. In some urban locations, that may be more important than transmitting a "death-ray" signal! There is something to be said about the age-old expression, "If you can't hear 'em, You can't work 'em."

An 80-meter Delta loop can be used on 160 meters by adopting one of two simple methods (Figure 4). A closed loop does not, however, offer good results when the overall length is a half wavelength. Either of the techniques in Figure 4 will work, but the method at A requires a ground radial system for best results.

Notes

[1]D. DeMaw, "Beat the Noise with a Scoop Loop," *QST*, July 1977, and "Maverick Trackdown," *QST*, July 1979.

[2]H. H. Beverage and D. DeMaw, "The Classic Beverage, Revisited," *QST*, Jan. 1982.

A Triband 75/40/30-Meter Delta Loop

WB4DFW wanted to use his Delta Loop on other bands besides 75 meters, so he installed some coaxial traps. He also reveals an effective technique to prevent breakage when support trees sway in the wind.

I've been using horizontally mounted, coax-fed Delta Loop antennas on 75 meters for many years. I've found them to be superior to dipoles—they are generally quieter on reception (perhaps because they are dc-grounded), exhibit broader bandwidth (typically less than 2:1 SWR over more than 200 kHz using a coaxial matching section), and are often more easily adaptable to available tree supports. They also don't fall down very often when installed correctly!

Outside of using an antenna tuner, a 265-foot-long 75-meter loop doesn't lend itself to working other bands–and I don't have a tuner. So while listening to a virtually dead 75-meter phone band one day and wishing I had a way to hit 30 meters for a long-overdue dose of CW, I had a brainstorm. Why not use traps? I'd homebrewed coaxial traps based on Sommer's article[1] several times in the past, and I knew they were easy to make and adjust.

The Traps

A quick examination of the latest incarnation of my 75-meter loop found it to be a reasonably equilateral triangle (a little over 85 feet per leg). I calculated that a 30-meter dipole should be something over 46 feet in total length (465/frequency in MHz). It appeared to be a simple enough matter to station some homebrewed coaxial traps 23 feet on each side of the feedpoint. A little more calculating found that it might even be practical to put in a pair of 40-meter traps as well.

Sommer gives a useful set of nomographs for computing trap dimensions on PVC pipe forms, using either RG-58 or RG-174 coax (the former capable of handling a kW with few problems). I opted for the larger cable. The 30-meter traps each require 46.8 inches of coax on a 1.62-inch OD section of PVC (6.75 turns). The 40-meter versions each use 61 inches of coax on a 2.25-inch OD PVC section (6.0 turns). Both PVC sections can be any convenient length. I used four-inch long pieces.

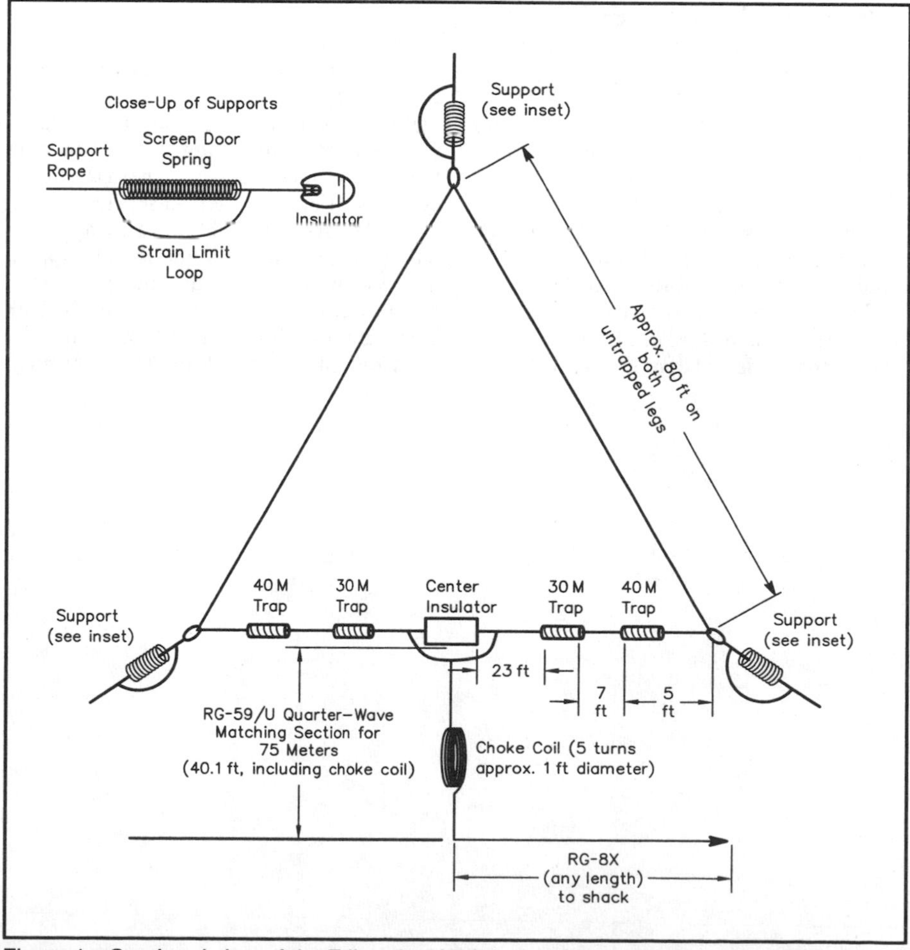

Figure 1—Overhead view of the Triband 75/40/30-meter Delta Loop, which WB4DFW mounted 40 feet off the ground. The traps were constructed with RG-58 coax. See text for winding details.

I wound the 30-meter traps in less than an hour and inserted them in the loop antenna 23 feet either side of the feedpoint. A quick SWR check found them resonant well below the bottom of the 30-meter band, but spacing the turns slightly apart and securing them with electrical tape raised the resonance easily enough, with the SWR bottoming out at 1.2:1 at 10.125 MHz.

The 40-meter traps went together with similar ease. Placement in the existing antenna was a little more challenging than with the first set of traps. The loading caused by the 30-meter traps made determining the physical location for the 40-meter traps uncertain. Grebenkemper notes that traps in a half-wave antenna tend to "pull" an antenna toward resonance,[2] even if they are not placed at exactly the optimum location. So, I simply took the formula length for 40-meter resonance, generously subtracted the physical length of the coax in each of the 30-meter traps and installed the 40-meter traps accordingly.

The approach was unscientific, but SWR was 1.3:1 at 7.1 MHz on the first try. Dumb luck, I guess, and I opted to make no further adjustments. The final antenna layout is shown in Figure 1, and a close-up photo of a trap is shown in Figure 2.

Figure 2—Close-up photo of one of the 40-meter traps. The ends of the coax cable used in each trap have been sealed against moisture with silicone caulking. See text for construction details.

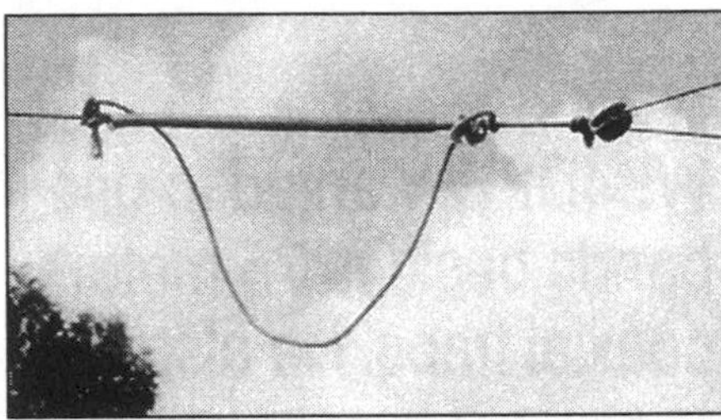

Figure 3—Photo of one of the support system springs, with rope used to limit total stretch. The rope is twice the length of the unstretched spring.

The Matching Section

The ARRL Handbook notes that the typical impedance of a full-wave loop antenna is on the order of 100 Ω, or about a 2:1 SWR presented to a 50-Ω feed line at resonance. The original WB4DFW 75-meter Delta Loop used a quarter-wavelength coaxial matching section (also described in the *Handbook*) between the feedpoint and the coax going to the shack. The matching section is merely 40.1 feet of RG-59 75-Ω coax [(465/3.85)/2 ×.66, the velocity factor] and transforms the load of the antenna very close to the 50-Ω characteristic impedance of the RG-8X feed line I use.

I also use a five-turn loop of the matching section coax to form an RF choke near the antenna feedpoint to minimize feed-line radiation. The matching section has no perceptible effect on the new 40 and 30-meter antenna SWRs (checked empirically by temporarily removing the section and feeding the antenna directly with the 50-Ω coax), so it has been left in service.

What effect did all this have on my original 75-meter loop? As one might expect, the resonant frequency dropped as a result of the traps' loading effects (nearly 200 kHz). But pruning about 8 feet off the far apex end of the loop brought my normal 1.2:1 resonant SWR back to my regular 3.842-MHz stomping grounds.

Keeping It Up in the Air

Here in South Carolina, pines are a great source of antenna support but their swaying during summertime thundershowers can wreak havoc on long, unprotected wire antennas. Over the years, I've found a simple shock-absorber system using a run-of-the-mill screen-door spring placed between a support rope and each antenna insulator can save a lot of grief. See Figure 3 for a closer look at the details.

A special aspect of my system is the employment of a "strain limit rope" in parallel with each spring. This piece of rope, approximately twice the length of the unstretched spring, keeps swaying pines from overly taxing the spring and gives a nice visual reference for gauging how much tension the antenna is operating under. It also provides a fail-safe for the inevitable corrosion-caused failure of the spring after several years aloft. It makes a fine bird perch, too.

So, How Does It Play?

Performance on all three bands has been most satisfactory. The 2:1 SWR bandwidth of the 30-meter section far exceeds the band edges. Similar 40-meter bandwidth is about 100 kHz (noticeably narrower than dipoles I've used, and I suspect partially attributable to my guesswork on the 40-meter trap placement). The 75-meter bandwidth is now approximately 150 kHz (slightly narrower than before the traps were installed). Signal reports both ways on all three bands are comparable to any other non-directional HF antennas I've used.

Notes and References

[1]Robert C. Sommer, "Optimizing Coaxial Cable Traps," *QST*, Dec 1984, pp 38-42.

[2]John Grebenkemper, "Multiband Trap and Parallel HF Dipoles-A Comparison," *QST*, May 1985, pp 29-30.

A Two-Band Loop for 30 and 40 Meters

After trying to find a way to place a 30-m delta loop inside an existing 40-m loop, I remembered an article in *All About Cubical Quad Antennas*[1] describing a $1^1/_2$-λ, or "Mini X-Q," loop. The gain of this antenna was said to be about 1 dB more than a 1-λ loop. I installed a large, ceramic SPST knife switch in the center of the delta-loop's bottom leg (see Fig 11). With this switch open, the full-wave, 40-m loop becomes a $1^1/_2$-λ, 30-m loop! The resonant frequency of this arrangement was 10.5 MHz. By adding 18-inch wires to the loop at both sides of the switch, I obtained resonance at 10.125 MHz.

Since the bottom of the loop is only 12 ft above ground, it's a simple matter to reach the band switch from ground level. (Caution: High RF voltage appears at the switch when the antenna is used for transmitting on 30 m.) Incidentally, the loop also works well on 15 m (SWR under 2:1 across the band) when set for 40 m, and I have used the 30-m configuration successfully on 80 m with the help of an antenna tuner.—*James Brenner, NT4B*

[1]W Orr and S. Cowan, *All About Cubical Quad Antennas* (Wilton, CT: Radio Publications, 1970).

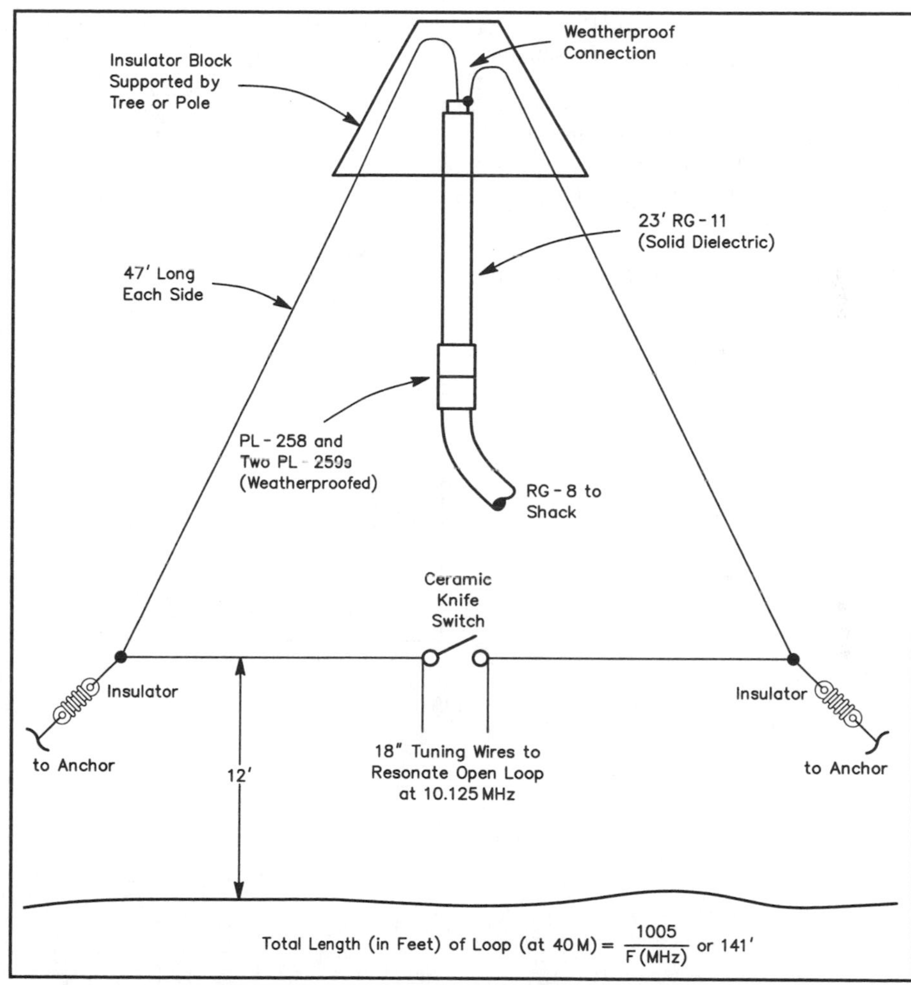

Fig 11—Jim Brenner's 30- and 40-meter loop. Note the 18-inch tuning wires used to lower the antenna's 30-m resonance from 10.5 to 10.125 MHz. The antenna is top-fed via a $^1/_4$-λ 40-m matching section. See text.

Feeding an 80-Meter Delta Loop at 160 Meters

After reading "The Full-Wave Delta Loop at Low Height,"[1] I found a satisfactory method of feeding an 80-meter loop at 160 meters. See Figure 1. C1 tunes the antenna to act as a $^3/_4$-λ resonator and allows the SWR at the feed point to be no more than 1.1 to 1 across the 160-meter band. *—Roy C. Koeppe, K6XK*

[1]D. DeMaw and L. Aurick, *QST* (Oct 1984, pp 24-25).

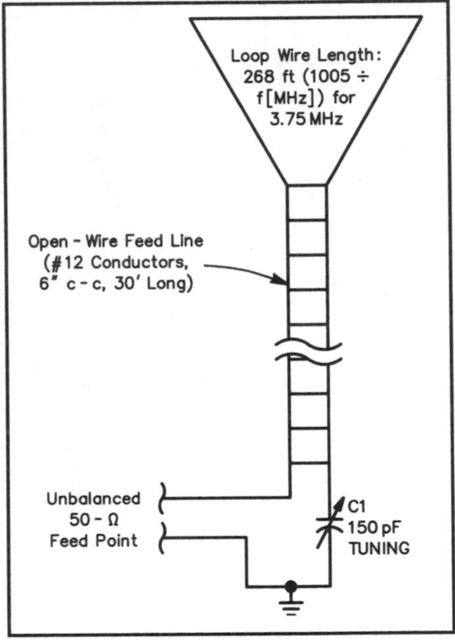

Figure 1—Roy Koeppe enjoys satisfactory 160-meter operation with an 80-meter delta loop by feeding the loop as shown here. On 80 through 10 meters, Roy takes C1 out of the circuit and feeds the loop via its open-wire feed line and a balanced tuner.

CHAPTER FIVE

COLLINEAR ANTENNAS

By Hugo Romander, W2NB

From *QST*, June 1938

The Extended Double-Zepp Antenna

Simple Antenna Structures Having Improved Gain and Horizontal Diversity

The questions of antenna directivity and antenna "gain" are becoming increasingly popular in discussions on antenna systems among amateurs, and it would therefore seem proper to preface an article dealing with directive antenna arrays with a warning, or perhaps a reminder, that one cannot have high gain and radiate in all directions at the same time. To most amateurs this is an obvious fact, but it may not be as well known that this principle is almost equally important in the *vertical* plane. Those of us fortunate enough to have available large open spaces in which to hang wires will find long-wire antennas, such as the "V" and the horizontal diamond, the easiest way to obtain high gain, but the radiation from such antennas is restricted in the vertical plane fully as much as in the horizontal plane. The result may well be that in the very direction such an antenna is supposed to work best, a simple horizontal doublet will put in a far better signal at certain distances.

It would seem, therefore, that the most practical antenna for a variety of distances is one with a fairly wide radiation pattern in the vertical plane.[1] The "stacking" of elements or the use of long wires is not recommended for distances under 1000 miles, except for the very short distances normally reached by the ground wave. The most universal high-gain antenna must restrict its radiation in the horizontal plane only and its height above ground must also be considered to obtain the best compromise in the vertical plane. It is the purpose of this article to discuss the merits of a simple antenna array which combines these desirable features.

THE DOUBLET

Let us discuss, first, the simple doublet, since this antenna will serve admirably as our basis of reference or comparison. To be more specific, consider this doublet as suspended horizontally and fed at one end in the time-honored fashion of the Zepp feeder, as in Figure 1A. Ignoring, for the moment, the fact that the open-ended feeder wire will be at somewhat higher potential than the other feeder wire and will therefore radiate, the horizontal radiation pattern about this doublet, if you're lucky, will be about as shown on Figure 2. Of course, it is assumed the antenna is sufficiently remote from power-line wires and house plumbing to be unaffected by such linear conductors, since our problem is complicated enough without having to consider the mutual impedance between our doublet and the neighbor's b.c.l. antenna!

This pattern of the horizontal doublet is, no doubt, familiar to most amateurs. Less familiar, perhaps, is the vertical radiation pattern (in the direction of maximum horizontal radiation) as shown in Figure 3. Here it is assumed that the height of the doublet above ground is one-half wavelength and that the earth has perfect conductivity. Fortunately, even the relatively low-conductivity soil and sand on Long Island reflects a high percentage of the horizontally polarized waves radiated at angles less than 50° to the earth's surface. Such reflected

[1] Although this viewpoint appears to be opposed to that expressed in the article "Simple Directional Arrays Using Half-Wave Elements," by N. C. Stavrou, in May, 1938 *QST* there is no actual conflict. As the present author points out, ihe broad vertical characteristic is to be preferred when the antenna is to give optimum results over short as well as long distances; the former article was concerned with long-distance transmissions, where the lowest possible angle gives best results under nearly all conditions. The type of work to be carried on naturally will be a determining factor. In any event, the simpler structures are not likely to be too sharp in either plane for satisfactory general work. —EDITOR

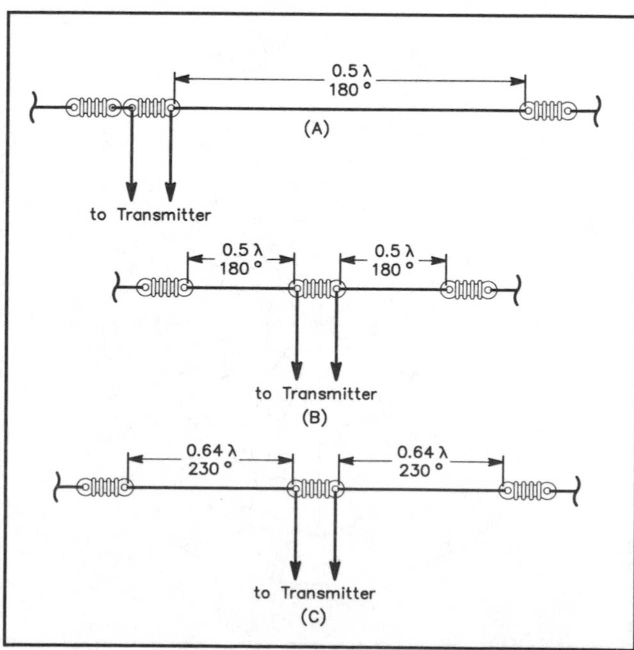

Figure 1—Basic Antenna Configurations (A) The half-wave Zepp; (B) Double Zepp or "two half-waves in phase"; (C) Extended double-Zepp.

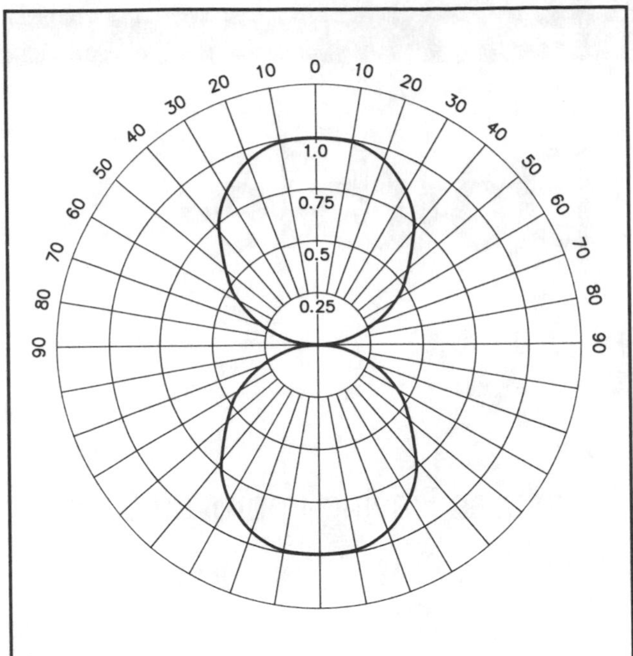

Figure 2—Horizontal pattern (relative field strength) of a doublet half-wave antenna.

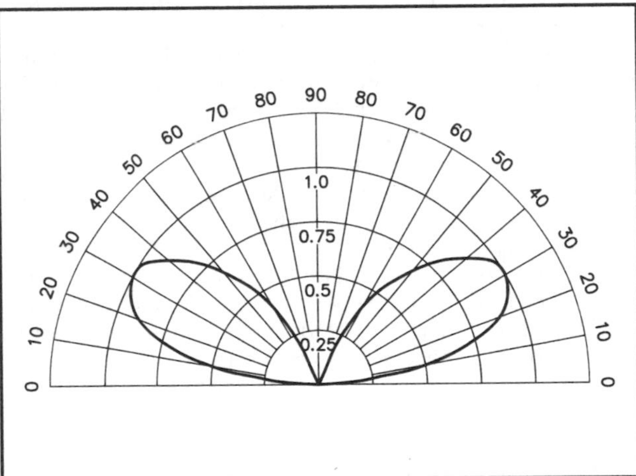

Figure 3—Vertical-plane pattern for a horizontal antenna, 1/2-wave above perfect earth in direction of maximum radiation.

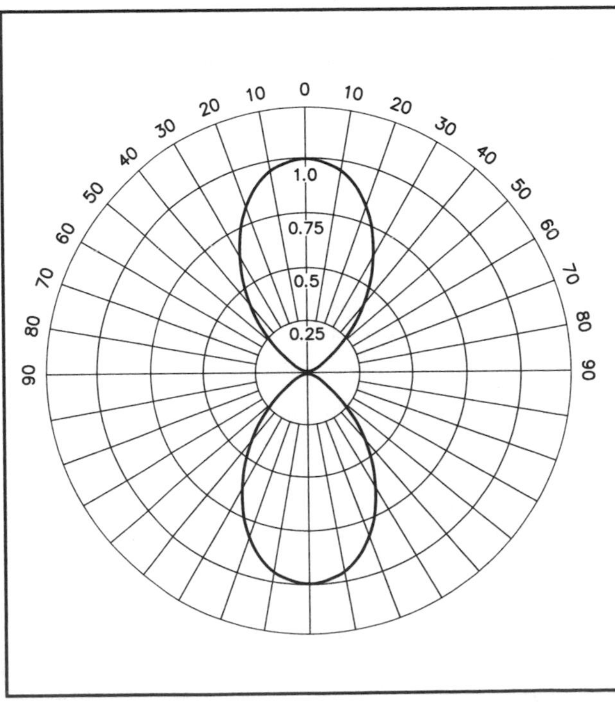

Figure 4—Horizontal pattern of a double-Zepp or two half-waves in phase.

waves, combining with the direct radiation from the antenna within the range of vertical angles in which we are most interested, are responsible for a maximum gain of nearly 6 db over the same doublet in "free space"; that is, without the presence of the earth.

This gift of 6 db must not be taken too much for granted, however. Consider the fact that most amateur communications at high frequencies utilize vertical angles ranging from 10 degrees to 50 degrees. That 6 db must, therefore, come from radiations reflected from the earth's surface quite some distance from the doublet, and in the desired direction of transmission. If the ground slopes sharply upward, or houses or wires are in this area, it becomes questionable, indeed, whether any great portion of the available 6 db is realized. This is especially true for the lower angles of radiation, but usually nothing can be done about this situation, so let us see if we can make up for our ground reflection losses by increasing the horizontal directivity.

THE DOUBLE-ZEPP

By the simple expedient of attaching another doublet to the open-ended terminal of the Zepp feeder, as in Figure 1B, and hanging this doublet parallel and coaxial with the original doublet, an appreciable gain may be obtained. The horizontal radiation pattern will be as shown in Figure 4. The gain should measure about 1.9 db, corresponding to a 55 per cent increase in power. This antenna is widely used among amateurs and is popularly known as the "double-Zepp" antenna, or "two half-waves in phase." A fair amount of gain and a reduction in interfering signals from end-wise directions is easily realized, but the gain is rather disappointing, since it would seem one is entitled to 100 per cent gain in power when doubling the number of radiating members.

The reason why only 1.9 db is realized with the double-Zepp is made clear only after a mathematical study of the situation. Briefly, the close proximity of the two doublets causes a mutual coupling between them, and this coupling (mutual impedance) has an adverse effect on the radiation resistance insofar as gain in the broadside direction is concerned. Obviously, the thing to do is to move the doublets further apart, but this complicates the method of feed. A much simpler way of obtaining increased gain was evolved by Mr. A. A. Alford, of Mackay Radio and Telegraph Company, from the principles discussed in Mr. G. H. Brown's article on broadcast antennas in the Proceedings of the I.R.E. for January, 1936. Mr. Alford presented this idea in a paper delivered at an I.R.E. meeting in Washington.

EXTENDED DOUBLE ZEPP

The gain of the double-Zepp may be increased from 1.9 to 3.0 db by the simple expedient of increasing the length of each doublet until its electrical length is 0.64

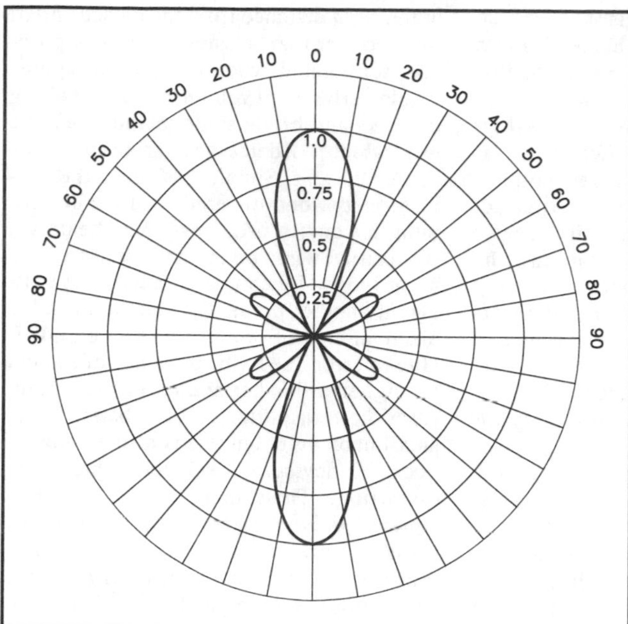

Figure 5—Horizontal pattern of double 230-degree Zepp

wavelength instead of 0.5 wavelength. In electrical degrees the double-Zepp consists of two 180-degree elements; the *extended* double-Zepp for maximum gain should consist of two 230-degree elements. See Figure 1C. In this way the power gain may be increased from 55 per cent to 100 per cent. The gain decreases rapidly for extensions beyond 230 degrees, and therefore, when operating over a band of frequencies, each of the two elements should not exceed 240 degrees for the highest frequency. The horizontal pattern for the double 230-degree Zepp is shown in Figure 5. The vertical pattern in a plane perpendicular to the antenna will be the same as for the simple doublet.

ANTENNA IMPEDANCE

The impedance of this antenna at the termination of the transmission line is of interest, since it has an important bearing on the standing-wave ratio of current or voltage in the line, and it will be compared with that of the ordinary double-Zepp. The double 180-degree antenna presents an impedance of approximately 4400 ohms of almost pure resistance as a termination for the transmission line. This value will be slightly affected by the size of wire used in the antenna and, to a moderate extent, by the height above ground or the influence of nearby conductors, and so the "free space" value is given for No. 14 wire. With this antenna and a 600-ohm surge-impedance line the ratio of maximum to minimum current along the line will be 4400 divided by 600, or 7.3. Incidentally, the terminating resistance of the simple Zepp-fed doublet is about 12,000 ohms, resulting in a standing wave ratio of 20 on our 600-ohm line. For calculation of the surge impedance of the line the reader is referred to *The Radio Amateur's Handbook*.

The impedance at the center of the double 230-degree antenna is not a pure resistance, and hence its effect upon the transmission line be such that maxima or minima of voltage and current along the line will not be odd multiples of a quarter wave from the antenna, as with the ordinary Zepp or double-Zepp antennas. As might be expected, the current or voltage maxima will be shifted towards the antenna, since the two doublets are longer than normal, and this shift is approximately 0.13 times the wavelength. At any rate, the antenna impedance is such that the equivalent pure resistance at any voltage maximum will be about 6000 ohms; that is, the standing-wave ratio will be 10 on a 600-ohm line.

Knowing the standing-wave ratio, it becomes an easy matter to calculate the input resistance to the transmission line if it was cut, let us say, at any current maximum. Thus, for the simple Zepp-fed doublet, this resistance would be 600 divided by 20, or 30 ohms. For the ordinary double-Zepp antenna this resistance would be 600 divided by 7.3, or 92 ohms. For the double 230 degree antenna this resistance would be 600 divided by 10, or 60 ohms. Those who do not use correcting stubs or other methods to obtain a "flat" line (standing wave ratio of 1), but who cut their transmission lines at the maximum current point after the manner of the "tuned feeder," will find these figures useful in calculating the power output of their transniitters.

FEED LINES

However, the use of flat lines, in amateur circles usually referred to as untuned lines, is becoming increasingly popular as their merits are more generally known, and the figures given are essential in calculating the length and position, for example, of a correcting stub, the latter being one of the simplest devices for reducing the standing-wave ratio to unity. A discussion of the use of the correcting stub may be found on page 307 of the *Handbook* (15th edition) where a table indicates the lengths and positions of the stub for various standing-wave ratios. The shorted stub or "loop" will generally be most practical, and some idea of dimensions may be obtained from the fact that when using the double 230-degree Zepp on 14,200 kHz the stub should be 3.5 feet long; that is, it should be composed of the same kind of wire as is used in the transmission line, this wire being 7 feet long and bent to form a U-shaped rectangle with a width equal to the transmission line. At the frequency mentioned, where each extended doublet is 43 feet long, the correct location for the stub will be about 8 feet from the antenna.

Ordinarily the best place to introduce power into an antenna is at a point symmetrically located with respect to the opposite ends of the system. This will not always be practical, because of space limitations, and hence it may become necessary to feed the extended double-Zepp antenna at one end in true Zepp fashion. Some sacrifice in gain will result, due, for the most part, to unequal distribution of current between the two doublets and to radiation from the feeder because it has an unbalanced load. Physical dimensions of feeder, extended doublets, and phasing stub are shown in Figure 6. Note that the phasing-stub length is shorter than a quarter wavelength, but by an amount not exactly

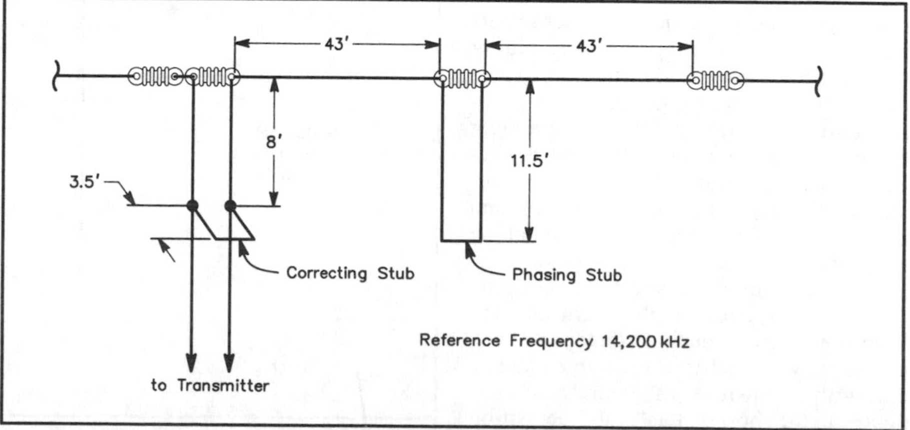

Figure 6—End-fed double 230-degree Zepp with 600-ohm non-resonant transmission line.

equal to the extension of each doublet beyond a half wavelength. This discrepancy is due to interaction between the doublet wires and the wires of the phasing stub. Again a reference frequency is given to simplify calculation of lengths at other frequencies by inverse proportion. The figures given should be regarded as approximations, and tests made individually to determine dimensions and positions of important elements.

ANTENNA ADJUSTMENTS

One of the most effective methods of adjusting an antenna system is to use the transmitter to excite, at the desired frequency, another antenna stretched out, perhaps temporarily, at least 8 or 10 feet from the ground. This "exciter" antenna should preferably be at least one-half wavelength away from the antenna to be adjusted. If the two are parallel to each other, so much the better. Using the end-fed extended double-Zepp antenna as an example, the first consideration is the length of each doublet, and this brings up the problem of "end-effect." We have approximately solved this by cutting our doublets 5 per cent short of the theoretical length. Actually the end effect, which normally exists only at the ends of the antenna, depends on the wire size and the size of the insulator fitting (if metal caps are used), as well as the frequency. Moreover, the end-effect does not vary quite as rapidly as a direct function of the wavelength, so that if an end-effect of 1 foot is correct at 14,200 kHz, it is more nearly 1.5 feet than 2 feet at 7100 kHz.

The theoretical "uncorrected" length of each half of an extended double-Zepp antenna at 14,200 kHz is 44.3 feet. The end-effect may be no more than 1 foot, but since it is better that the antenna be a little too short than too long, let's make the correction 1.3 feet, so that each half will be exactly 43 feet long. At the center insulator which separates the two halves a "stub" must be connected, if the system is to be end-fed. Now, this stub will, take the form of a transmission line shorted at the end opposite from the antenna, and for the frequency under consideration its shortest length will be about 12 feet. Since it is our intention to make this transmission line somewhat longer than the anticipated proper position of the short in order to permit locating the correct position by sliding the short up or down, and furthermore since 12 feet will ordinarily not bring the probable correct shorting point at a convenient distance from ground, it will be more practical to add any multiple of a half wavelength (34 feet) to the transmission line and do the shorting experiment at a point along the line accessible from the ground.

With all this done and with the Zepp feeder *not* attached to the main antenna (why not use it to feed the exciter antenna?), power may be fed to the exciter antenna and with a sensitive r.f. instrument connected in the short on the "stub," a position of the short may easily be found where maximum current flows. This assumes that the main antenna has been hauled up into its normal operating position. Note this maximum-current position of the short and measure in exact multiples of one-half wavelength from this point towards the antenna, thus arriving at the nearest point to the antenna at which a short may be placed. This will give the shortest possible center stub. Of course, it is not necessary that the center stub be made that short, since very little loss will be incurred if it is left a half-wave or even a wavelength longer–that is, if the wire is not smaller than No. 14. The Zepp feeders may now be attached to either end of the antenna, as it is now ready for operation.

The method of adjusting the center stub just outlined is also useful where it is desired to feed any of the antenna forms discussed above with an untuned transmission line—that is, a transmission line terminated by its surge impedance. Using the example cited in the previous paragraph, let us assume the point of maximum current through the short has been found and, after soldering a piece of wire across the line at this point, the extra length of transmission line is chopped off. The two-wire line from the transmitter may now be tapped in above this shorted end of the stub, and it is only a question of position of the feeder tap from the shorted end to terminate the line properly. See Figure 7. This point will be approximately 3.3 feet from the short for a 600-ohm transmission line. Of course, when all this is done, we will have the same transmission line with correcting stub discussed earlier in this article.

LINE-CURRENT MEASURING DEVICES

The same principle holds true for the double Zepp and for center feeding the simple doublet, the only ambiguous point being that of knowing when the transmission line from the transinitter is connected at the right distance from the closed end of the stub. The transmission line will be properly terminated when there is no appreciable variation in voltage or current along the line, and hence some method must be used which will detect any variation in voltage or current. Simplest of all is the old-fashioned wood-covered lead pencil, for with this device bright arcs may be drawn from either wire and the voltage at any one point judged by comparison with the voltage at another point. This method has the disadvantage of being both crude and liable to error due to the presence of second or higher-order harmonics in the output from the transmitter. A little better is the small neon bulb, but this is also a voltage-operated device and subject to harmonic distortion. Then there is the current-squared galvanometer with a few small turns of wire connected to its terminals, these turns being coupled to either line. This device is quite good if one is careful to use constant coupling to the wire, but its disadvantages are that coupling between the meter and line is both inductive and capacitive, causing the meter to read differently according to how it is held up to the line, and that the meter will also be influenced by harmonics.

The most satisfactory device seems to be a sensitive galvanometer in a miniature tuned circuit coupled to the line. Such a device combines sensitivity with freedom from harmonic distortion, but the effects of capacity are still there to some extent. The tuned circuit should therefore be carried along the line with its position relative to the plane of the two wires maintained constant. See Figure 8. Checking along one of the two wires only is usually sufficient, but if one suspects an unbalance exists, the other line should also be checked. If the antenna system is reasonably symmetrical, unbalanced currents will most likely be due to improper coupling to the transmitter, but that is another story. At any rate, if a cur-

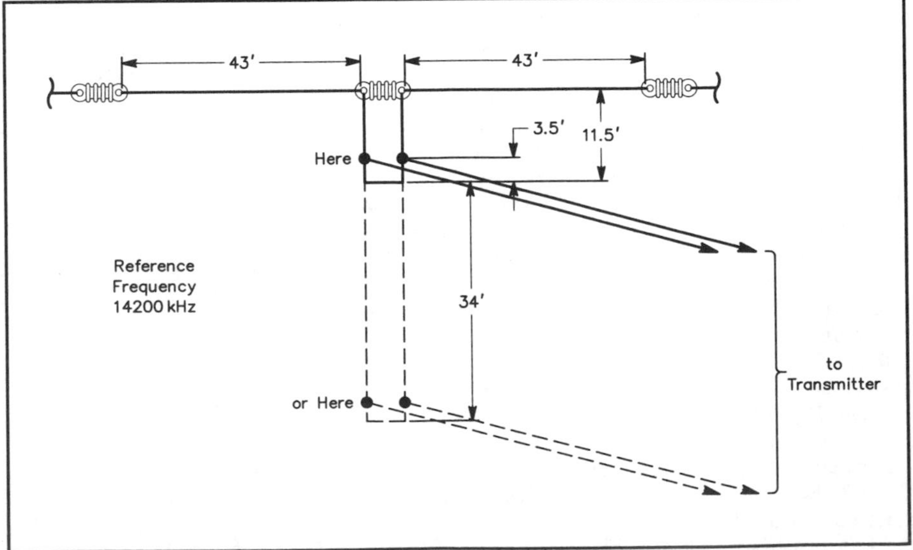

Figure 7—Center-fed double 230-degree Zepp with 600-ohm transmission line.

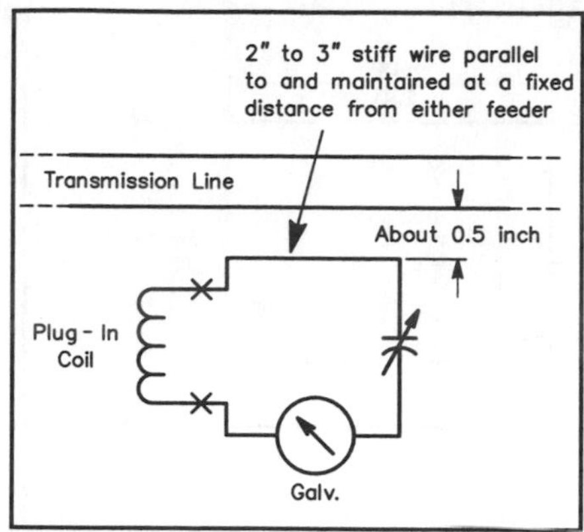

Figure 8—Portable line-current indicator.

rent maximum (voltage minimum) appears at the junction of transmission line and stub, the connection to the stub is too near the shorted end, and if the current is a minimum the opposite, of course, will be true. If the current maximum or minimum (assuming standing waves actually exist) does not occur near the stub, there has been a slip-up somewhere in the procedure, for this will be evidence that the stub itself did not tune the antenna to resonance.

PARASITIC ELEMENTS

The use of spreaders several feet in length to support each end of a "flat-top" antenna composed of two wires is again becoming popular, and for a good reason. Gains of around 5.5. db for a bi-directional array and around 7 db for a unidirectional array are thus made possible. One of the most promising arrangements is that of two double-230-degree antennas supported parallel to each other only 0.2 wavelength apart, as illustrated in Figure 9. Antenna A is excited by the transmitter using the same technique as when exciting a single double-230-degree antenna. Antenna B may be tuned by proper positioning of a short on its stub to become either a director or a reflector. In fact, this stub might be extended by multiples of a half wavelength to enter the operating room where, by means of a single switch the shorting position may be changed from that corresponding to a director to the position corresponding to a reflector, thus reversing the directivity of the antenna system. Information as to the gain of such an antenna system when one antenna is used as a director is not available, but when it is used as a reflector and adjusted to give a *minimum* signal to the rear, the signal forward will be 7 db better than a simple doublet, and the signal backwards will be 7 db less than from a doublet, resulting in a front-to-back ratio of 14 db. Adjustment of the reflector shorting bar should be on the basis of minimum backward signal from the antenna, or minimum signal received from a station in that direction when using the antenna system for reception, since this adjustment can be made more accurately than one resulting in maximum front signal. This will minimize QRM for both transmitting and receiving. The same principles hold true when adjusting the auxiliary antenna for use as a director.

If adjustments of the auxiliary antenna are to be made when transmitting and an ammeter is temporarily inserted in the shorting bar, minimum backwards signal will occur for the reflector when the short is moved slightly further from the antenna than that position corresponding to maximum current through the short. For the director, optimum conditions will be obtained when the short is slightly closer to the antenna than the maximum shorting current position. From this it can be seen that the proper adjustment of the auxiliary antenna is quite critical and certainly very important. Since field strength instruments are not generally available, a receiver fitted with a signal strength indicator of some sort located within a few miles of the antenna and in the desired direction should prove to be the next best thing. Connecting your own receiver to the antenna, with the other fellow transmitting, is perhaps even more practical in search for that minimum-signal adjustment.

The presence of the auxiliary antenna only 0.2 wavelength away from the driven antenna and adjusted properly for maximum forward or backward radiation will obviously affect the radiation resistance and impedance of the driven antenna. As a result, if the auxiliary antenna is left open-circuited and the stub on the driven antenna is adjusted in accordance with the method previously given, this adjustment will not be correct when the auxiliary antenna is, in turn, properly tuned. Moreover, the impedance of the driven antenna will differ slightly according to whether the auxiliary antenna is tuned as a director or as a reflector. If quick change from reflector to director is contemplated, some sort of compromise is indicated in the adjustment of the stub on the driven antenna. From a practical standpoint, however, the adjustment of the auxiliary antenna should be made first, be it reflector or director, with the driven antenna excited by the two-wire transmission line without a correcting stub. This may require temporary adjustment of the transmission-line length, either physically or by means of series coils or condensers, to bring a low-impedance point at the coupling coil to the transmitter in order to load the latter satisfactorily. In fact, if the line length from antenna to transmitter is not more than a wavelength or so, the reduction in line losses resulting from the use of a correcting stub is hardly worth the trouble of installing it. The point is, if a "flat" line is desired, adjustments of the stub at the driven antenna should not be made until after the auxiliary antenna has been tuned to give the desired radiation pattern.

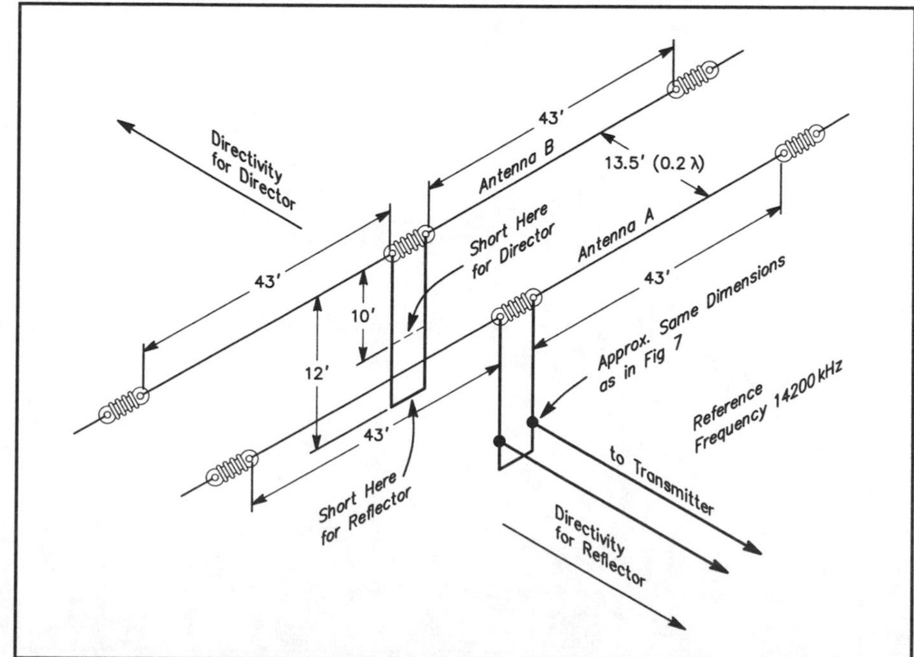

Figure 9—Double 230-degree Zepp with parasitic reflector-director.

LARGER COLINEAR ARRAYS

Now let us suppose our backyard is big enough to hang up more than two doublets end to end. An interesting possibility would be a four-element array of 230-degree elements. But here the principles of the extended double-Zepp must be carefully considered in designing the length of each element and the length of the phasing stub separating them. The actual arrangement will be as shown in Figure 10. Both degrees and dimensional designations are used to indicate the electrical and physical length of each element. The length of the stubs indicated is approximate only and must be adjusted for best results. The lengths of the various antenna elements as shown, however, may be assumed correct. The transmission line can be tapped on to any of the stubs, but connection to the center one will give the greatest gain.

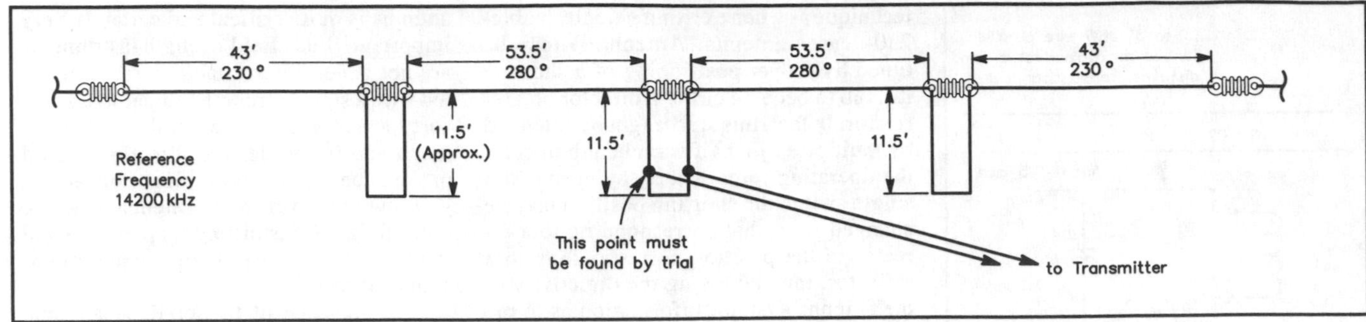

Figure 10—Four-element array with 230-degree elements.

The principle of the design shown for four elements is to provide the same separation in space between each doublet center as is provided between the two doublet centers in the double-230-degree antenna. Each phasing stub will, of course, be considerably shorter than a quarter wavelength, and when adjusting their length to tune the whole antenna to resonance, all three stubs must be made the same length; that is, if one is shortened two inches, the other two must also be shortened by two inches, assuming they were all the same length to start with.

This may prove to be an awkward and tedious method of adjustment, but the procedure may be simplified by adjusting the length of the center stub only as a first approximation. The use of an "exciter" antenna temporarily rigged nearby is assumed. If it is found the center stub must be shorter than the other two stubs by 3 feet, for instance, in order to obtain maximum current through its shorting bar, the other two stubs should be shortened by 1 foot and a new position found for maximum current through the center stub. This new position should correspond closely to equal length for all three stubs. This procedure of tuning the four-element array should be followed even though a stub at the center will not be used, finally, its place being taken by the transmission line itself. Such a connection will result in standing waves along the line and it will have to be "tuned" to permit easy coupling to the transmitter or receiver.

The horizontal pattern of the four-element array just described will have two major lobes at right angles to the antenna and several minor lobes of small amplitude. The gain in actual practice has proved to be greater than anticipated, and probably is more than 7 db. The major lobes will be much narrower than for the double-230-degree antenna, thus requiring more careful "aim" in erecting the array, or provisions for swinging it about. It is assumed this array, as well as the others described, is hung horizontally. It may be hung vertically, of course, but the possible restrictions due to sharp directivity in the vertical plane should then be considered. In general, the horizontally polarized antenna will prove most practical, chiefly because the ground reflection usually encountered with vertically polarized radiations is inferior, and because high masts are not often available.

The adjustment of the antennas described in this article may seem a bit involved as compared with the simple doublet or the double-Zepp antenna, but the extra gain thus made available should make their use worthwhile. For that matter, the double-230-degree antenna need not be tuned at all, its construction being made in accordance with the dimensions given and the transmission line tuned to fit the transmitter. In this respect the simplest of all directive antennas, size and gain also taken into consideration, is the extended double-Zepp.

Try an Extended Double-Zepp Antenna

Although the extended double Zepp (EDZ) antenna (Figure 1) has been in just every antenna handbook since the year one, hams seldom use it. Its overall length is 1.28 wavelengths (1.28 λ), and it's bidirectional broadside. Fed with open-wire line and a balanced antenna tuner, an EDZ also makes a fine multiband antenna. Let's look at an extended double Zepp for 17 meters. We can calculate the overall length of its two wire elements with the formula

$$\frac{984}{f\,(\text{MHz})} \times 1.28 = \text{length in feet} \qquad \text{(Eq 1)}$$

Using this formula, an 18.15-MHz EDZ works out to be 69.4 feet long. At this frequency, the EDZ exhibits, 3 dBd gain [*AB* page 8-34][1] in a figure-8 pattern with two major and four minor lobes. It still performs usefully, when operated on several bands lower in frequency than 17, however. At 20 meters, a 17-meter EDZ acts as two slightly long half waves in phase, exhibiting between 1.6 and 2 dBd gain [*AB* page 8-32]. At 40 meters, it's a slightly long ½-λ dipole [*AB* page 2-16,2-17,3-11 and 3-12]. All of these modes are directional broadside if the EDZ is positioned at least ½-λ high at 40 meters [*AB* pages 3-8 and 3-9]. At 15 meters, it exhibits a four-leaved-clover pattern, with minor lobes broadside; at 10 and 12 meters, it's close to two full waves in phase and produces a pattern similar to that at 15 meters. It can even be used as a short 75-meter dipole–not bad for a 70-foot piece of wire.

Scaled for 28.7 MHz–43 feet, 10 inches long—the extended double Zepp gives a 3-dBd-gain, figure-8 pattern at 10 meters and a similar pattern with a bit less gain at 12 meters. It acts as two halfwaves in phase at 15 meters, with about 1.6 dBd gain and a figure-8-pattern. At 17 and 20 meters, it's somewhat long for a half-wave dipole; a tuner can make it work at 40. On all of these bands, assuming that it's at least ½ λ high at its lowest band of operation, the 10-meter EDZ is directional broadside.

There's nothing magic about the extended double Zepp. It's a tried-and-true dipole that offers useful gain at its design frequency and good multiband performance. *–Bob Baird, W7CSD*

[1] This and the other AB references in this item refer to pages in G. Hall, ed, *The ARRL Antenna Book*, 16th ed (Newington: ARRL, 1991).

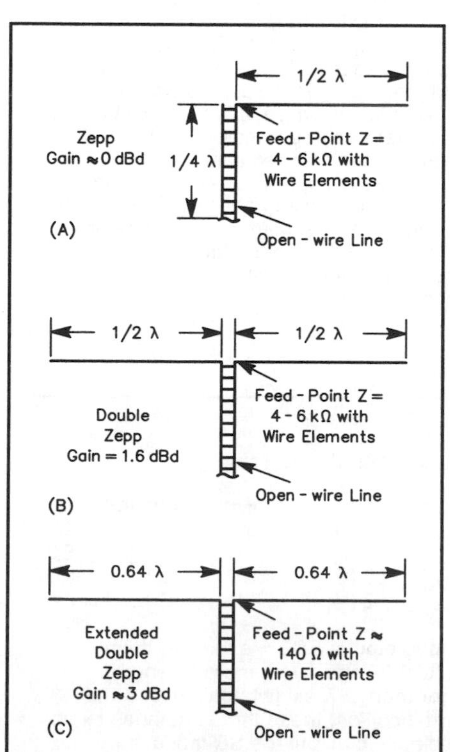

Figure 1—Evolution of the extended double Zepp antenna. At A, the classic Zepp—a ½-λ wire end-fed via an open-wire feeder operated with one of its two wires unterminated. (Antenna lore tells us that this feed method received its name from its use in zeppelin radio installations.) Commonly, Zepp systems use a feeder length of ¼ λ (or an odd multiple thereof) to transform the antenna's high feed-point impedance to a low impedance at the feeder terminals. Assuming efficient feed, this system exhibits 0 dBd gain—that is, *no gain* relative to a half-wave dipole cut for the same frequency and erected in the same position.

B shows a double Zepp, also known as *two half waves in phase* or (incorrectly) a center-fed Zepp. This is a form of collinear antenna because each ½-λ wire acts as an element, and because both elements lie along one line. It's also a *dipole*—a *full-wave* dipole that exhibits about 1.6 dB gain relative to a half-wave dipole in the same position. As with the Zepp, feeders cut to an odd multiple of ¼ λ make for least hassle feed with the double Zepp, but a good balanced antenna tuner should be able to handle whatever impedance appears at the feed-line input terminals.

A double Zepp becomes an *extended* double Zepp when you make its wires 0.64 λ long instead of 0.5 λ. Its gain increases, too—to about 3 dB relative to a half-wave dipole in the same position and cut for the same frequency.

The rough feed-point impedances shown here are for wire elements (fatter elements, such as those made of aluminum tubing, will lower these values somewhat) and convey only the resistive portion of the antennas' impedance. For more on the extended double Zepp antenna, see *The ARRL Antenna Book* and J. Reh, "An Extended Double Zepp Antenna for 17 Meters," *QST*, December 1987, pp 25-27.

Notes on Wire-Antenna Construction

Robert J. Zavrel, W7SX, seconds last month's extended-double-Zepp hint as he reports on how he keeps three EDZs high in trees for world-encompassing coverage (Figure 1) at 80, 40 and 20 meters—*Ed.*

What is a terminal DX hound to do when he suddenly moves to a rural area and has access to the stuff dreams are made of: several 100-foot-plus ponderosa pine trees atop a gentle rise in a reasonably clear area, only 500 feet from his house? The answer: *Wire antenna arrays!* A few months spent confronting the practical problems of installing three extended-double-Zepp antennas in these trees has yielded some solutions that I'd like to share.

Counterweights. Although my pines are wonderful supports for wire antennas, they have a few minor disadvantages. Among these is the fact that they sway in the wind. Because too much tension can snap antenna wires and supporting cords in strong winds, a wire-antenna engineer must compensate for variable tension in a horizontal wire antenna strung between two of these majestic plants. The textbook solution to this problem involves installing pulleys where the wires meet the trees. Ropes attached to the wire-end insulators pass over the pulleys and connect to counterweights hanging near the ground. Installed with antenna-to-counterweight lines of sufficient length, such a system also allows the antenna to be raised and lowered without further climbing.

My system includes three extended double Zepps, each fed with 450-Ω, low-loss TV twinlead. Including its supporting lines, each EDZ spans 165 feet. Counterweighting the ends of these antennas serves two purposes: It keeps tree movement from snapping them, and it keeps their wires reasonably horizontal. But how much weight is required, and what should the weights consist of? Charts exist for computing tension necessary to keep a singlewire span taut, but the presence of a twinlead feeder at the span's center renders such charts useless.

Enter experimentation! I strung the extended double Zepp, minus feed line and plus pulleys, between the two trees at about 15 feet above ground. Next, I weighed the feed line. Then I located an object of simi-

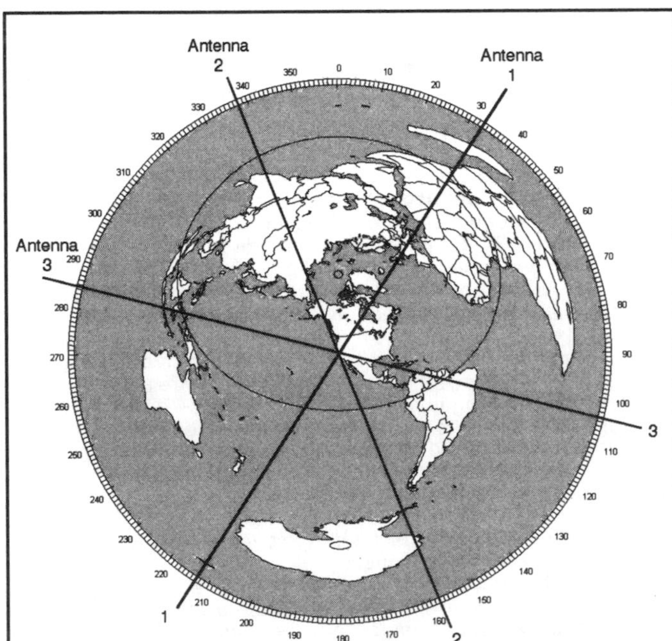

Figure 1—Proving the multiband effectiveness of the extended-double-Zepp antenna, "The W7SX Wire Farm" consists of three center-fed wires supported by three ponderosa pines that form a nearly equilateral triangle atop a small rise. (Total cost for the system, including all hardware, wire, feeders, control lines and $150 fee for the tree climber, approximately $700.) Because the ground slopes to the south at about a 10° angle, this system works best for VK, ZL and long paths into Europe and Asia. To the north, the horizon is essentially level. Table 1 shows how the antennas function on the system's three bands.

You're not a second-class citizen just because your best antenna is a dipole. You can easily set the pattern and gain of center-fed horizontal wire antennas by adjusting their length. Get the wires as high as possible, keeping them away from nearby objects, use a tuner and open-wire line for multiband feed, and have a blast! Surrounding terrain may greatly influence your antenna's DX performance.

Table 1
The W7SX Extended Double Zepp Trio

			Band/Antenna Mode		
	Length	*Height*	*80*	*40*	*20*
Antenna 1	82 ft	98 ft	SD	LD	EDZ
Antenna 2	82 ft	98 ft	SD	LD	EDZ
Antenna 3	164 ft	82 ft	LD	EDZ	LW

SD = short dipole; LD = long dipole; EDZ = extended double Zepp, LW = long wire (dipole legs 1 λ or more in length). Because a dipole longer than ½ λ exhibits gain over a half-wave dipole in the same position, these three antennas exhibit gain over a half-wave dipole in all but the SD mode.

Figure 2—Bob Zavrel uses a triangular relay box to switch feeders between antennas mounted in a triangular pattern; a square box in this situation twists feeders, en-couraging antenna sway to break feeder wires. *(W7SX photo)*

Figure 3—A combination feedthrough insulator/strain relief made of fasteners and two plexiglass plates.

lar weight, put it in a plastic bag, tied the bag at the antenna feed point, and experimented with various weights. Two 15-lb weights kept the antenna wire horizontal without overtension.

Next, I needed to construct multiple 15-lb weights at minimum cost—weights that had to be long and narrow to avoid collisions with the supporting trees and other nearby objects. I decided that slumpblock bricks inside a 2 × 4 wood frame would be satisfactory. (Slump blocks are long, narrow and about 12 lbs each; some are just as wide as a standard 2 × 4.) Including bricks, wood and nails, each weight cost me about $1.50, so the price was right—and 2 × 4s can be easily drilled for attaching ropes.

The relay box. A squarc relay box centered between three antennas twists the antennas' open-wire feeders at the feed points. I solved this problem for a triangular, relay-switched vertical array[1] and my three-EDZ system by building triangular relay boxes (Figure 2). (Use redwood, cedar or cypress and weatherproof the box with sealant.[2]) Plexiglass plates and screws insulate the feeder wires from the box, continue the feeders to the relays and provide strain relief (Fiureg 3).

Results with the system. Using a 600-W amplifier, I often get through pileups first on long path. Short path to Europe is more difficult because the ground slopes slightly upward in that direction. My blissful encounters with nature's gift to the RF enthusiast continue!–*W7SX*

[1]See Volume 3 of *The ARRL Antenna Compendium*.

[2]In my work with wire antennas, I discovered a reference on wood that may be of interest to amateurs: *Wood Handbook: Wood as an Engineering Material* (Washington, DC: US Department of Agriculture, rev 1987), 466 pages (for sale by the Superintendent of Documents, US Government Printing Office, Washington, DC 20402).This text describes, among many mechanical and structural considerations, electrical qualities of wood that come into play when wood supports are used in RF fields.

Collinear Arrays

Collinear arrays are always operated with the elements in phase. (If alternate elements in such an array are out of phase, the system simply becomes a harmonic type of antenna.) A collinear array is a broadside radiator, the direction of maximum radiation being at right angles to the line of the antenna.

Power Gain

Because of the nature of the mutual impedance between collinear elements, the feed point resistance is increased as shown earlier in this chapter (Figure 9). For this reason the power gain does not increase in direct proportion to the number of elements. The gain with two elements, as the spacing between them is varied, is shown by Figure 38. Although the gain is greatest when the end-to-end spacing is in the region of 0.4 to 0.6 λ, the use of spacings of this order is inconvenient constructionally and introduces problems in feeding the two elements. As a result, collinear elements are almost always operated with their ends quite close together–in wire antennas, usually with just a strain insulator between.

With very small spacing between the ends of adjacent elements the theoretical power gain of collinear arrays is approximately as follows:

2 collinear elements—1.6 dB
3 collinear elements—3.1 dB
4 collinear elements—4.2 dB

More than four elements are rarely used.

Directivity

The directivity of a collinear array, in a plane containing the axis of the array, increases with its length. Small secondary lobes appear in the pattern when more than two elements are used, but the amplitudes of these lobes are low enough so that they are not important. In a plane at right angles to the array the directive diagram is a circle, no matter what the number of elements. Collinear operation, therefore, affects only E-plane directivity, the plane containing the antenna. At right angles to the wire the pattern is the same as that of the individual 1/2-λ elements of which it is composed.

When a collinear array is mounted with the elements vertical, the antenna radiates equally well in all geographical directions. An array of such "stacked" collinear elements tends to confine the radiation to low vertical angles.

If a collinear array is mounted horizontally, the directive pattern in the vertical plane at right angles to the array is the same as the vertical pattern of a simple 1/2-λ antenna at the same height (Chapter 2).

TWO-ELEMENT ARRAY

The simplest and most popular collinear array is one using two elements, as shown in Figure 39. This system is commonly known as "two half-waves in phase." The manner in which the desired current distribution is obtained is described in Chapter 26. The directive pattern in a plane containing the wire axis is shown in Figure 40.

Depending on the conductor size, height, and similar factors, the impedance at the feed point can be expected to be in the range from about 4 to 6 kΩ, for wire antennas. If the elements are made of tubing having a low λ/dia (wavelength to diameter) ratio, values as low as 1 kΩ are representative. The system can be fed through an open-wire tuned line with negligible loss for ordinary line lengths, or a matching section may be used if desired.

THREE- AND FOUR-ELEMENT ARRAYS

When more than two collinear elements are used it is necessary to connect "phasing" stubs between adjacent elements in order to bring the currents in all elements in phase. It will be recalled from Chapter 2 that in a long wire the direction of current flow reverses in each 1/2-λ section. Consequently, collinear elements cannot simply

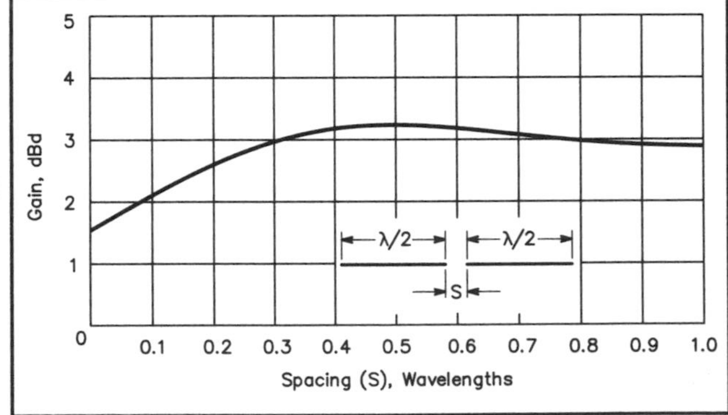

Figure 38—Gain of two collinear 1/2-λ elements as a function of spacing between the adjacent ends.

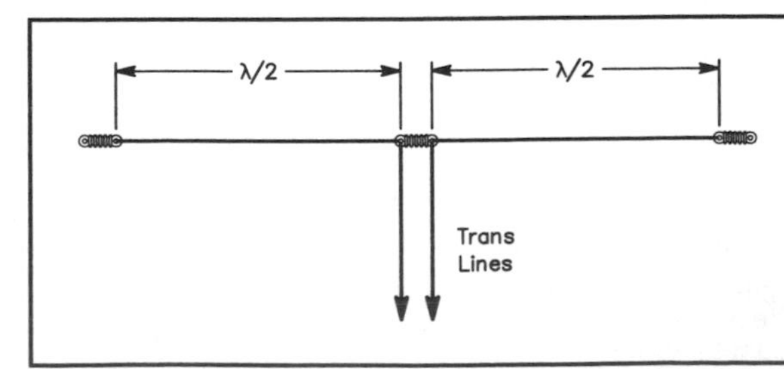

Figure 39—A two-element collinear array (two half-waves in phase). The transmission line shown would operate as a tuned line. A matching section can be substituted and a nonresonant line used if desired.

be connected end to end; there must be some means for making the current now in the same direction in all elements. In Figure 41A the direction of current flow is correct in the two left-hand elements because the transmission line is connected between them. The phasing stub between the second and third elements makes the instantaneous current direction correct in the third element. This stub may be looked upon simply as the alternate $^1/_2$-λ section of a long-wire antenna folded back on itself to cancel its radiation. In Figure 41A the part to the right of the transmission line has a total length of three half wavelengths, the center half wave being folded back to form a $^1/_4$ λ phase-reversing stub. No data are available on the impedance at the feed point in this arrangement, but various considerations indicate that it should be over 1 kΩ.

An alternative method of feeding three collinear elements is shown in Figure 41B. In this case power is applied at the center of the middle element and phase-reversing

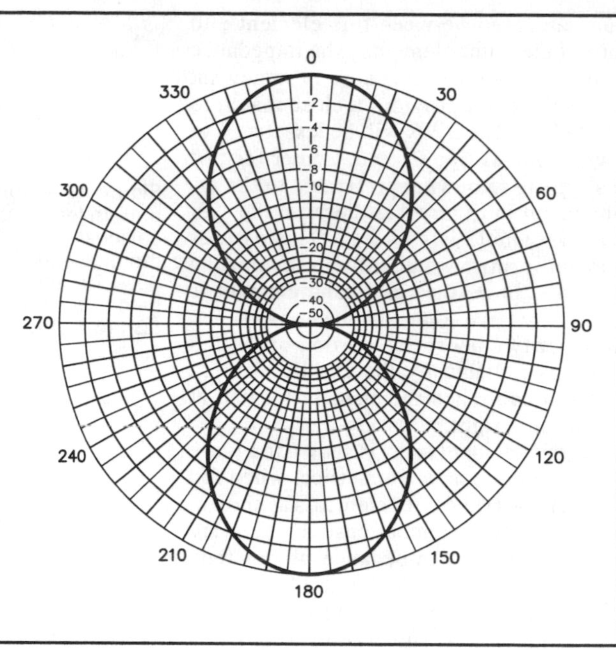

Figure 40—Free-space E-plane directive diagram for the two-element collinear array of Fig 39. The axis of the elements lies along the 90°-270° line. This is the horizontal pattern at low wave angles when the array is horizontal. The array gain is approximately 1.6 dBd (3.8 dBi).

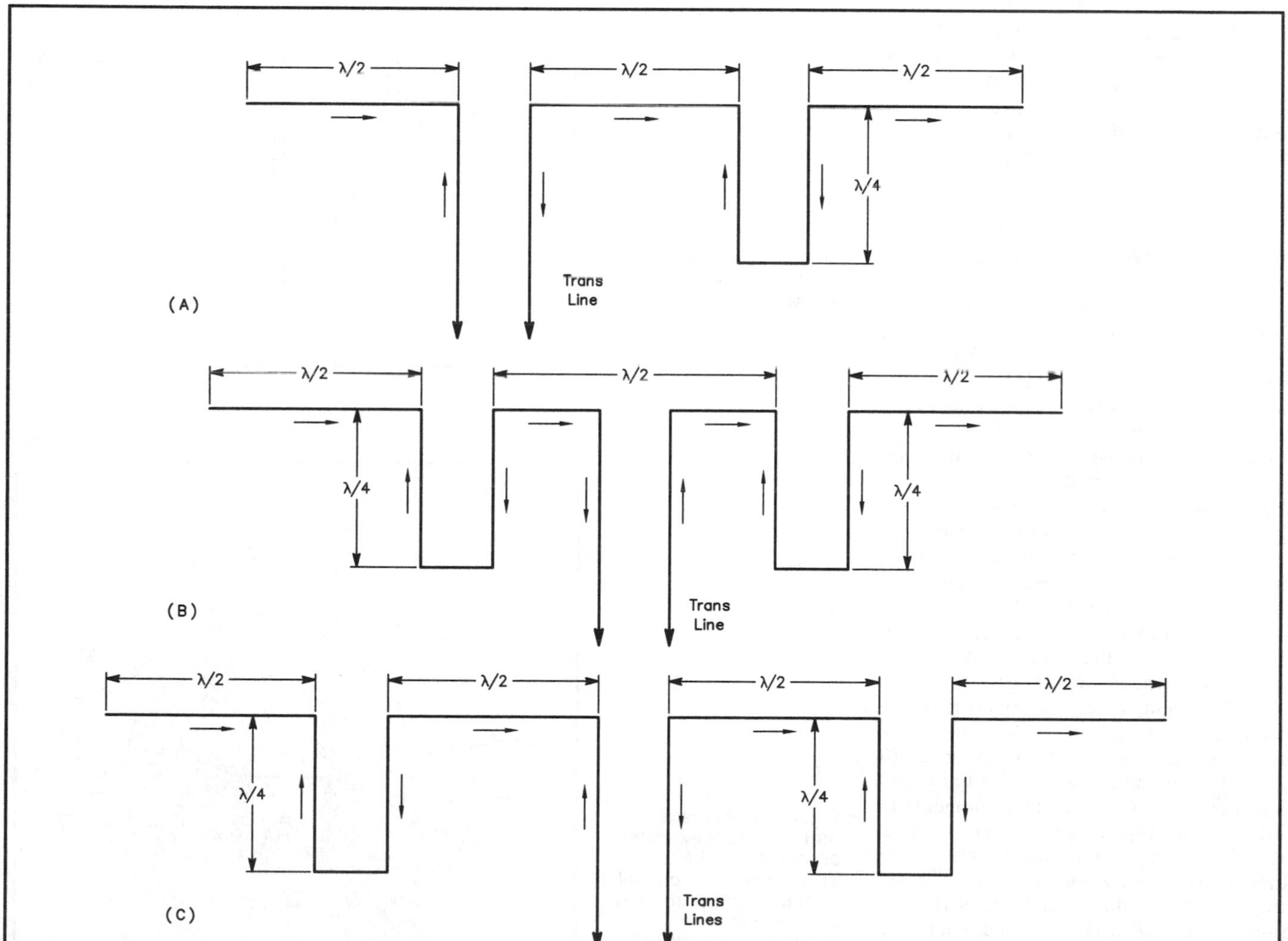

Figure 41—Three and four-element collinear arrays. Alternative methods of feeding a three-element array are shown at A and B. These drawings also show the current distribution on the antenna elements and phasing stubs. A matched transmission line can be substituted for the tuned line by using a suitable matching section.

stubs are used between this element and both of the outer elements. The impedance at the feed point in this case is somewhat over 300 Ω and provides a close match to 300-Ω line. The SWR will be less than 2 to 1 when 600-Ω line is used. Center feed of this type is somewhat preferable to the arrangement in Figure 41A because the system as a whole is balanced. This assures more uniform power distribution among the elements. In A, the right-hand element is likely to receive somewhat less power than the other two because a portion of the fed power is radiated by the middle element before it can reach the element located at the extreme right.

A four-element array is shown in Figure 41C. The system is symmetrical when fed between the two center elements as shown. As in the three-element case, no data are available on the impedance at the feed point. However, the SWR with a 600-Ω line should not be much over 2 to 1. Figure 42 shows the directive pattern of a four-element array. The sharpness of the three-element pattern is intermediate between Figures 40 and 42, with four small minor lobes at 30° off the array axis.

Collinear arrays can be extended to more than four elements. However, the simple two-element collinear array is the type most used, since it lends itself well to multiband operation. More than two collinear elements are seldom used because more gain can be obtained from other types of arrays.

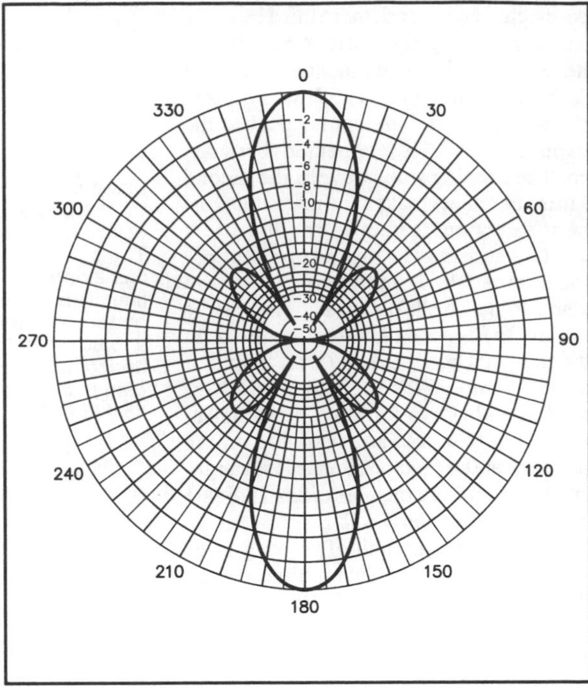

Figure 42—E-plane pattern for a four-element collinear array. The axis of the elements lies along the 90°-270° line. The array gain is approximately 4.2 dBd (6.4 dBi).

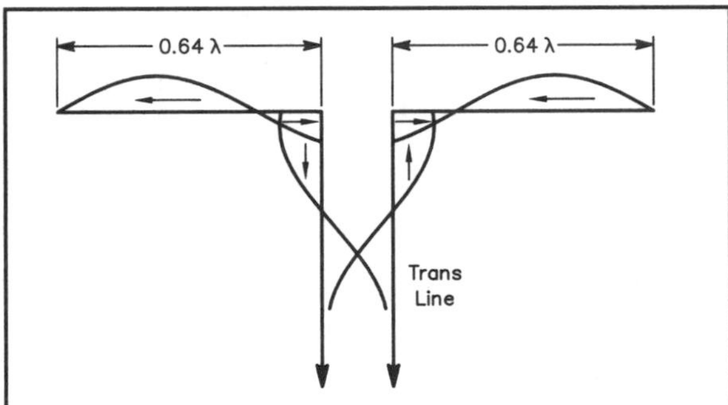

Figure 43—The extended double Zepp. This system gives somewhat more gain than two 1/2-λ collinear elements.

Adjustment

In any of the collinear systems described the lengths of the radiating elements in feet can be found from the formula $468/f_{MHz}$. The lengths of the phasing stubs can be found from the equations given in Chapter 26 for the type of line used. If the stub is open-wire line (500 to 600 Ω impedance) it is satisfactory to use a velocity factor of 0.975 in the formula for a 1/4-λ line. On-the-ground adjustment is, in general, an unnecessary refinement. If desired, however, the following procedure may be used when the system has more than two elements.

Disconnect all stubs and all elements except those directly connected to the transmission line (in the case of feed such as is shown in Figure 41B leave only the center element connected to the line). Adjust the elements to resonance, using the still-connected element. When the proper length is determined, cut all other elements to the same length. Make the phasing stubs slightly long and use a shorting bar to adjust their length. Connect the elements to the stubs and adjust the stubs to resonance, as indicated by maximum current in the shorting bars or by the SWR on the transmission line. If more than three or four elements are used it is best to add elements two at a time (one at each end of the array), resonating the system each time before a new pair is added.

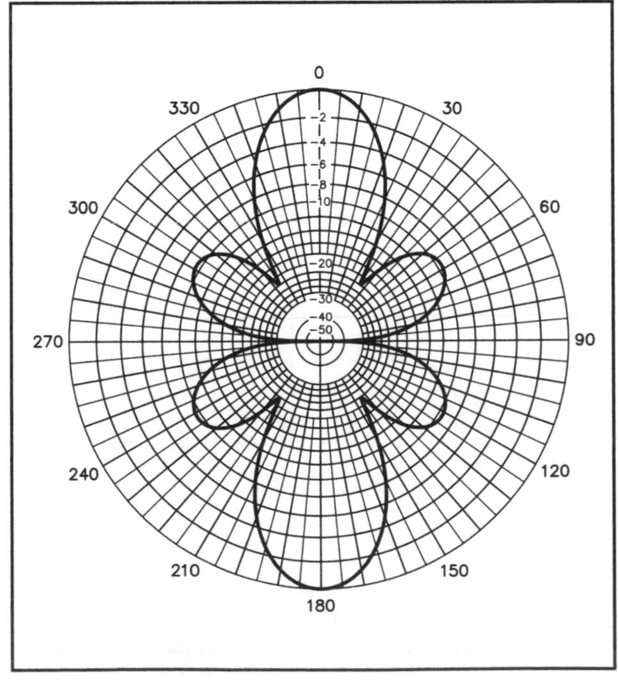

Figure 44—E-plane pattern for the extended double Zepp of Fig 43. This is also the horizontal directional pattern when the elements are horizontal. The axis of the elements lies along the 90°-270° line. The array gain is approximately 3 dBd.

THE EXTENDED DOUBLE ZEPP

An expedient that may be adopted to

obtain the higher gain that goes with wider spacing in a simple system of two collinear elements is to make the elements somewhat longer than 1/2 λ. As shown in Figure 43, this increases the spacing between the two in-phase 1/2-λ sections at the ends of the wires The section in the center carries a current of opposite phase but if this section is short the current will be small; it represents only the outer ends of a 1/2-λ antenna section. Because of the small current and short length, the radiation from the center is small. The optimum length for each element is 0.64 λ. At greater lengths the system tends to act as a long-wire antenna, and the gain decreases.

This system is known as the "extended double Zepp." The gain over a 1/2-λ dipole is approximately 3 dB, as compared with approximately 1.6 dB for two collinear 1/2-λ dipoles. The directional pattern in the plane containing the axis of the antenna is shown in Figure 44. As in the case of all other collinear arrays, the free-space pattern in the plane at right angles to the antenna elements is the same as that of a 1/2-λ antenna—circular.

CHAPTER SIX

WIRE BEAMS

The K6NA 80-Meter Wire Beam

Ever get crunched in a DX pileup on 80/75 meters by K6NA? Lots of folks have! K6NA reveals the secrets behind his two-element "killer" Yagi.

This article describes a practical design for a two-element, horizontal wire Yagi for the 75/80-meter band using just two support towers. Designed with DXing and contesting in mind, this radiator-and-reflector combination features moderate gain and good front-to-back ratio, combined with instant beam reversal and band-segment (mode) switching. It uses tuned feeders and a balanced-output antenna tuner at the operating position for matching at the transmitter. Remote relays at ground level handle reversal and parasitic-tuning requirements, taking advantage of the convenient properties of half-wave transmission lines.

I designed this antenna empirically, before PC-modeling programs were readily available. I've confirmed that it really works, through years of successful 80-meter operation. Later, I verified this with computer modeling. Figure 1 shows the general layout of my system.

Background

At my former location I had two towers, 70 and 100-feet high, fairly close together on a half-acre of property. There was an assortment of quads and Yagis for 10-40 meters,[1] but as is often the case with all-band DXers, the 80-meter antenna was an afterthought. Through the years I experimented with many types, looking for the "silver bullet" antenna for 80 that would finally make this challenging band a pleasure to operate.

Each season numerous magazine articles and endless discussions by experts extolled the virtues of this or that magic 80-meter DX antenna. I even tried some of them: a diamond quad loop, a delta loop (bottom-fed or corner-fed), a pair of parasitic deltas, phased quarter-wave slopers (tilted ground planes), a λ/4 vertical with lots of radials. I always came back to a basic truth: with the tower heights available, and faced with the reality of rough, rocky, lossy earth, my 95 foot high inverted-V dipole was usually as good as–or better than—any of the "trick" antennas I had spent hundreds of hours building. As a bonus, usually the dipole was quieter on receive.

Even Jim Lawson, W2PV, in an early *QST* article about broadbanding an 80-meter antenna, mentioned almost in passing that his high (110-foot) inverted-V was superior to his four-element phased vertical array, at least on transmit.[2] What I observed about 80-meter DXing in the 1970s was that most attempts using vertically polarized systems to achieve low-angle radiation were not satisfactory. This was due to excessive ground losses (both near-field return-current losses and far-field reflective losses); in addition, verticals were generally noisy on receive.

Neither vertically polarized antennas nor relatively low horizontal antennas were really getting thejob done. I resolved to construct my next station thinking about the 80-meter band from the start."High and horizontal" would be the goal.

Design Criteria

My new station would have two main towers, each 140 feet high. Because I had all-band contesting in mind, there would be numerous other antennas on the towers. I wanted to put a three-element 40-meter beam on top of one tower and a large 20-meter Yagi on the top of the other. Though it likely would be an excellent performer, a rotating beam of some type for 80 meters was ruled out because I did not want to dedicate the top of one tower to it. Also, there was the near-certainty of difficult maintenance problems with such a gigantic antenna. Instead, I settled on a design for a fixed-wire beam. These would be the main criteria for reliable, day-to-day operation:

- Oriented properly for best results in both DX and domestic contests
- Moderate gain, with a wide lobe for good azimuth coverage
- Instant beam reversal with a decent front to-back ratio
- Near-instant band-segment (phone or CW mode) switching, especially for casual, daily DXing
- Easy, low-loss matching anywhere in the band
- Horizontally polarized to minimize both ground-reflection losses and noise on receive
- High enough to produce reasonably low angle radiation
- Reliable and easy to maintain without disrupting other antennas on the towers.

Design Discussion

Some discussion of these interrelated design criteria will aid in understanding the trade-offs and choices I made while planning this antenna. For example, from San Diego the typical short path to Europe on 80 meters *is not* the expected true Great-Circle path of about 25°. Rather, the short path toward Europe is almost always a "bent" or skewed path, where signals pass to the south or southeast of the highly absorptive auroral oval, propagating by means of a scatter mode that uses ionized patches over the central Atlantic Ocean. Consequently, our pseudo short-path heading toward Europe is commonly 50° to 90° in azimuth. Occasionally, European signals on 80 meters arrive in California from the southeasterly direction. Rarely, around the time of the Equinox, they may even arrive by scatter path from straight south, as reported by some rotary beam users in California.

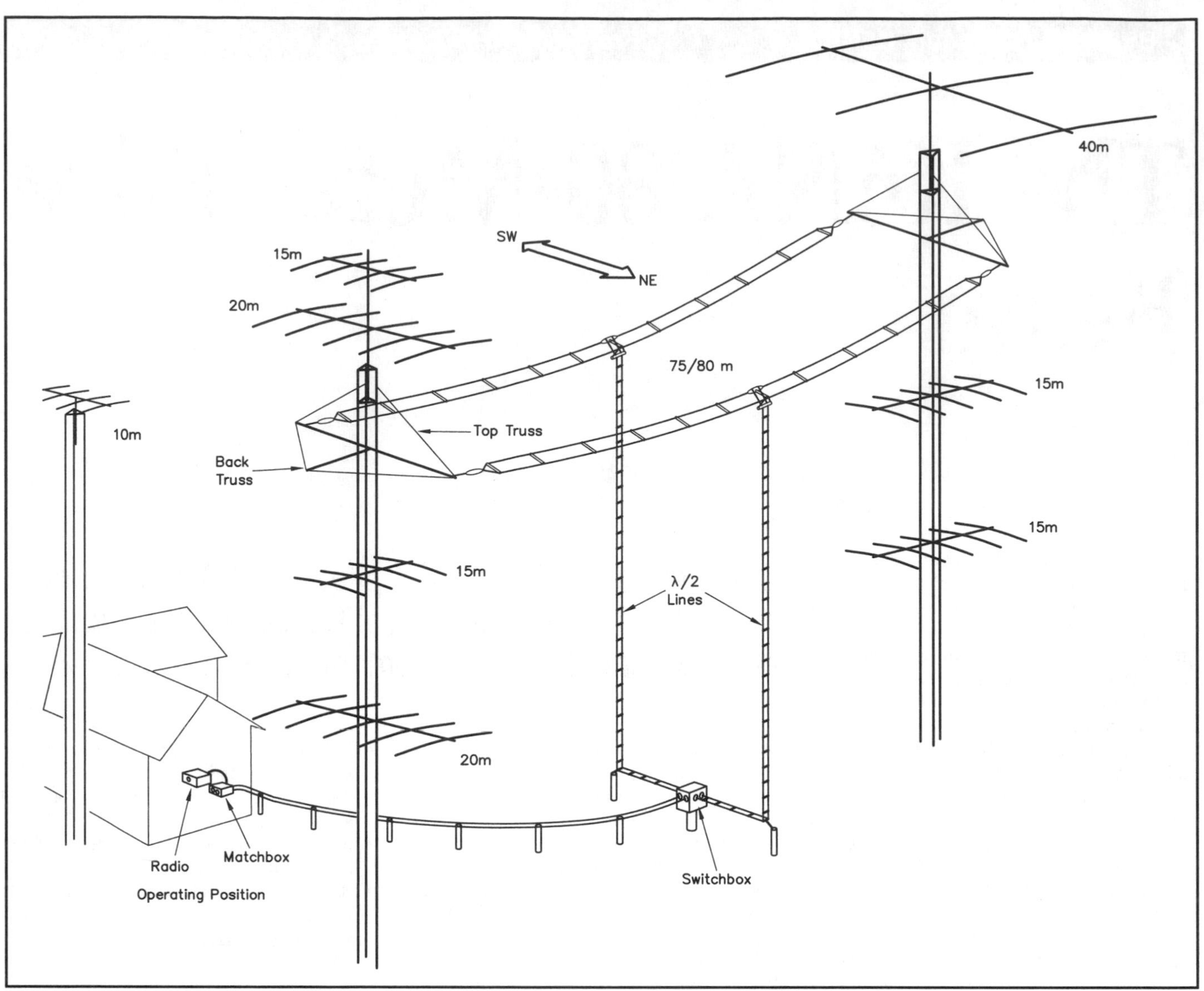

Figure 1—The antenna farm at K6NA, showing the 80-meter wire beam installed on the pair of 140-foot towers, using 35-foot long crossarms made of aluminum tubing and fiberglass. The rearward truss and upper truss wires make each crossarm rigid. For clarity, the tower guy wires are not shown.

I decided to orient the two support towers so that the reversible wire beam, slung between the towers on a pair of horizontal crossarms, would point at 60°/240°. This would provide good coverage across the USA and into Europe (both short and long path), Africa, the Caribbean and South Pacific areas, which together make up about 85% of my annual QSOs on 80 meters.

I could build a simple two-element parasitic design, using λ/2 elements on a reasonable boomlength of about 35 feet, or λ/8.[3] Figure 2B shows that such a design has a half-power azimuth beamwidth of about 74° in the azimuth plane. I was willing to accept the slightly lower gain and front-to-back ratio compared to a three-element design (which would be more difficult mechanically), or a two-wire phased array using double-extended Zepp elements. The Zepp would require the towers to be placed about 350 feet apart. Either of these antennas conceivably would provide more gain, but at the price of reduced azimuthal coverage. A reversible, all-driven array has the additional disadvantage that its phasing and matching networks become overly complicated to make it work properly at both ends of the band.

To achieve the twin requirements of easy mode-switching and beam reversal, I decided to construct a symmetrical, two-element Yagi beam. The elements are identical dipoles, self-resonant near 3.675 MHz. This frequency is about 3% to 4% lower than the phone operating area around 3.800 MHz, meaning that each element could be used as a reflector in the phone band with no additional loading. Each dipole is fed with λ/2 of open-wire line. These lines come together in a relay box at ground level. Power is applied to one or the other driven element, and a short is applied across the base of the remaining feeder to force the remaining element to act as a parasitic reflector.

To operate near 3.5 MHz, the reflector simply is retuned to about 3.4 MHz, using an inductor in place of the direct short in the relay box. Supplying RF to the relay box is accomplished using low-loss, open-wire feeders. A balanced-output tuner in the shack is simply touched up to move between phone and CW subbands, and to accommodate the change in SWR. There is no requirement for an exact match at the line input to the driven element because of the low-loss transmission lines.

With the crossarm supports mounted at 130 ft, the dipoles are about λ/2 high and thus produce fairly low-angle radiation. A horizontally polarized antenna is less responsive to man-made noise sources, which have a large vertical component. The height allows λ/2 lines to reach down to the switchbox near ground level. All switching relays are in this weatherproof box, so the system is reliable and easy to maintain. No inaccessible relays or matching networks are up at the elements, where failures might occur.

Construction

No doubt, the 80-meter wire beam can be built and integrated into many existing

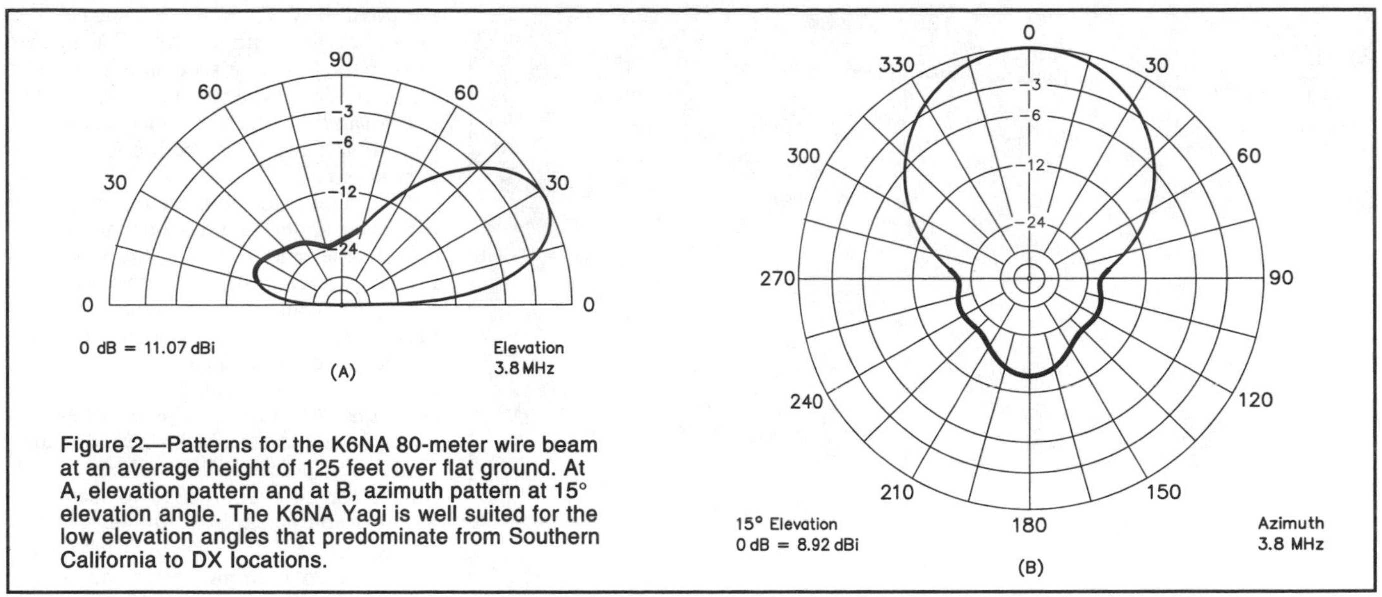

Figure 2—Patterns for the K6NA 80-meter wire beam at an average height of 125 feet over flat ground. At A, elevation pattern and at B, azimuth pattern at 15° elevation angle. The K6NA Yagi is well suited for the low elevation angles that predominate from Southern California to DX locations.

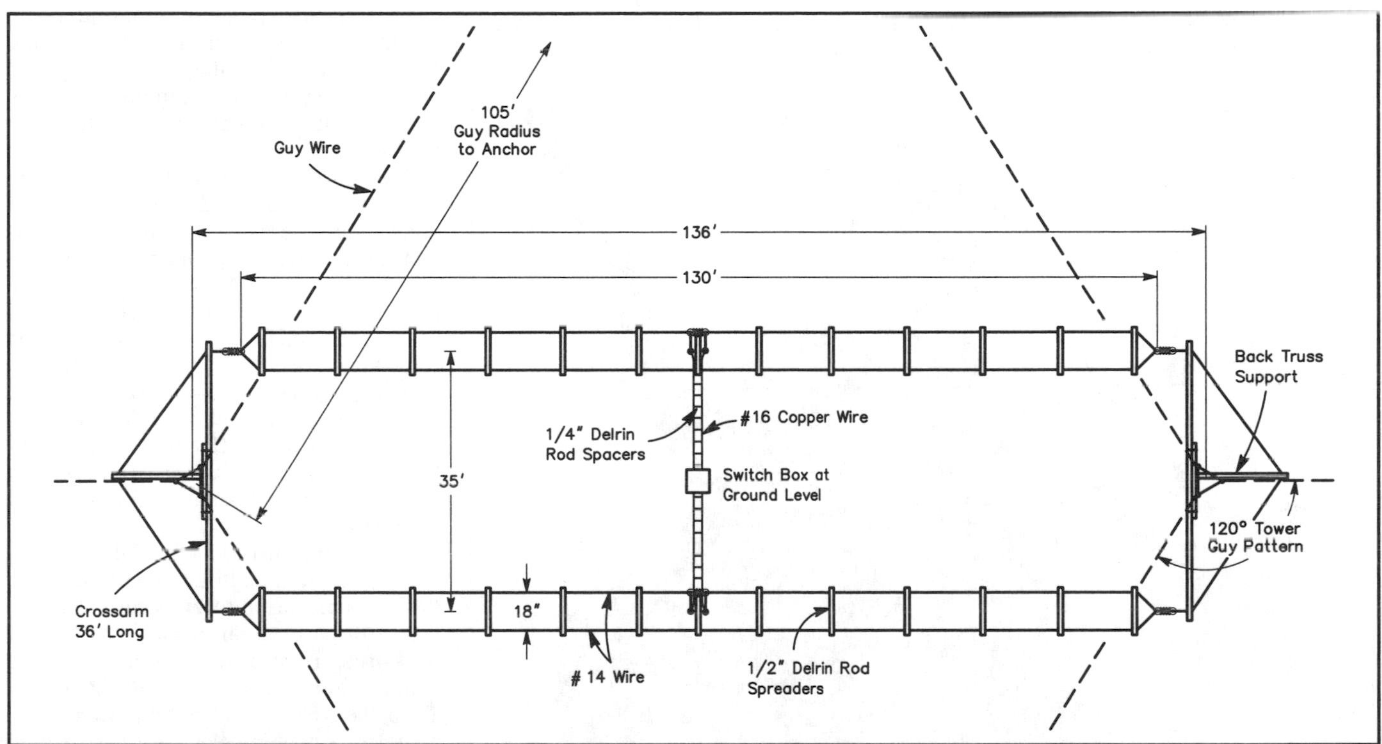

Figure 3—Top view of the 80-meter wire beam mounted on the towers, with details of the crossarms and rearward truss system that prevents the tensioned antenna wires from bending the crossarms inward. The tower-guy placement geometry provides several advantages for antenna-farm management! See text.

multi-tower systems. I had the luxury of planning the tower and guy arrangement well in advance of building the antenna. This helped make erection and maintenance simpler. Such planning contributes to convenient erection of other antennas too!

Figure 3 shows the recommended two-tower layout that has served very well for this and other projects. Through symmetry of guy-anchor placement, any forces due to loading by wire antennas or temporary trams are balanced out. Note that each tower has an opposing, backward-facing guy exactly in line with its counterpart and parallel with the centerline of the towers.

When the 36-foot crossarm is mounted about 10 feet below the top of the tower and the topmost guy bracket, the ends extend out beyond and above the sloping, upper guys. This arrangement keeps the dipoles in the clear, allowing them to be raised and lowered using ropes and pulleys with minimal guy-wire interference, without having to remove or move the crossarms.

This tower/guy layout also presents two parallel tower faces on which to mount the booms and to correctly orient the wire antenna. Non-parallel tower faces can work, but the mounting brackets would be more complex and some advantages of symmetry would be lost. For safety, significant sideloads should be distributed across *at least two* tower legs. In addition, guys passing through the central area between the towers would complicate the raising and lowering of the wire antennas, so this should be avoided if possible.

Another advantage of the layout in Figure 3 concerns the overall use and maintenance of other antennas in the system. Note that the 140-foot towers are 136 feet apart, and the guy radius (distance from tower to each of its guy anchors) is 105 feet. These facts, together with the generally uncluttered work area between the towers, make

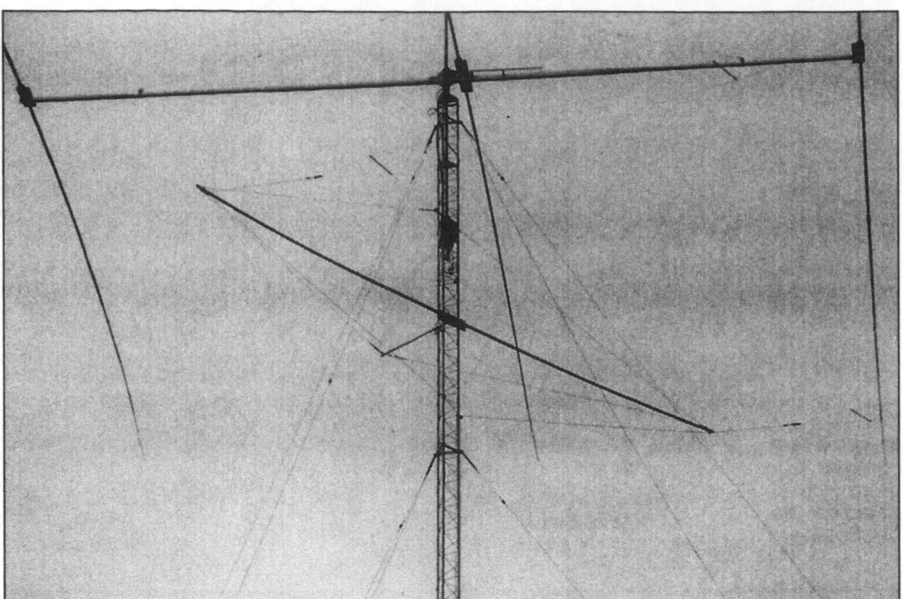

Figure 4—Photo of one of the crossarms (booms) mounted on the tower. The truss system and 80-meter folded dipoles are visible. Note that the truss, dipoles and pull-down ropes provide a stable truss/guy system for each crossarm. The large Yagi at the top of the tower is a full-sized three-element 40-meter beam.

Figure 5—Photo showing mounting method for installing the 80-meter crossarm and associated horizontal back-truss pipe. The horizontal riser prevents the crossarm from collapsing when the side load (antenna tension) is applied. Note how a plate with U-bolts is used to secure the back-truss pipe to one of the tower legs, while the other end is secured with an L-shaped bracket and U-bolts to the mounting plate for the crossarm.

it convenient to run a temporary tram wire anchored at the base of one tower to the top (or anywhere along the length) of the other tower. The tram can be used to raise the 80-meter crossarms themselves, or to move other HF Yagis. If an upper-mast-mounted beam is to be erected, the 80-meter wires first can be lowered easily to the ground. If the tram has to reach only to the area of a side-mounted beam, the wire antenna may stay in place above it. I can detach the 80-meter feeders from the feed-through insulators at the switchbox, and pull them away from the vicinity of the tram wire, if necessary.

Crossarms and Trusses

I made each crossarm (boom) using aluminum and fiberglass. In order to eliminate the possibility of "boom resonance" that might interfere with nearby HF Yagis and to keep the 80-meter wire ends away from metal conductors, only the middle 14 feet of crossarm is made of 2-inch OD, 1/4-inch wall aluminum. This section is stiff and very strong.

The outer extensions are fiberglass tubing, each about 11 feet long, for a total crossarm length of 36 feet. Surplus vaulting poles were used but other types of fiberglass would also be suitable. The fiberglass should be wrapped with high-quality black tape (such as Scotch 33 or 88) to prevent decomposition due to ultraviolet radiation. At the end of each fiberglass pole, I mounted a six-inch sleeve of aluminum tubing, before drilling a clearance hole for the forged eyebolt used to secure the pulley used to hoist each dipole.

Each crossarm is clamped to the tower face using an aluminum plate and muffler clamps. See Figures 4 and Fig 5. Once in place, the crossarm is made rigid by means of a dual truss system. When the dipoles are erected and tension is applied, the crossarms cannot collapse. The top and back truss system and dipole pull-ropes together transfer all the balanced forces to the tower.

Layout of the truss system can be seen in Figures 1, 4 and 5. The horizontal riser behind the tower is the key to the truss system. This rigid support is a six-foot length of $1^1/_2$-inch OD, 1/4-inch wall aluminum pipe, extending horizontally about five feet beyond the back leg of the tower. This "sideways riser" is anchored to the crossarm mounting plate with a right-angle bracket, and to the back tower leg by means of U clamps and a flat plate. The back-truss lines connect to eyebolts on the outer end of the riser using turnbuckles and shackles.

The various truss lines and dipole pull-ropes are coiled and individually taped to the crossarm prior to erection. After the crossarm has been raised and installed, the back-truss riser pipe is brought up and clamped into the tower from the back side. Then, the crossarm is first trimmed to horizontal by attaching the top-truss lines to the tower legs and adjusting the turnbuckles until the crossarm looks good, much like a Yagi boom truss is adjusted. Then the backtruss lines are untaped from the crossarm, one at a time, and connected to the far end of the horizontal back-truss pipe. This is an interesting maneuver, which should be accomplished only by an experienced climber/rigger! The turnbuckles are adjusted initially for moderate tension. Properly adjusted, the result is a crossarm that doesn't sag downward or fold inward when antenna tension is applied.

After the truss system has been rigged, the installer untapes one of the coiled-up dipole pull-ropes, which has been prerigged to extend along the arm to the outer crossarm pulley and back again to the vicinity of the tower. By attaching either a weight or tag line to the thimble, the dipole-end of the rope is released to hang vertically from the pulley. The rigger then pays out the coiled rope until a ground man can reach the thimble end to attach the dipole insulator. The pull-ropes ultimately are brought to the tower legs through pulleys and thence brought down to ground level. The angle of the pull-ropes completes what is essentially a four-point guy on each end of the crossarm (antenna tension, back-truss, up-truss, and angled pull-down rope). This configuration presents a stable structure when the dipoles are raised and tension applied.

I must emphasize that all subassemblies should first be fabricated and fitted to the tower at ground level. Make sure that all plates, clamps and hardware mate to the tower, referenced to diagonal braces, welds, or leg bolts. Premark the tower legs up at the target location ahead of time to preclude surprises on the tower!

Construction of the Dipoles

Two-wire folded dipoles were selected to help bring the feedpoint impedance up closer to the characteristic impedance of the feeders and to improve bandwidth. This helps minimize tuner adjustments when moving within a subband, but there is no need to go overboard trying to effect a perfect match at the load.

Each folded dipole is built with #14 stranded, non-stretch copper wire. The two parallel wires are held apart by seven horizontal spreaders located about every 20 feet along the wire length. These spreaders of black Delrin are 1/2-inch diameter rod, 18 inches long (See Figure 6). They are held in position by tie wires, as are the ribbed end insulators. I used an unbroken length (about 260 feet) of wire for each two-wire dipole, making the free ends come together at the center-feed insulator. The free ends of the wires and feeders are tied to the center insulator for strain-relief, and then the loose ends are soldered. This technique prevents breakage, because soldered connections (which are weaker due to wire heating) are not under strain, or flexing. After each dipole is assembled on flat ground, it should be stretched temporarily at low level between the towers in order to check the symmetry and mechanical behavior. When adjustments are complete, the feeder can be soldered to the antenna.

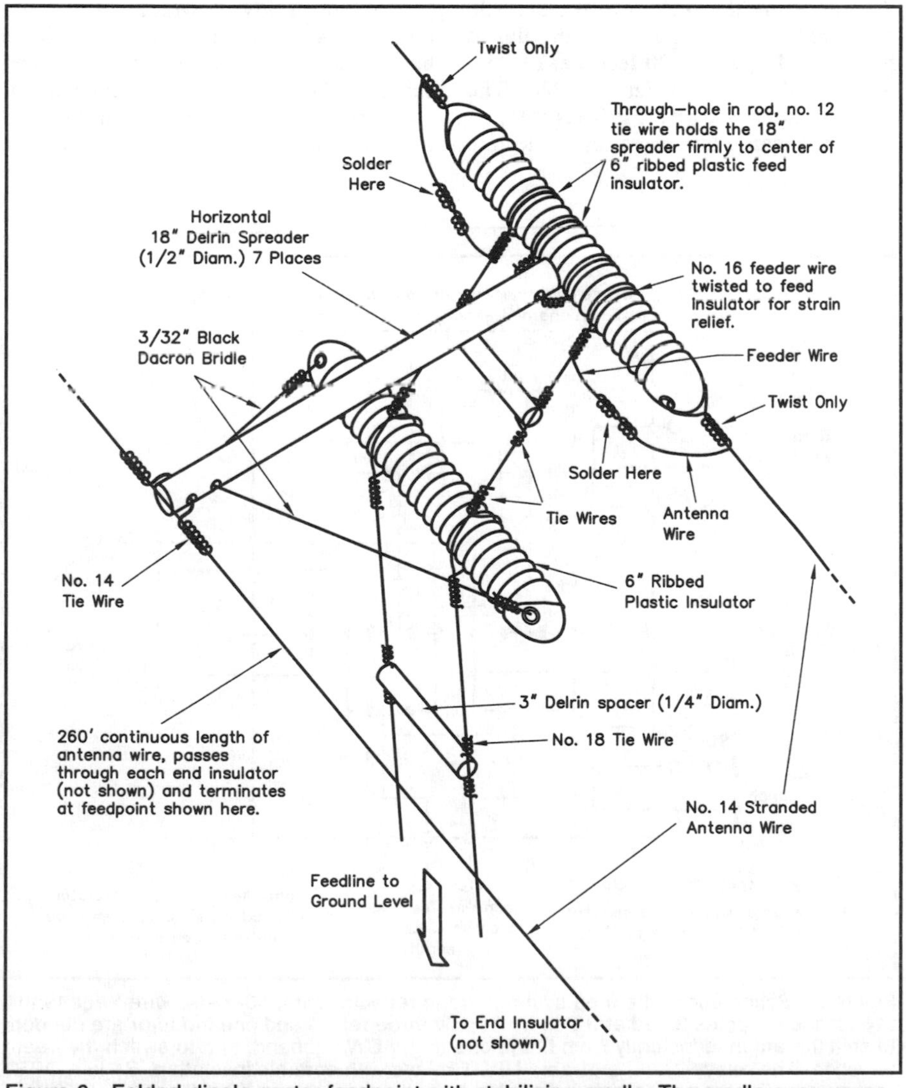

Figure 6—Folded-dipole center feedpoint with stabilizing cradle. The cradle encourages stability in the large, two-wire dipoles, preventing twist. See text.

There is little point in making any sort of analyzer measurements at five feet off the ground because the results will be misleading. When the dipole is installed in the clear at a half-wavelength in height, it will perform as predicted. However, just to be sure, I grid-dipped one dipole at its final height through the half-wave line. I had first built the $\lambda/2$ transmission line and had checked for resonance near 3.650 MHz with a grid-dip meter. When the first dipole and its feeder are on frequency, the second dipole and feeder was cloned and the array was erected. The dipoles at K6NA are about 130 feet long.

Feed Line Construction

Each feed line was constructed using old techniques but modern materials. Minimum weight is very important. The lines were made about 130 feet long, using #16, prestretched solid copper wire. Spacers are 1/4-inch Delrin rods, three inches long and spaced every four feet along the feeder. Spacers were notched and drilled, and held in position using #18 copper tie wires through the holes. The result is a very lightweight, low-loss feeder with a characteristic impedance of about 575 Ω.[4] To minimize sag in the dipole, avoid using heavy ceramic spacers.

A good technique for constructing the long, open-wire feeder is to stretch two wires tightly between the towers about four feet above the ground. Space the wires three inches apart in a plane parallel to the earth. Precut all the tie wires and make a simple, four-foot long, interval-measuring rod from a piece of wood or tubing. With all the spacers, tie wires, rod and long-nose pliers in a tool bag, the installer begins at one end, moving easily along the line, installing spacers at the correct intervals. The feeder wires sit in the notched ends, while the tie wires pass through the holes and are twisted in place over the feeder wires. When complete, the suspended line is in a good position for checking resonance with a grid-dip meter, assuming the line is insulated from the towers. The final length of each feeder described here is about 128 feet and each weighs less than three pounds.

Some additional comments about feedpoint construction will be helpful to those planning assembly of such a large, wide-spaced folded dipole and associated feeder. Initially, the dipole was built with upper and lower wires, as depicted in most antenna books. An 18-inch Delrin rod connected the lower feed insulator to the upper wire in order to distribute the weight of the feeder onto both wires and to maintain correct spacing. This worked, but with unforeseen side effects. When the dipole was raised under tension, the result was a half-twist in the parallel wires about halfway back toward the end insulator. This twist was not due to "live" or "kinky" wire! The twist was due to a catenary sort of effect.

The solution to the twist problem is to avoid the upper wire/lower wire configuration, and allow all the spreaders (including the center spreader) to lay over into the horizontal plane. To encourage this, a simple cradle was constructed to distribute the weight of the feeder at the center of the two antenna wires. This forces the wires (and all the other balanced spacers) to form a plane parallel to the earth. Fig 6 shows the centering cradle below the Delrin rod. The feeder passes over the cradle insulator (tie wires take the load) and continues up to the feedpoint. Small Dacron lines tie the cradle insulator to the other end of the rod, keeping the load centered. These large folded dipoles have been well behaved, even in high winds.

Remote Switching

A plywood "doghouse" was constructed under the antenna center, and mounted about four feet above ground for convenience. Fig 7 is a schematic of the switching network housed in the box. Two DPDT relays control the direction of the beam pattern. The main feeder connects through K1 and K2 to one dipole, while the other dipole becomes a phone-band reflector due to the short across K3. In the shack, switch S1 activates K1/K2 for reversal. S2 is a phone/CW selector switch that allows K3 to retune the reflector element to about 3.4 MHz by means of an inductor. Relays are large, ceramic-base open-frame types.

The center-loading inductor for a CW reflector is best chosen experimentally. A coil is connected across K3, S2 is set for CW, and the tuner in the shack is adjusted for a match low in the CW band. From the tuner, a coax cable is routed temporarily back to the switchbox. At the box, the temporary coax feeds a receiver, which is used to monitor signals arriving off the back of the antenna around 3.515 MHz. A variable tap is moved along the coil until the best front-to-back ratio is obtained. After a period of testing, the tap can be soldered. With the dipoles self-resonant around 3.650 MHz, a CW reflector coil of about 17 µH was required. This consisted of a 2-inch OD coil, 2³/₄-inches long, on which were wound about 25 turns of #16 wire.

Operation and Maintenance

The 80-meter wire beam performs as expected. Typical F/B is 11 to 14 dB on both phone and CW modes. For high-band operators accustomed to quiet bands with low atmospherics, this level of F/B may not seem like much. However, on a consistently noisy band like 80 meters, this kind of rejection on receive is very important. The high wire beam makes it possible to work long-path Europeans (southwest heading, before sunrise in San Diego) on a regular basis in the winter months, due to effective rejection of storm noise over North America. The antenna performance is roughly comparable to that of a nearby, two-element rotary beam (195 feet high) at N6ND—we hear and work many of the same weak stations during DX sessions. In domestic contests, reports from the East Coast are excellent on both modes.

The wire beam has been very reliable for over 10 years. The first version of the antenna used # 16 solid-copper wire for the elements, and there was a broken wire or two after perhaps six years. The antenna was rebuilt in 1992 with #14 stranded wire and has been perfect since then. The original dipole pull-ropes of black, 1/4-inch polypropylene lasted about seven years before they were replaced. Black Dacron, readily available from several *QST* advertisers, would be an excellent replacement for the pull-ropes, but it is more expensive than polypropylene. Do not use any color except black, and do not use polyethylene lines for permanent antenna supports.

Antenna Comparisons and Modeling

Recently, I wanted to investigate this 80-meter wire beam using computer modeling. By comparing the predicted performance of this antenna with that of several other common types of 80-meter DX antennas, the relative merits of each can be described briefly. Example antennas were modeled over average earth, with a dielectric constant of 13 and a ground conductivity of 5 mS/m. Gain figures listed are referenced to an isotropic antenna in free space.

Fig 8A shows the elevation pattern of the wire beam 125 feet high over real earth, compared to single horizontal dipoles at heights of 125 and at 70 feet. Peak radiation for the Yagi occurs at 28°. The azimuth pattern shown in Fig 8B is at 15° elevation, because low elevation angles are required for success on most DX paths from Southern California. For example, the W6 short path to Europe supports signals between 8° and 18° elevation, essentially 100% of the time.[5] The Yagi has about 5 dB of gain over the high dipole.

Note that a single high dipole is an excellent DX performer, and likely will beat any sort of beam antenna on a 70 to 80-foot tower. In addition, a high dipole has better rejection of local (high-angle) signals, so it will be a superior DX receiving antenna. Inspection of Figure 8A indicates that the high wire beam probably will outperform a 70-foot high dipole by nearly 10 dB at low elevation angles.

Some people might say: "With 140-foot towers, you should be using quad loops!" Not so fast. Figure 9A shows that with the top of a diamond-like, full-wave loop at 135 feet, the λ/2 dipole at 125 feet is superior by at least 1 dB at all angles lower than 40°, in spite of the small stacking gain expected from the loop. The expected gain from the bigger loop cannot be realized fully because the upper and lower wires are at drastically different heights (in terms of fractional wavelengths) relative to each other and the earth. This prevents proper phase addition. Further, the average height of the loop is lower, which results in a higher peak angle of radiation. The dipole has better high-angle rejection, too, and should be better on receive. If both antennas were up a couple of wavelengths on 80 meters, the quad loop would beat the

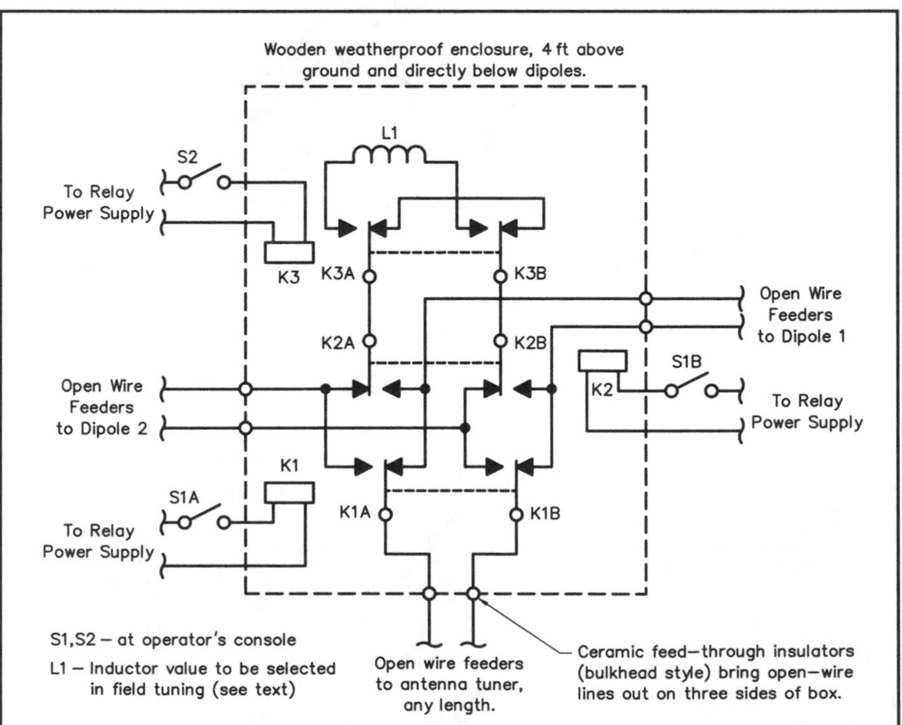

Figure 7—Schematic of the feed system for the reflector-type, 80-meter wire Yagi. With two identical dipoles tuned at 3.650 MHz, only three relays and one inductor are needed to shift the antenna instantly from the phone to the CW subband, and to switch the beam direction. The relays are open-frame DPDT relays with ceramic insulation. S1 is a DPST switch used to reverse the beam direction by activating K1 and K2. S2 activates K3 to tune the parasitic element as a CW reflector.

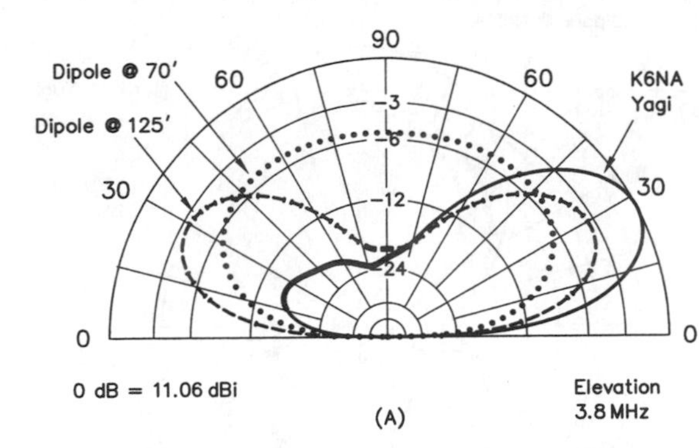

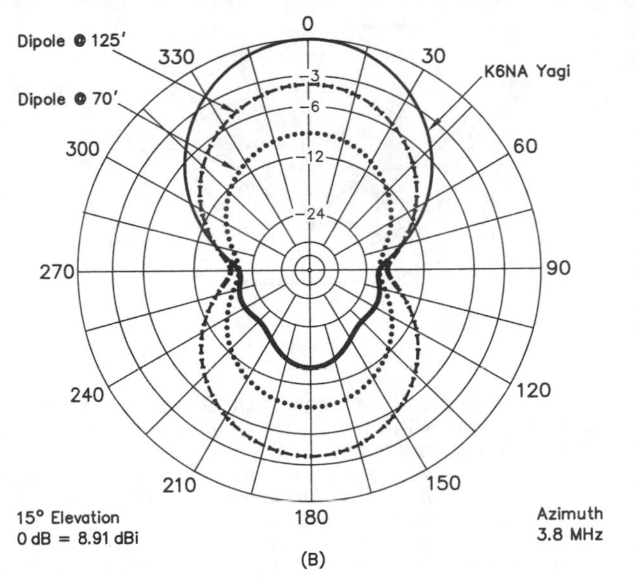

Figure 8—Comparison of patterns for K6NA two-element Yagi at 125 feet with two dipoles, at 125 and at 70 feet heights. At A, elevation pattern comparisons and at B, azimuth pattern comparisons.

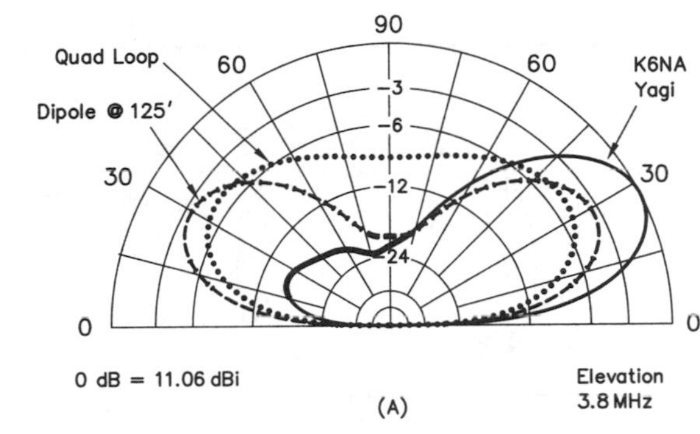

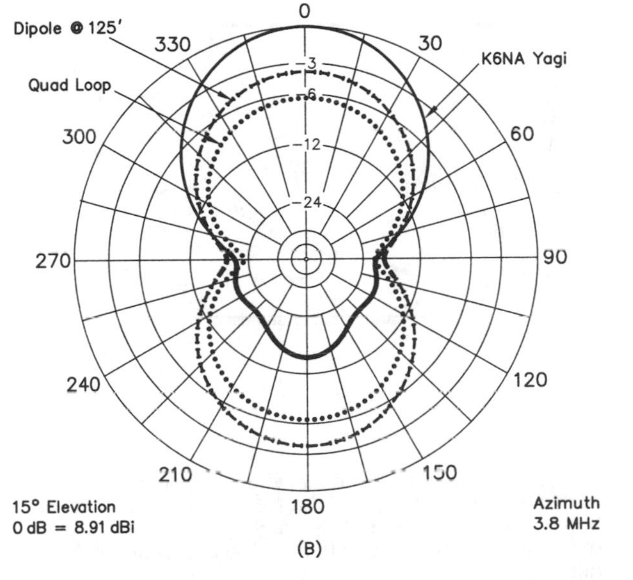

Figure 9—Comparison of elevation responses for full-wave diamond-shaped loop, K6NA two-element Yagi and a single horizontal dipole. The Yagi and dipole are at a height of 125 feet and the loop's apex is at 135 feet.

dipole...but who has a 550-foot tower?

How about a pair of half-wave slopers (that is, tilted vertical dipoles) tied from the tops of the 140-foot towers one $\lambda/2$ apart, and fed in phase for broadside gain? See Figure 10A. All the antenna books show the phased-sloper array to be a 4 dB gain, low-angle monster. The author had this exact antenna up for a few months in 1980. It performed fairly well, but was very responsive to powerline noise. For comparison, a single horizontal dipole was hung between the towers a few feet above the top ends of the slopers. With this single dipole, my local noise level dropped by 8 to 10 dB, and stations in New Zealand reported a consistent transmit advantage over the sloper array. This was a real eye-opener. As seen in Fig 10A, the high wire beam *buries* the phased slopers by 7 dB at the peak angle of 28°, and by 5 dB at 15° elevation!

The famous W8JK is a bi-directional, all-driven wire beam made from two dipoles fed 180° out of phase. At 125 feet in the air, this antenna has excellent rejection of high-angle signals, and is only slightly down from the peak gain of the Yagi configuration. See Figure 11A. It requires no relay box. For DXing, it is inferior to the Yagi because there is no way to reject signals (noise) off the back. Years ago, the author fed the two new folded dipoles as a W8JK for a few months prior to building the switchbox. It performed much like the patterns show, but once the switchbox was available to configure the array as a Yagi with good F/B ratio, the W8JK configuration was abandoned. However, a W8JK antenna 100-feet high in Kansas City would be an outstanding Sweepstakes antenna.

Figure 12 compares the horizontal wire beam to a similar antenna with inverted-V elements with a 90° included angle. The V-style gain is down a bit and the pattern is not as clean. But if you have a single tall tower and install this antenna (only one crossarm is required), you will be in the 95th percentile of effective 80-meter DXing antennas.

In Figure 13 the K6NA Yagi is compared to a quarter-wave vertical mounted over average ground. Since there are less losses due to ground reflection characteristics, horizontal polarization allows the Yagi to outperform substantially the verti-

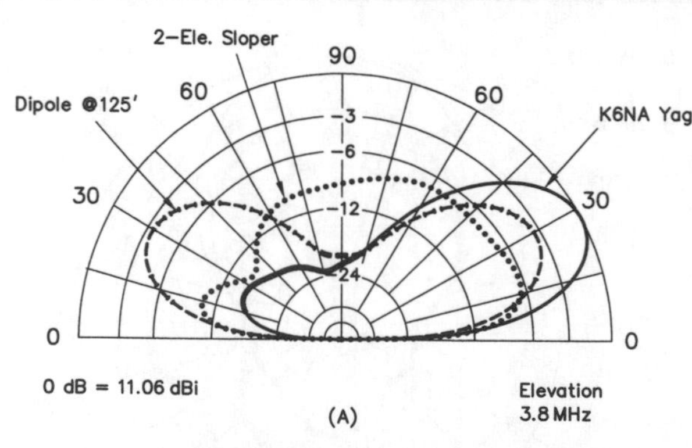

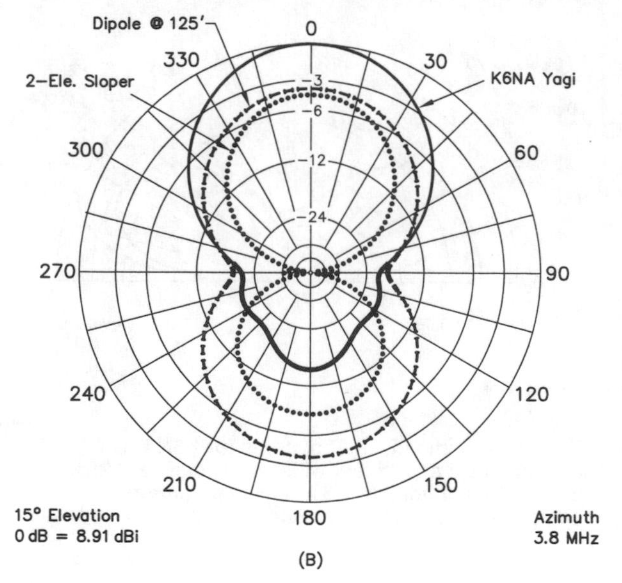

Figure 10—Patterns for a pair of λ/2 sloping dipoles fed in phase, compared with K6NA two-element Yagi and a single horizontal dipole. The slopers both start at a height of 140 feet and are mounted over average ground, with a dielectric constant of 13 and a conductivity of 5 mS/m. The high horizontal Yagi is a decidedly superior performer. Surprisingly enough, even the simple horizontal dipole is a better performer than the phased-sloper system. This is mainly due to far-field losses experienced by the mainly vertically polarized sloper system.

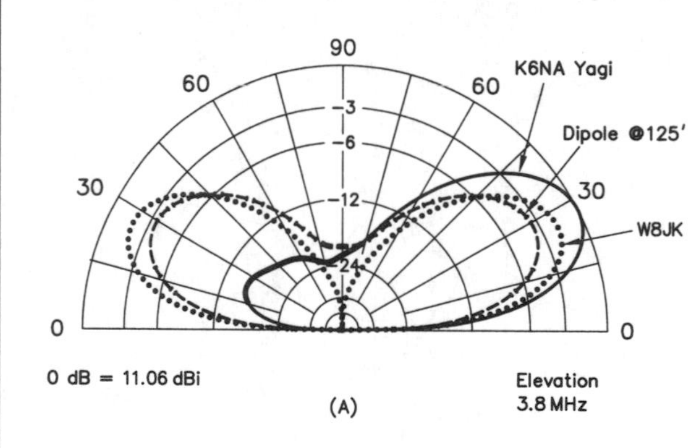

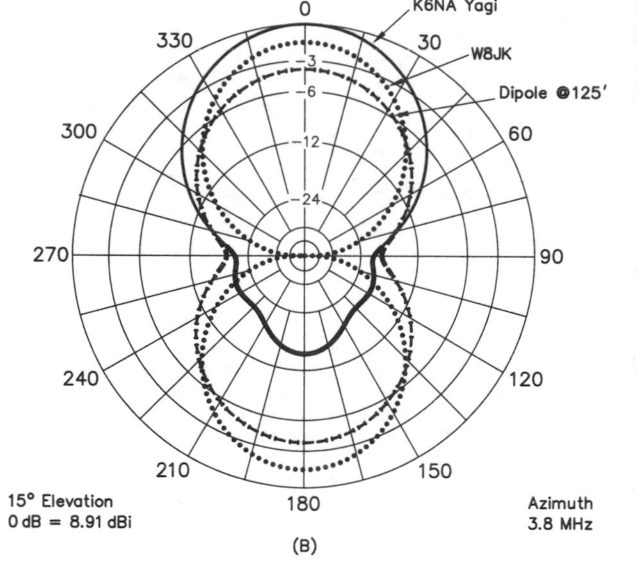

Figure 11—Comparison of elevation responses for a horizontal W8JK array and the K6NA two-element Yagi, both at height of 125 feet. As might be expected, the unidirectional Yagi has gain over the bidirectional W8JK design.

cal at all angles of interest, even down to 10°. In most locations, the vertical will be noisy on receive. *An extensive local ground screen (or elevated radials) will not make any significant change in the pattern relationship depicted here.* Only by installing the vertical in, say, a saltwater marsh extending perhaps 100 λ would its low-angle performance be improved substantially.[6]

The "Four-Square" has become popular in recent years. Figure 13 also shows the beautiful pattern of this array, along with that of the two-element beam and a single vertical. In a quiet location this vertical system with a switching matrix makes a terrific receiving antenna, allowing full azimuth coverage in four directions. However, the horizontal wire Yagi still has superior gain at all elevation angles of interest—note the 4 dB advantage at 15°—and the Yagi likely will be quieter on receive in all but the most remote locations. The remarks in the paragraph above about attempts to improve the lowangle performance of the λ/4 vertical also apply to the Four-Square array.

Conclusions

Someone once said: "All antennas work... some more so." The sometimes-large differences seen during model comparisons can be misleading, as we know from our day-to-day operating. Propagation, pileup dynamics, and operator skills (one skill is picking a quiet location ...) are important factors that determine 80-meter DXing success. Virtually every antenna discussed in the comparison section of this report is a "good" antenna. The author's first 180 countries on 80 meters were worked with an inverted-V dipole at 70 feet.

If *tall* supports are available, a horizontally polarized, gain antenna will certainly provide a statistical increase in performance over the others. In lieu of really tall

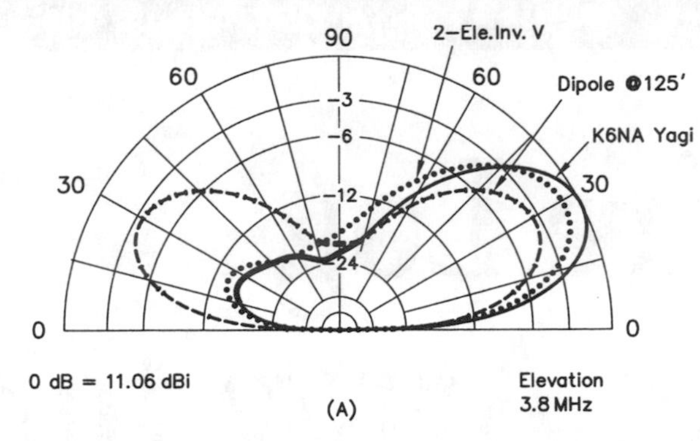

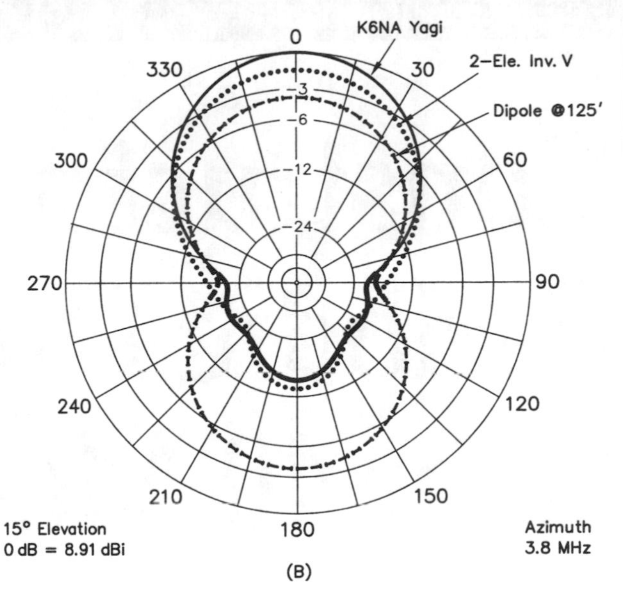

Figure 12—Comparison of the horizontal K6NA wire beam to a similar antenna using inverted-V dipole elements with a 90° included angle. The inverted-V configuration loses gain and directivity compared to the fully horizontal dipole Yagi, but the loss is only about 1 dB. The inverted-V style may be quite practical for those with a single, high tower.

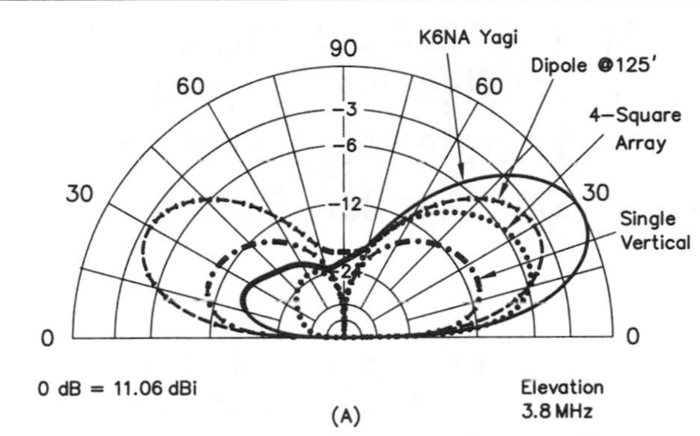

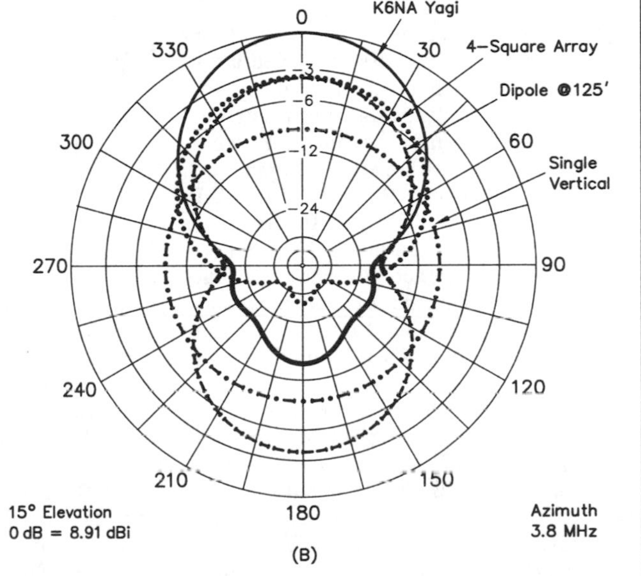

Figure 13—Comparison of K6NA Yagi to both a single ground-mounted λ/4 vertical on average ground (with a dielectric constant of 13 and a conductivity of 5 mS/m), and a "Four Square" phased vertical array. Each vertical is assumed to have 120 radials, an "optimal" ground system. Again, far-field ground losses are detrimental to verticals over poor ground.

towers, a well-installed vertical system can be an excellent antenna, especially in a quiet receiving location. As the computer models show, height above ground is the most significant variable when designing a lowband station using horizontally polarized antennas. [Hills help too!–*Ed.*]

It is hoped that some of the construction techniques for the crossarms, wire antennas, and feed lines described in this report will stimulate others to build even better horizontal arrays. You will enjoy special satisfaction using homemade open-wire lines, and a 40-year old Johnson Matchbox tuner makes a nice addition to your modern shack.

The author would like to thank H. Shepherd, W6US, for his valuable advice regarding wire antennas and open-wire feeder construction; R. Craig, N6ND, for mechanical suggestions; E. Andress, W6KUT, for manuscript review; and N6ND and J. McCook, W6YA, for modeling assistance.

Notes and References

[1]Wayne Overbeck, N6NB, "Quads vs. Yagis Revisited," *Ham Radio*, May 1979, p 12.

[2]James L. Lawson, W2PV, "160/80/75-Meter Broad-Band Inverted-V Antenna," *QST*, Nov 1970, p 17.

[3]James L. Lawson, W2PV, *Yagi Antenna Design*, W. Myers, K1GQ, editor (Newington: ARRL, 1986), p 2-2.

[4]Doug DeMaw, W1FB, editor, *ARRL Electronics Data Book* (Newington: ARRL, 1976), p 82.

[5]Frank Witt, AI1H, "Optimizing the 80-Meter Dipole," *The ARRL Antenna Compendium, Vol 4* (Newington: ARRL, 1995), p 39.

[6]Charles J. Michaels, W7XC, "Some Reflections on Vertical Antennas," *QST*, Jul 1987, p 18.

Construct a Wire Log-Periodic Dipole Array for 80 or 40 Meters

These log-periodic dipole arrays are simple and easy to build. They are also lightweight, strong and inexpensive. The design parameters can be used to construct antennas for the other ham bands.

My desire to work DX and obtain DXCC certification caused me to build my first antenna in the early 1960s. I needed a directional antenna that had reasonable gain, was inexpensive, lightweight and rotatable, and could be assembled with stock items found in large hardware stores. My choice of antennas then was the cubical quad. I had much success DXing with different quads, and I quickly earned DXCC certification. Quads are excellent antennas, but the ones I built lacked the mechanical stability needed in southern Louisiana. I soon learned this when they were ruined by hurricanes.

After my fourth quad was destroyed some years later, I purchased a triband Yagi and forgot about building antennas ... until the day I had a QSO with Ansyl Eckols, YV5DLT. What started as a normal QSO that day in the late 1970s led to a full-fledged experiment with the design, construction, erection and use of log-periodic dipole arrays made of wire. At that time, YV5DLT was using a triband log-periodic dipole array (LPDA) for 20, 15 and 10 meters. What immediately piqued my interest was that his beam was made of wire, and that his signal had outstanding quality and strength.

During the QSO, I asked Ansyl for construction details of his antenna. His response was generous. He mailed me diagrams, schematics and photographs of the LPDA that he had named Telerana. (He subsequently published his design in *QST*.[1]) After reading and studying all of his data, I was convinced that his design had the mechanical stability to withstand hurricanes, and I began plans in my mind to build a copy of Telerana.

I began a search of the literature, reading all of the LPDA articles that I could find.[2-8] By the time I gathered and read several references, three years had passed. Sunspot activity had diminished and band conditions weren't as good. Openings on 10 and 15 meters were few and of short duration, and future conditions would be worse. I did not duplicate the Telerana for these reasons, but decided instead to apply the LPDA theory to the design of wire LPDAs for use on the 160, 80 and 40-meter bands. By making some preliminary calculations I found that an LPDA for 160 meters would be too large to fit my lot size, but LPDAs for 80 and 40 would fit. However, it would not be possible for me to rotate these LPDAs.

LPDAs for 80 and 40 Meters

I placed the same criteria on the LPDAs that I had placed on the quads–that they have reasonable gain, be inexpensive and lightweight, and that they could be assembled with stock items found in large hardware stores. This article is written to detail the design, construction, erection and use of wire LPDAs for the lower frequency bands. Figure 1 shows one method of installation. You can use the information presented here as a guide and point of reference for building similar LPDAs.

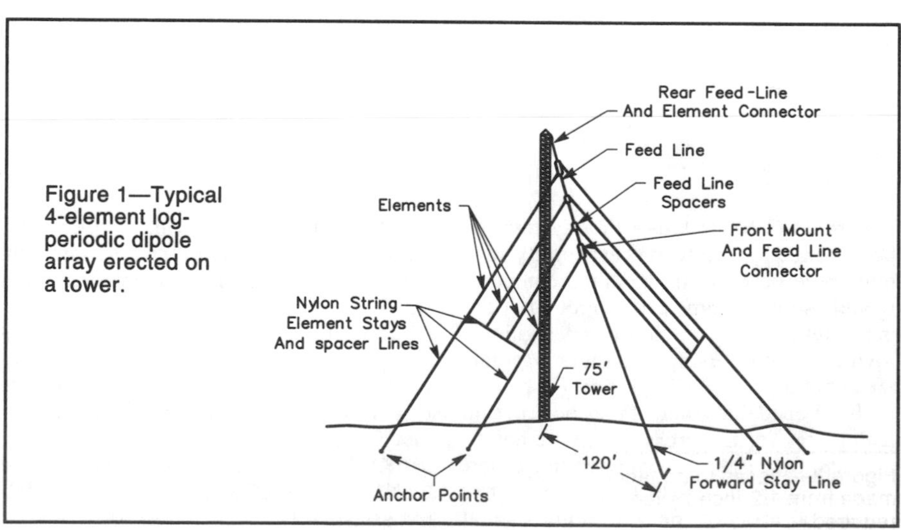

Figure 1—Typical 4-element log-periodic dipole array erected on a tower.

If space is available, the antennas can be "rotated" or repositioned in azimuth after they are completed. A 75-ft tower and a clear turning radius of 120 ft around the base of the tower are needed. The task is simplified if only three anchor points are used, instead of the five shown in Figure 1. Omit the two anchor points on the forward element, and extend the two nylon strings used for element stays all the way to the forward stay line.

For the design procedure, refer to *The ARRL Antenna Book*.[9] By using the formulas given there and other data in the text, all of the dimensions can be obtained and the LPDAs will take shape on paper. The design results are summarized in the concluding section of this article. The next step is to fabricate the fittings; see Figure 2 for details. Cut the wire elements and feed lines to the proper sizes and mark them for identification. After the wires are cut and placed aside, it will be difficult to remember which is which unless they are marked. When you have finished fabricating the connectors and cutting all of the wires, the antenna can be assembled. Use your ingenuity when building one of these antennas; it isn't necessary to duplicate my LPDAs exactly.

The elements are made of standard no. 14 stranded copper wire. The two parallel feed lines are made of no. 12 solid copper-coated steel wire, such as Copperweld. This will not stretch when placed under tension. The front and rear connectors are cut from 1/2-in-thick Lexan sheeting, and the feed-line spacers from 1/4-in Plexiglas sheeting.

Study the plans carefully and be familiar with the way the wire elements are connected to the two feed lines, through the front, rear and spacer connectors. Details are sketched in Figure 4. Connections made this way prevent the wire from breaking. All of the rope, string and connectors must be made of materials that can withstand the effects of tension and weathering. Use nylon rope and strings, the type that yachtsmen use. Figure 1 shows the front stay rope coming down to ground level at a point 120 ft from the base of a 75-ft tower. It may not be possible to do this in all cases. In my installation I put a pulley 40 ft up in a tree and ran the front stay rope through the pulley and down to ground level at the base of the tree. The front stay rope will have to be tightened with a block and tackle at ground level.

Putting an LPDA together is not difficult if it is assembled in an orderly manner. It is easier to connect the elements to the feeder lines when the feed-line assembly is stretched between two points. Use the tower and a block and tackle. Attaching the rear connector to the tower and assembling the LPDA at the base of the tower makes raising the antenna into place a much simpler task. Tie the rear connector securely to the base of the tower and attach the two feeder lines to it. Then thread the two feed-line spacers onto the feed line. The spacers will be loose at this time, but will be positioned properly when the elements are connected. Now connect the front connector to the feed lines. A word of caution: Measure accurately and carefully! Double-check all measurements before you make permanent connections.

Connect the elements to the feeder lines through their respective plastic connectors, beginning with element 1, then element 2, and so on. Keep all of the element wires securely coiled. If they unravel, you will have a tangled mess of kinked wire. Check that the element-to-feeder connections have been made properly. (See Figure 4.) Once you have completed all of the element connections, attach the 4:1 balun to the underside of the front connector. Connect the feeder lines and the coaxial cable to the balun.

You will need a separate piece of rope and a pulley to raise the completed LPDA into position. First secure the eight element ends with nylon string, referring to Figures 1 and 3. The string must be long enough to reach the tie-down points. Connect the front stay rope to the front connector, and the completed LPDA is now ready to be raised into position. While raising the antenna, uncoil the element wires to prevent their getting away and balling up into a mess. Use care! Raise the rear connector to the proper height and attach it securely to the tower, then pull the front stay rope tight and secure it. Move the elements so that they form a 60-degree angle with the feed lines, in the direction of the front, and space them properly relative to one another. By adjusting the end positions of the elements as you walk back and forth, you will be able to align all the elements properly. Now it is time to hook your rig to the system and make some QSOs.

Performance

The reports I received using the LPDAs were compared with an inverted-V dipole. All of the antennas are fixed; the LPDAs radiate to the northeast and the dipole to the northeast and southwest. The apex of the

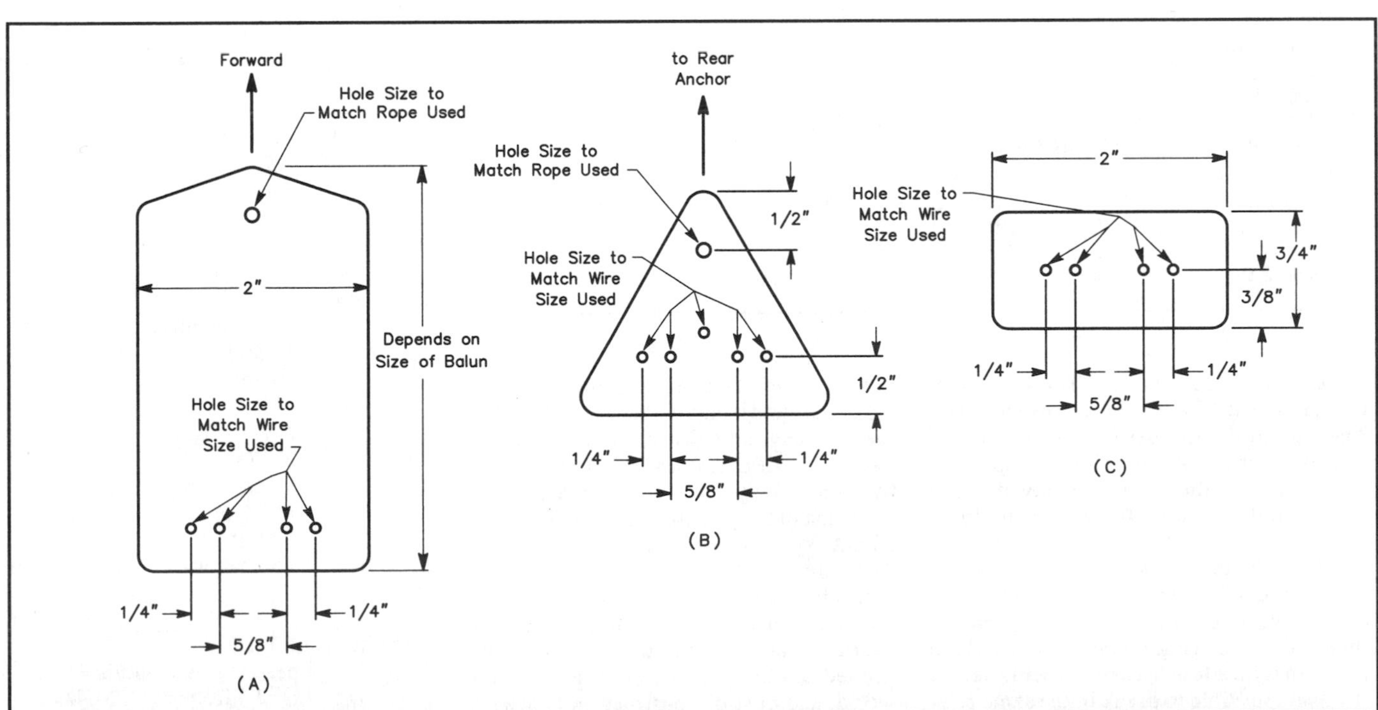

Figure 2—Pieces to be fabricated for the LPDA. At A, the forward connector, made from 1/2-inch Lexan. At B, the rear connector, also made from 1/2-inch Lexan. At C is the pattern for the feed-line spacers, made from 1/4-inch Plexiglas. Two of these spacers are required.

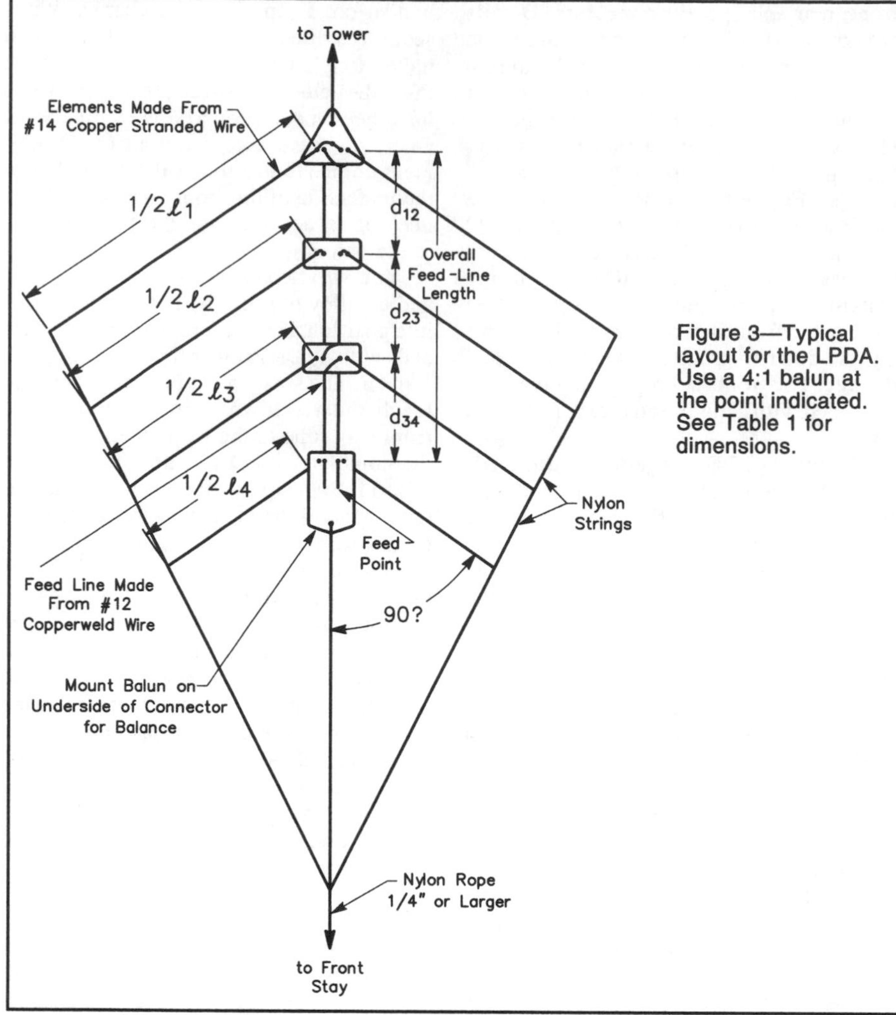

Figure 3—Typical layout for the LPDA. Use a 4:1 balun at the point indicated. See Table 1 for dimensions.

Table 1
Calculated Array Dimensions

Element Length	*Half Length*	*Element Spacing*
80-Meter Array		
ℓ1 = 149.09 ft	½ ℓ1 = 74.55 ft	d_{12} = 17.89 ft
ℓ2= 125.98 ft	½ ℓ2 = 62.99 ft	d_{23} = 15.12 ft
ℓ3 = 106.45 ft	½ ℓ3 = 53.23 ft	d_{34} = 12.77 ft
ℓ4 = 89.95 ft	½ ℓ4 = 44.98 ft	
40-Meter Array		
ℓl = 71.30 ft	1/2 ℓl = 35.65 ft	d_{12} = 8.56 ft
ℓ2 = 60.25 ft	1/2 ℓ2 = 30.13 ft	d_{23} = 7.23 ft
ℓ3 = 50.91 ft	1/2 ℓ3 = 25.46 ft	d_{34} = 6.11 ft
ℓ4 = 43.02 ft	1/2 ℓ4 = 21.51 ft	

dipole is at 70 feet, and the 40-and 80-meter LPDAs are at 60 and 50 feet, respectively. The gain of the LPDAs is in the range of 7 to 9 dB over the dipole. This was apparent from some of the reports received: "The quality of the audio on the log is superior to the inverted V." "The signal on the log is much stronger and steadier than the V, about 10 dB." "The LPDA does not fade, but fading conditions are present on the inverted V." During pileups, I was able to break in with a few tries on the LPDAs, yet it was impossible to break in the same pileups using the dipole.

During the CQ WW DX Contest I was able to break into some *big* pileups after a few calls with the LPDAs. Switching to the dipole, I found it impossible to break in after many, many calls. Then, after I switched back to the LPDA, it was easy to break into the same pileup and make the QSO.

Think of the possibilities that these wire LPDA systems offer hams worldwide. They are easy to design and to construct, real advantages in countries where commercially built antennas and parts are not available at reasonable cost. The wire needed can be obtained in all parts of the world, and cost of construction is low! If damaged, the LPDAs can be repaired easily with pliers and solder. For those who travel on DXpeditions where space and weight are large considerations, LPDAs are lightweight but sturdy, and they perform well. They'll even withstand a hurricane!

Calculations for Log-Periodic Dipole Arrays

Design constants and the results of design procedures follow. (Terms are defined at the end of this section.)

τ = 0.845
σ = 0.06
σ_{opt} = 0.152
cot α = 1.548
α = 32.86°
Gain = 7.5 dBi (5.35 dBd). [By sloping the elements forward, the gain may be increased 3 to 5 dB over this figure. *–Ed.*]
σ' = 0.065
B_{ar} = 1.39; see note 10.
R_0 = 70

Fed with 50-ohm coaxial cable and a 4:1 balun

For the 80-meter antenna,
fn = 4.1
f1 = 3.3
B = 1.24
B_S = 1.72
λ_{max} = 298.18 ft
L = 48.42 ft
N = 4.23 (rounded to 4)
ℓ1 = 149.09 ft
R_0 = 70 ohms
h = 62.4
a = 2.667×10^{-3}
h/a = 23400
Z_{av} = 937.26 ohms
Z_0 = 80.72 ohms
d_{12} = 17.89 ft
See Table 1 for calculated array dimensions.

For the 40-meter antenna,
fn = 7.5
fl = 6.9
B = 1.09
B_s = 1.51
λ_{max} = 142.61 ft
L = 18.57 ft
N = 3.44 (rounded to 4)
ℓ1 = 71.30 ft
R_0 = 70 ohms
h = 32.727
a = 2.667×10^{-3}
h/a = 12273
Z_{av} = 859.82 ohms
Z_0 = 81.76 ohms
d_{12} = 8.56 ft
See Table 1 for calculated array dimensions.

Definitions of Terms
B = operating bandwidth = fn/f1
fn = highest frequency, MHz
fl = lowest frequency, MHz
τ = design constant
σ = relative spacing constant

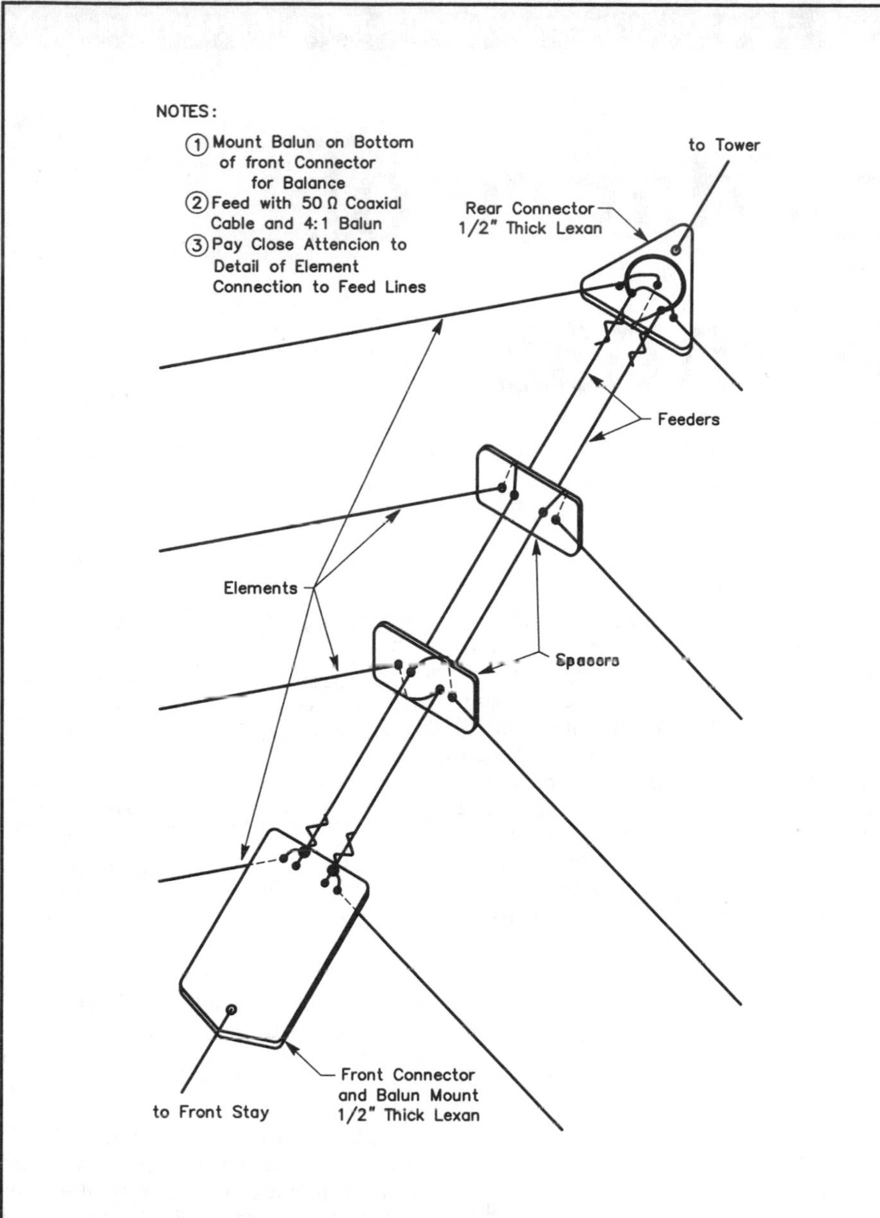

Figure 4—Details of electrical and mechanical connections of the elements to the feed line. Knots in the nylon stay lines are not shown.

σ_{opt} = value of σ for optimum gain
σ' = mean spacing factor
α = apex half-angle
B_{ar} = bandwidth of the active group. See note 10.
B_s = structure (array) bandwidth
L = boom length for N elements
N = number of elements
$\ell 1$ = longest element = 492/f1
λ_{max} = longest free-space wavelength = 984/f1
Z_0 = characteristic impedance of feeder
R_0 = mean radiation resistance level of required input impedance of active region
Z_{av} = average characteristic impedance of a dipole

$$= 120\left(\ln\frac{h}{a} - 2.25\right)$$

h = element half length
a = radius of element
ℓ = length of elements
d = spacing between elements

Notes

[1]A. Eckols, "The Telerana—A Broadband 13- to 30-MHz Directional Antenna," *QST*, Jul 1981, pp 24-27.
[2]C. T. Milner, "Log Periodic Antennas," *QST*, Nov 1959, p 11.
[3]D. E. Isbell, "Log-Periodic Dipole Arrays," *IRE Transactions on Antennas and Propagation*, Vol AP-8, No. 3, May 1960, pp 260-267.
[4]P. D. Rhodes, "The Log-Periodic Dipole Array," *QST*, Nov 1973, pp. 19-22.
[5]*The GIANT Book of Amateur Radio Antennas* (Blue Ridge Summit, PA: Tab Books, 1979), pp 55-85.
[6]C. A. Balanis, *Antenna Theory, Analysis and Design* (New York: Harper and Row, 1982), pp 427-439.
[7]G. L. Hall, ed., *The ARRL Antenna Book* (Newington, CT: The American Radio Relay League, Inc, 1982), pp 6-24 to 6-26, 9-12 to 9-14.
[8]D. A. Mack, "A Second-Generation Spiderweb Antenna," *The ARRL Antenna Compendium Vol 1* (Newington, CT: The American Radio Relay League, Inc, 1985), pp 55-59.
[9]See note 7.
[10]See Ref 6, p 435. B_{ar} is found from the following equation:

$$B_{ar} = 1.1 + 7.7(1 - \tau)^2 \cot \alpha$$

Input Impedance of LPDA Antennas

The log-periodic dipole array (LPDA) has intriguing possibilities, some of which have been explored by Uhl.[1] Unfortunately, some misleading information in *The ARRL Antenna Book* has kept the subject from being more accessible to the amateur.[2] The design procedure called out by *The Antenna Book* and by Rhodes refers to graphs of R_0, the mean antenna input resistance, versus the log-periodic design parameters τ and α.[3] It would appear that R_0 is dependent solely on these parameters; such is not the case. As shown by Eq 2 and 4, R_0 is dependent on τ, σ, Z_{av} and Z_0. This relationship is demonstrated by Carrel.[4] The graphs in question are from the seminal experimental study of the LPDA by Isbell, and are used to demonstrate the effect on R_0 of changing τ and σ; Z_{av} and Z_0 were constant.[5] Since τ and σ are usually chosen to provide the desired antenna geometry and directivity (gain), R_0 should be controlled by either Z_{av} or Z_0. Z_{av} is usually constrained by mechanical considerations, which leaves Z_0 as the easiest parameter to adjust for desired R_0.

The erroneous procedure determines R_0 from a graph. The correct procedure is to calculate the necessary Z_0 based on the *desired* R_0. For most practical LPDA antennas, this results in a Z_0 of less than 100 ohms, if an R_0 of 50 ohms is chosen. Impedances in this range are difficult to achieve with open-wire line, the kind normally used in an LPDA, because thick conductors and/or close spacing are needed. For this reason, it is usual to design for a higher R_0 and use an impedance-transforming balun between 50-ohm coaxial cable and the antenna feed point. This is the method chosen by Uhl. Assuming a 4:1 balun, an antenna-input impedance of 200 ohms is required. Z_0 is then calculated (using Eq 9 and 11 from page 6-25 of *The ARRL Antenna Book*) as follows:

$$Z_0 = \frac{R_0^{\,2}}{8\sigma' Z_{av}} + R_0\sqrt{\left(\frac{R_0}{8\sigma' Z_{av}}\right)^2 + 1} \quad \text{(Eq 1)}$$

where

$$\sigma' = \frac{\sigma}{\sqrt{\tau}} \quad \text{(Eq 2)}$$

$$Z_0 = \frac{200^2}{8(0.065)\,(937.26)} + 200\sqrt{\left(\frac{200}{8(0.065)\,(937.26)}\right)^2 + 1}$$

$$= 298.26 \text{ ohms}$$

The actual impedance of the feed line used by Uhl can be computed from the formula on p 3-16 of *The ARRL Antenna Book*:

$$Z_0 = 276 \log\left(\frac{2S}{d}\right) \quad \text{(Eq 3)}$$

where S is the spacing between conductors, 0.75 inch, and d is the conductor diameter, 0.081 inch for no. 12 AWG wire. Therefore:

$$Z_0 = 276 \log\left(\frac{2(0.75)}{0.081}\right) = 349.86 \text{ ohms}$$

Although this appears to be considerably in error, recasting Eq 1 to solve for R_0 based on the true value of Z_0 gives:

$$R_0 = \frac{Z_0}{\sqrt{1 + \dfrac{Z_0}{4\sigma' Z_{av}}}} = \frac{349.86}{\sqrt{1 + \dfrac{349.86}{4(0.065)(937.26)}}}$$

$$= 224.17 \text{ ohms} \quad \text{(Eq 4)}$$

which, when transformed by the 4:1 balun, results in a load impedance of 56.04 ohms; this is close to the design impedance. From this, we can infer that the actual Z_0 used is not critical, although it isn't negligible, either.

The design Z_0 values listed in the Uhl article are incorrect, since they were produced using the erroneous procedure from *The ARRL Antenna Book*. The values should be 298.26 and 308.56 ohms for the 80- and 40-meter antennas, respectively. In addition, R_0 could be made closer to the design value of 200 ohms by using a different feeder spacing to change the feeder impedance. From Eq 2:

$$S = \left(\frac{d}{2}\right)10^{(Z_0/276)} = \left(\frac{0.081}{2}\right)10^{(298.26/276)}$$

$$= 0.488 \text{ inch}$$

Rounding this value to 0.5 inch results in a Z_0 of 301.26 ohms and an R_0 of 201.46 ohms for 80 meters and 196.62 ohms for 40 meters. These translate to load impedance of 50.37 and 49.15 ohms, respectively.—*Jon Bloom, KE3Z*

[1] J.J. Uhl, "Construct a Wire Log-Periodic Dipole Array for 80 or 40 Meters," *QST*, Aug 1986, p 21.

[2] G.L. Hall, ed., *The ARRL Antenna Book* (Newington: ARRL, 1982), pp 6-24 to 6-26.

[3] P.D. Rhodes, "The Log-Periodic Dipole Array," *QST*, Nov 1973, pp 16-22.

[4] R.L. Carrel, "The Design of Log-Periodic Dipole Antennas," *1961 IRE International Convention Record, Part 1, Antennas and Propagation*, pp 61-75; also PhD thesis, University of Illinois, Urbana, 1961.

[5] D. E. Isbell, "Log-Periodic Dipole Arrays," *IRE Transactions on Antennas and Propagation*, Vol AP-8, No. 3, May 1960, pp 260-267.

The WA1AKR 40- and 75-Meter Slopers

Several amateurs have suggested that I submit a description of my sloper antenna system for publication in "Hints and Kinks." Other amateurs may be interested in this adaptation of the 8JK beam. Construction information is shown for both the 75- and 40-meter bands.

As shown in the accompanying diagram, the array has two half-wave sloping elements joined by a 1/8-wave, 300-ohm phasing line. Transposing the phasing line should bring the element currents into phase. I find the antenna is broadbanded. There appears to be no need for a Transmatch.

If one desires to suspend an additional sloper from the tower for a directional change, installation of remote switching at the top of the tower will permit the use of a single transmission line. Otherwise, separate transmission lines will be required.

Ends of the antenna are suspended by ropes with the tops placed roughly 1 foot away from the tower. An angle of 45 degrees between the antenna and ground should be maintained. Do not use an angle greater than 50 degrees. Resonance with the dimensions shown should occur near 3.8 MHz for the 75-meter sloper and 7.150 for the 40-meter antenna.

How well do my antennas work? I have contacted stations "across the pond" while competing with the big boys who sport three and four-element beams. I have also experienced little difficulty in working VKs and ZLs. – *Carl Bissonnette, WA1AKR*

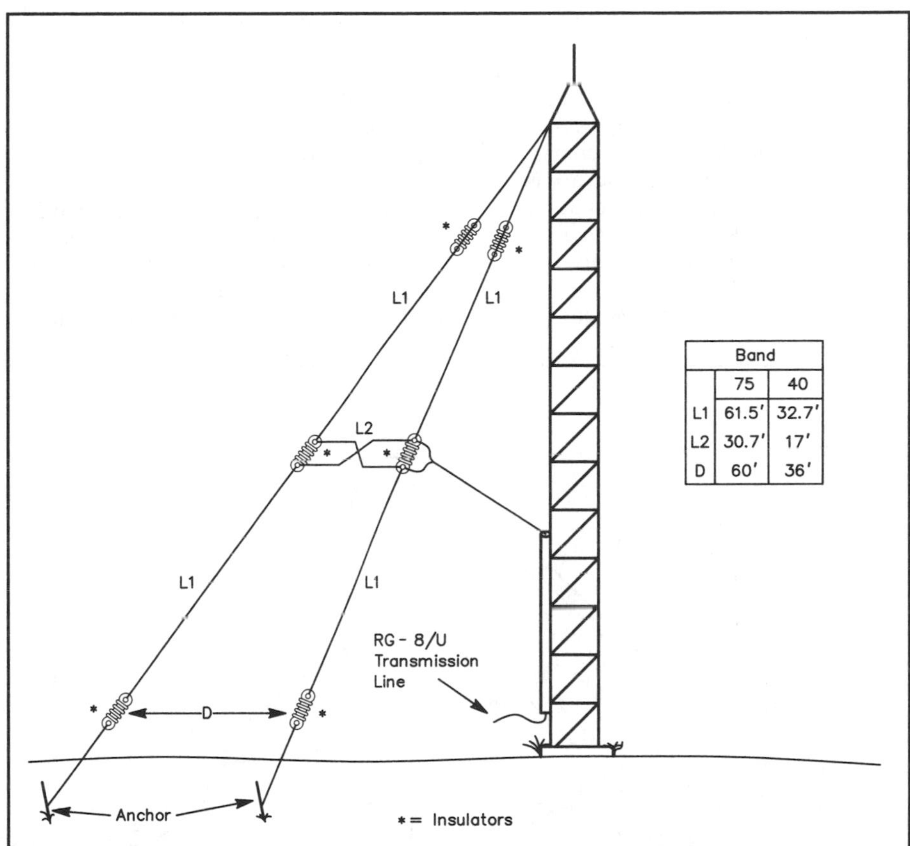

	Band	
	75	40
L1	61.5′	32.7′
L2	30.7′	17′
D	60′	36′

When Carl Bissonnette, WA1AKR, chases DX he uses a sloper like the one illustrated. Carl's arrangement is fashioned after the famous 8JK beam. The feed system resembles that of the ZL Special.

Inexpensive 30-Meter Beam Antenna

In about two months of operating on the new 30-meter band I have worked all 50 states and over 50 countries. My antenna is simple but effective. It is a rotatable inverted V beam. Figure 1 shows the construction details. The antenna boom is suspended from a tree branch about 50 feet in the air. The antenna can be rotated 360° simply by moving the two ground stakes. All of the materials to build this antenna cost me less than $25.

Eq. 1 gives the driven element length, Eq. 2 gives the director length, and Eq. 3 gives the element spacing that I used.

D.E. length = $476/f_{MHz}$ (Eq. 1)

Dir. length = $450/f_{MHz}$ (Eq. 2)

Spacing = $120/f_{MHz}$ (Eq. 3)

The feed-point impedance is around 30 ohms. I used a matching transformer made by connecting two 1/4-λ sections of RG-59/U coaxial cable in parallel. One end of the transformer connects to the antenna, and the other end goes to 50-ohm cable to the shack. Figure 1B shows how this is wired. Perhaps the easiest method to join the two pieces of 75-ohm cable is to use coaxial T connectors. You should use a balun at the antenna feed point to prevent rf from flowing on the outside of the shield braid.

I also built an antenna of this type for 40 meters, and it works great. I guess the key word is rotatable! —*Jon Ferrara, N9DWR, Chattanooga, Tennessee*

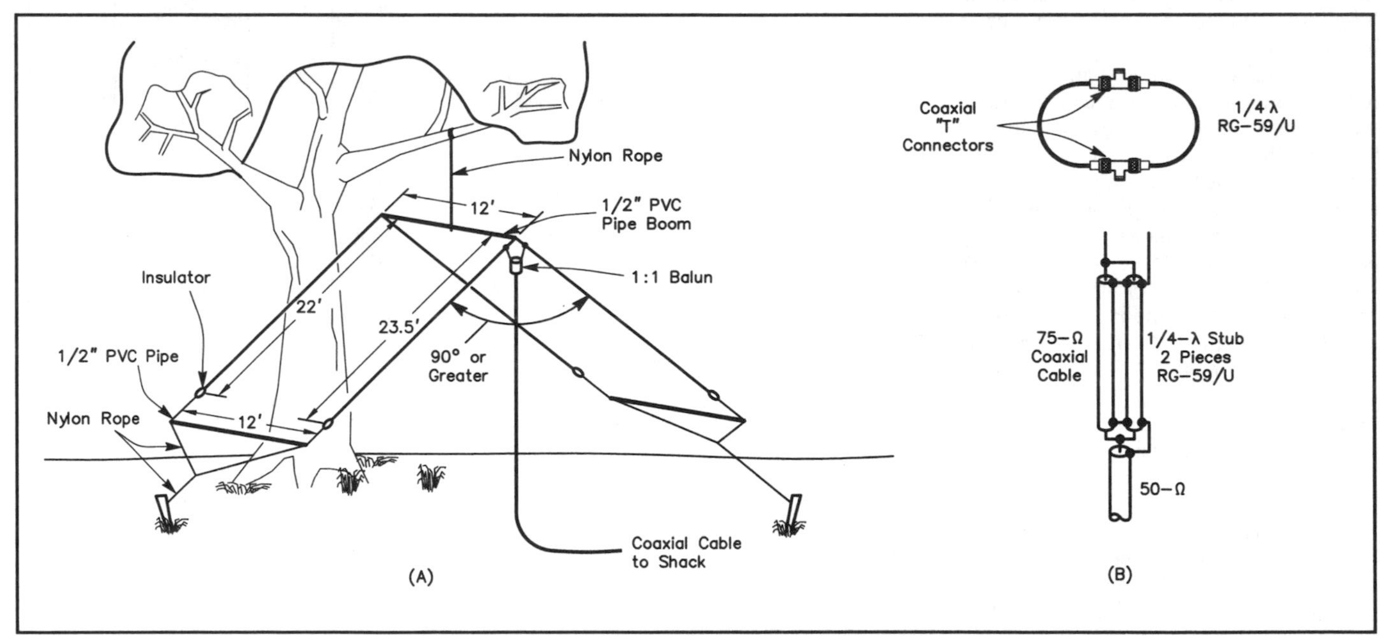

Figure 1—Construction details for a 30-meter inverted V beam are given at A. A coaxial-cable impedance-matching transformer is shown at B.

Try the "FD Special" Antenna

Looking for an antenna that's simple, inexpensive, lightweight and easy to install? Here's one that fits the description.

While the Field Day "juices" were flowing last year, my Field Day partners and I decided to become more competitive without compromising our general philosophy: Use all homemade gear, pack it in to the site, and don't let operating interfere with watching the scenery. Since most of our gear already ran the accepted QRP power limit of 10-W dc input and had been designed to provide high output efficiency, the antenna seemed like the reasonable point of attack (our high quality superheterodyne receivers have not been the limiting item). The constraints dictated that an antenna be lightweight, portable and easy to put up. It should also have substantial gain.

Because of our West Coast location, the front-to-back ratio was not a concern. But having a reasonably wide lobe toward the East was. We desired a low SWR because of the relatively high loss of our RG-58/U and/or RG-174/U feed line. We were interested only in the CW portion of the band, but this antenna works well over all of the band. It seemed that 20 meters would be our main "money-maker," so we designed the antenna for that band. It can be scaled for other bands, too.

The Research

Although a number of antenna types might have done the job, I settled quickly on a horizontal, close-spaced, driven array. Experience has shown that driven arrays are generally more tolerant of imperfect construction and erection than are parasitic arrays. Experience and much measurement have convinced me that horizontal arrays outperform vertical ones in the high-frequency bands, except perhaps from an exceptional location. In addition, we didn't want the nuisance of establishing a decent ground system — which most vertical arrays require.

The theoretical gain and front-to-back ratio of 2-element arrays with 1/8-wavelength spacing between the elements are shown in Figures 1 and 2. Note the lower curve of Figure 1. It shows the effect of losses on the gain (losses don't affect the front-to-back ratio, and change only the scaling of the pattern). Figure 3 shows the patterns of arrays with 135, 160 and 180-degree relative spacing. All are drawn to the same scale. The 1/8-wavelength spaced, 135-degree-fed array is frequently called the "ZL Special."[1] The close-spaced, 180-degree-fed array is known as an "8JK."[2] From 135 to 160 degrees, phasing was chosen because of the combination of relatively high insensitivity to loss, reasonable gain and wide forward lobes. Note that the gain stays about the same in this range, ensuring good performance if the phasing isn't exactly as predicted. Actually, it's much easier to generate and maintain precise 180-degree phasing than the angles I've chosen — particularly over a wide frequency range.

There's one major flaw (usually fatal) in a simple analysis like the one presented here: It assumes that equal-magnitude currents are flowing in the elements. This is not easy to realize, for even in arrays with elements spaced 1/2 wavelength or greater, mutual coupling has a profound effect on element impedances. This changes them dramatically and unequally, as a rule. This

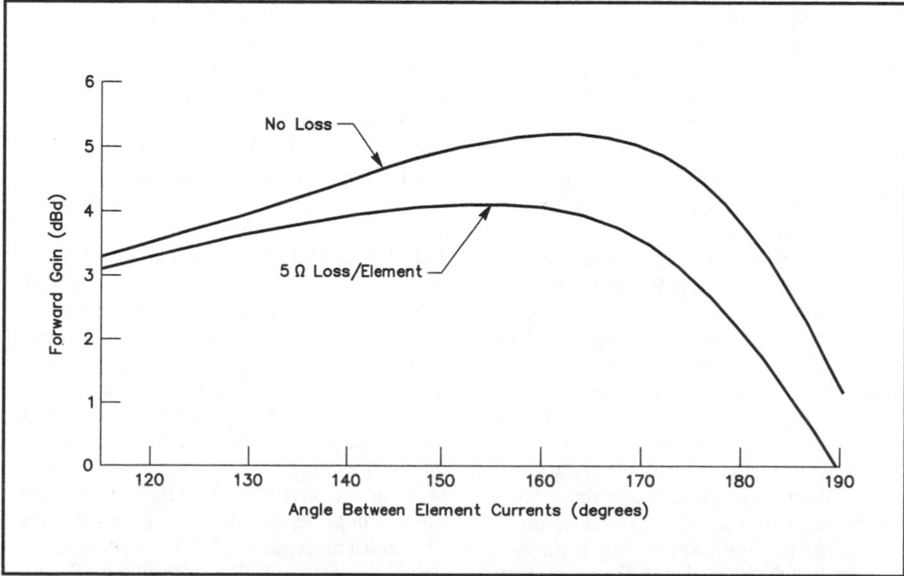

Figure 1—Curves that show gain versus phase angle for two-element arrays with 1/8-wavelength spacing.

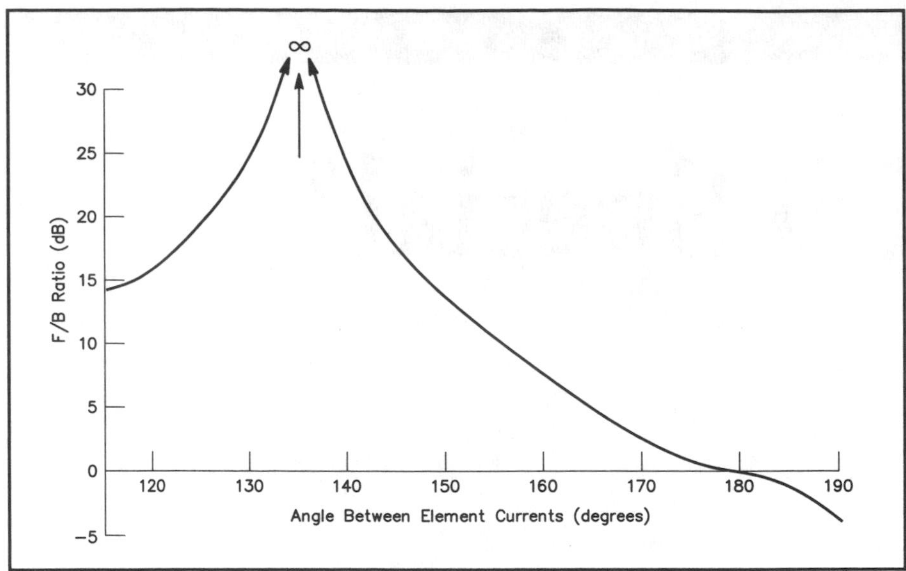

Figure 2—Theoretical F/B ratio for a two-element array with 1/8-wavelength spacing.

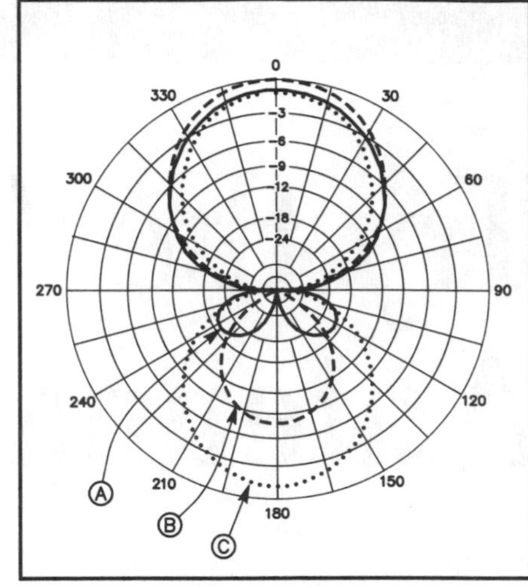

Figure 3—Dipole array patterns for 135, 160 and 180-degree relative phasing at λ/8 spacing. Curves A, B and C, respectively, represent these conditions. Add 5 dB for dBd. These curves are based on the array being fed with equal currents.

impedance change is a function of not only "mutual impedance," but also the relative magnitudes and phases of the currents flowing in the elements. In an array as closely spaced as these, coupling is so intimate that it could be argued that the term "driven array" is a misnomer. For example, the feed-point impedances of the elements in a 1/8-wavelength-spaced array, assuming equal currents can be made to flow, are

Phase Angle Between Currents (Degrees)	*Loading Element (Ohms)*	*Lagging Element (Ohms)*
135	28 – j46	28 + j46
160	13 – j22	13 + j22
180	9 + j0	9 + j0

This shows quite a change from the 74 + j0 ohms value that each element exhibits when it is not coupled to another element. The fact that the resistive parts of the two-element impedances are equal, and the reactances are equal in magnitude, is a peculiarity of the particular element spacing chosen. For other spacings they will be unequal, and the reactances can be different in magnitude, as well as in sign.

This mutual coupling isn't undesirable; in fact, it's essential for obtaining gain in the presence of rather severe pattern cancellation that is common in these closely spaced arrays. The lower impedances cause more element current to flow for a given power input, thereby increasing the fields from the elements. In these arrays, the increased field strength is sufficient to compensate for the fact that the fields from the elements don't add in phase in any direction. They partially or completely cancel instead. But, the lower feed-point impedances make them more sensitive to losses, and the low resistance with relatively high reactance makes them tricky to feed properly.

Why do these different and reactive feed impedances make feeding the arrays so difficult? The first problem is that, with few exceptions, the magnitude of current out of a line not terminated in its characteristic impedance won't be equal to the current into the line. In classic "If you can't fix it, feature it!" fashion, this impedance-transforming property is put to good use in the form of the 1/4-wavelength Q section.[3] The second (and almost always overlooked) difficulty is that, again with only a few exceptions, the phase delay of current in an imperfectly terminated transmission line doesn't equal the electrical length of the line. This effect isn't minor: The phasing of a casually designed array can easily be off by tens of degrees. In one design I investigated, an 80-degree line produced 139 degrees of phase shift.

The Solution

There are a number of approaches toward correct feeding of an array. My choice was to investigate some simple feed systems to see if any would yield results that came close to the desired characteristics. I wrote a computer program that would solve, iteratively, for element-current magnitude and phase angle, plus feed-point impedances for this one type of array, given the array specifics. Several configurations looked promising, and one of the simplest proved adequate. This was an array of two folded dipoles that were self-resonant, spaced 1/8 wavelength apart and connected by a taut piece of 300-ohm TV ribbon with one half twist. The feed impedance was close to 50 ohms resistive. There was some inductive reactance that could be corrected by adding two small-value capacitors at the feed point. The element current ratio was 1.13:1, with element phasing that was 154 degrees. This was not the 124 degrees one might expect from the 56 electrical degrees of line — assuming a velocity factor of 0.8 — minus the 180 degrees caused by the half twist.

It was this array that we used for Field Day, with very good results (see section on performance). However, when the array was reconstructed at the home QTH, a dramatic rise in SWR was noticed when operating the antenna at other than the low end of the band. Computer analysis showed that, above the design frequency, the phase angle increased. This caused a substantial lowering of the element feedpoint impedance, plus narrowing of the forward lobe. The analysis also showed the antenna to be well-behaved *below* the design frequency. Consequently, a similar array was designed (figuratively speaking) for 14.5 MHz. It gave good results over the 20-meter band.

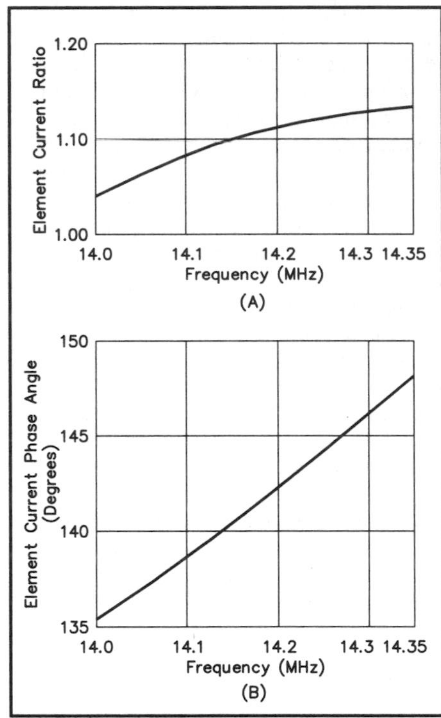

Figure 4—Element current ratio (A) and phase angle (B) as a function of frequency.

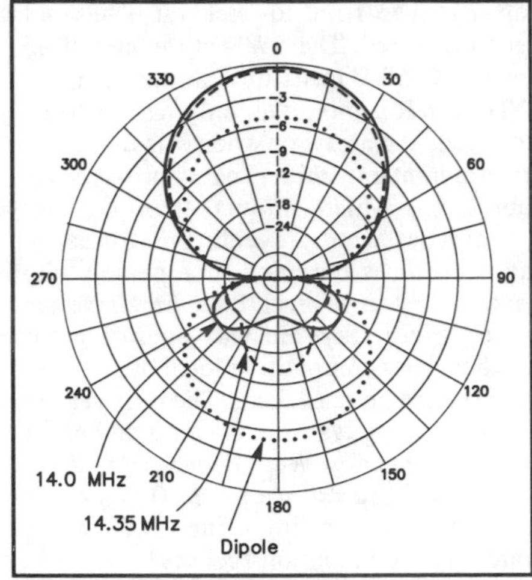

Figure 5—Calculated antenna patterns for the high and low ends of the 20-meter band. Add 5 dB for dBd.

Element phasing varies from 135 degrees at 14.0 MHz to 148 degrees at 14.35 MHz, with current ratios from 1.04 to 1.13:1 (see Figure 4). The gain can be calculated as fairly constant from 4.5 to 4.6 dBd across the band. Again, the array feed-point impedance can be corrected easily to provide a low SWR. The calculated patterns for the antenna at the top and bottom ends of the 20-meter band are shown in Figure 5. These take into account the changes in element phasing, spacing, current magnitude and element self-impedance with frequency.

Construction

The antenna is made from quality 300-ohm TV line to the dimensions given in Figure 6. Sketches of the insulators are provided in Figure 7. They are made from scrap pieces of epoxy-glass PC-board material. This results in ruggedness and minimum weight. The spreaders are readily available 10-foot lengths of "1 inch" ($1^5/_{16}$ in OD)[4] schedule 40 PVC pipe. The capacitors are used only to provide a good match to 50-ohm feed line: They don't otherwise affect the performance of the array. Small 500-V mica or monolithic ceramic units may be used for power levels up to a few hundred watts, since they are at a relatively low-voltage part of the system. Open-ended stubs could probably be substituted for the capacitors, if desired. I recommend that a balun transformer be used with this antenna. Attempts to measure the impedance of one element of this array resulted in a unique experience — the first substantial evidence of the need to use a balun transformer. The antenna-bridge readings varied greatly as the measuring equipment was moved, or as I placed my hand around the feed line. This ceased when I added a balun transformer. The phenomenon is explained by Maxwell in a recent paper.[5] Nearly any style of balun transformer will prevent the unwanted flow of current on the coaxial cable outer conductor. I use a choke type of balun transformer. It consists of 10 turns of small-diameter coaxial cable

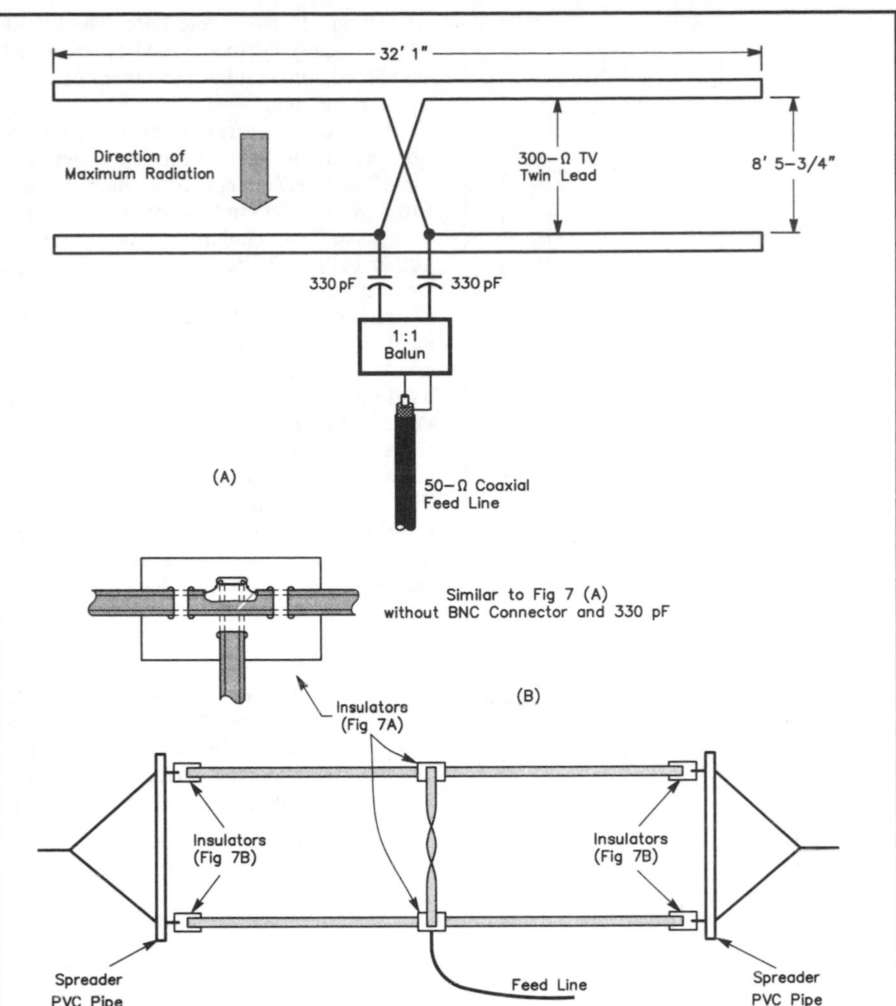

Figure 6—Electrical dimensions for the W7EL array (A). Illustration B shows how the antenna is assembled on spreaders of PVC pipe.

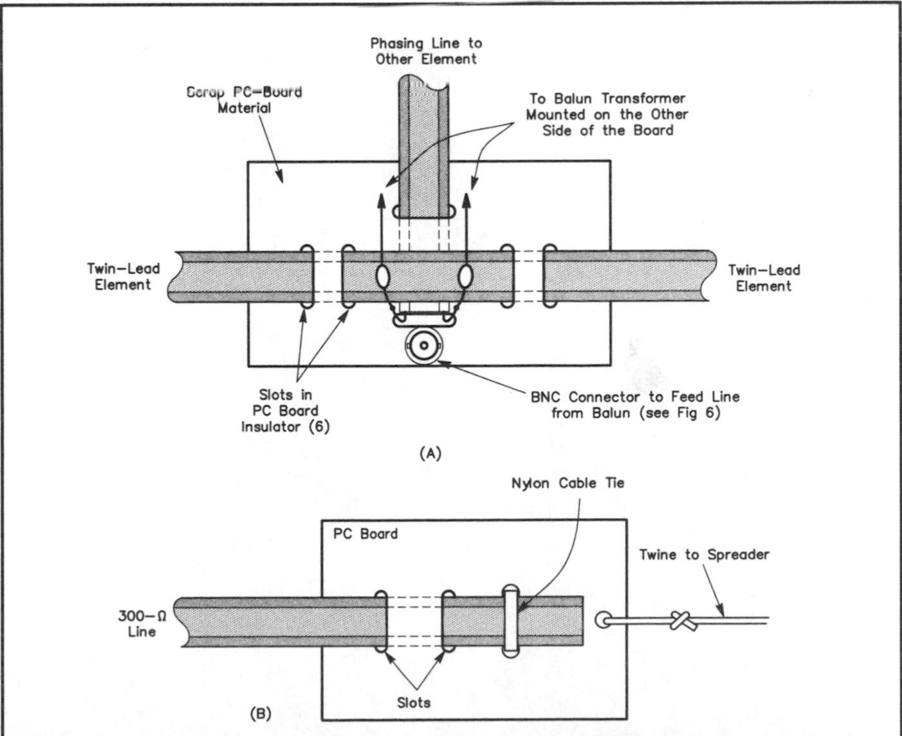

Figure 7—Details for the antenna insulators used in the two-element 20-meter array. The drawing at A shows how the TV ribbon is affixed to the feed-point/phasing-line insulating block. The example at B provides details for the end insulators.

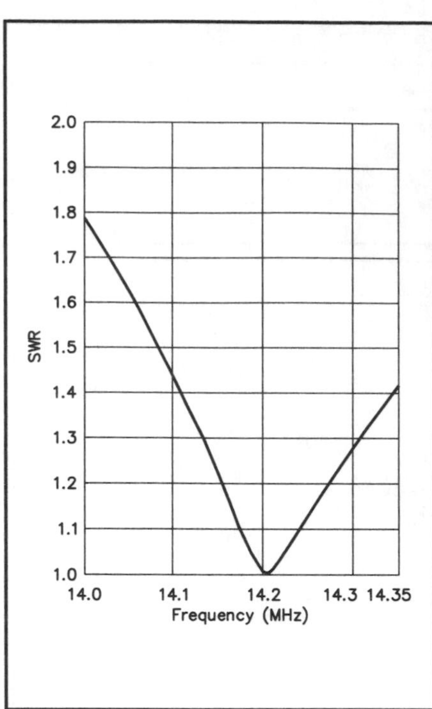

Figure 8—SWR curve obtained at the end of a 45-foot length of RG-58/U coaxial cable.

wound on a ferrite toroid core. The OD is approximately 1 1/4 inches, and it is mounted at the feedpoint insulator by means of small nylon cable ties.

If you use the array for portable operation, as we did, the 10-foot spreaders are out of the question, at least in their original form. We cut ours in half for "packing in," then used PVC cement to glue them together at the FD site. PVC pipe couplings were used to join the sections. The glue container was enclosed in polyethylene sandwich bags, just in case a leak developed. At the end of our FD exercise, we used the saw blade of KØED's Swiss army knife to cut the PVC pipes again for easy transport. The spreaders were glued together again for use at the home station. The antenna is held horizontal easily by attaching a piece of twine to the nondirectly driven element insulator. This counteracts the weight of the feed line that is connected to the other element.

Performance

The "FD Special" has been in use at W7EL for some time. Array gain has been compared to that of an inverted V at the same height. The calculated performance values appear correct within the measurement capability. The front-to-back ratio has not been measured. The SWR at the end of 45 feet of RG-58/U feed line is shown in Figure 8. The SWR is important only when a lossy line feeds the array, or when it is driven by an intolerant transmitter (with built-in SWR shut down), which is now the norm.

Perhaps the most revealing performance indication was provided by a person who encountered us several times on 20 meters during Field Day. He was operating for another, very competitive local club. After the exercise he remarked,"The only reason I believe you guys were running an honest 10 W is that I know Wes Hayward (W7ZOI) was there." Indeed, we used 10 W or less input while operating—and 0 W while watching the mountain scenery!

Notes

[1]My apologies to the first person who described or named this antenna. I don't know its history. (See L.A. Moxon, "Two-Element Driven Arrays," *QST*, July 1952, p 28.)

[2]Named after W8JK, Dr. John Kraus, "Antenna Arrays with Closely-Spaced Elements," *Proc. IRE*, Feb. 1940.

[3]G. Hall, ed., *The ARRL Antenna Book* (Newington: ARRL, 1982).

[4]mm = in x 25.4; m = ft x 0.0348.

[5]W. Maxwell, "Some Aspects of the Balun Problem," *QST*, March 1983.

By Riki Kline, 4X4NJ From *QST*, February 1985

Build a 4X Array for 160 Meters

Low-angle radiation and electrical rotation of directivity are the features of this vertically polarized top-band antenna. If you are interested in 160-meter DX, this system could be your secret weapon.

Is your lament a common one — no room for an effective DX antenna on 160 meters? This complaint is voiced frequently by amateurs who live in urban areas, or who are programmed toward horizontal wire antennas. But, a number of successful top-band operators have adopted the philosophy, "If you can't go out, go up!" It is no secret that a physically short vertical antenna is generally more effective than a horizontal antenna that is close to the ground electrically, at least for DX work.

My 4X array is electrically rotatable. It is compact and is effective as a low-angle radiator. Let's examine how my antenna evolved from some basic designs. I will also cover the practical details of construction and system performance.

The Tilted Ground-Plane Look

The tilted ground plane is almost identical to the usual vertical. The physical format of this antenna resembles a four-conductor ground plane. The major difference is that the radiating elements tilt up toward the supporting structure. The 4X array contains four sloping ground planes. Each of the slope wires is 100 feet long.[1] They are supported at the high end by an 80-foot tower. A four-element, 20-meter Yagi antenna is atop the tower.

Each of the sloping wires is fed separately near ground by means of a tapped-coil matching device (Figure 1) that is returned to radial wires and ground rods. In effect, each radiator is a ground-plane vertical antenna that is slightly less than 0.25 wavelength. The matching inductor provides resonance and effects an impedance match to the coaxial feed line.

I believe that the metal tower and 20-meter antenna may possibly be functioning as a reflector because the Yagi antenna and tower combined with ground wires are resonant slightly below 1.8 MHz.

Two Tilted Ground-Plane Verticals in Phase

I had excellent results with one sloping vertical. Next, I installed a second system in the opposite direction. Switching between the two antennas (north-south sloping radiators) showed considerable front-to-back ratio (a relative reading of 15-20 dB). Subsequently, I connected the two antennas in phase. This gave a bidirectional pattern, east and west. Although I did not gather extensive data on the performance, I observed a 6-dB signal improvement with stations about 700 miles to the

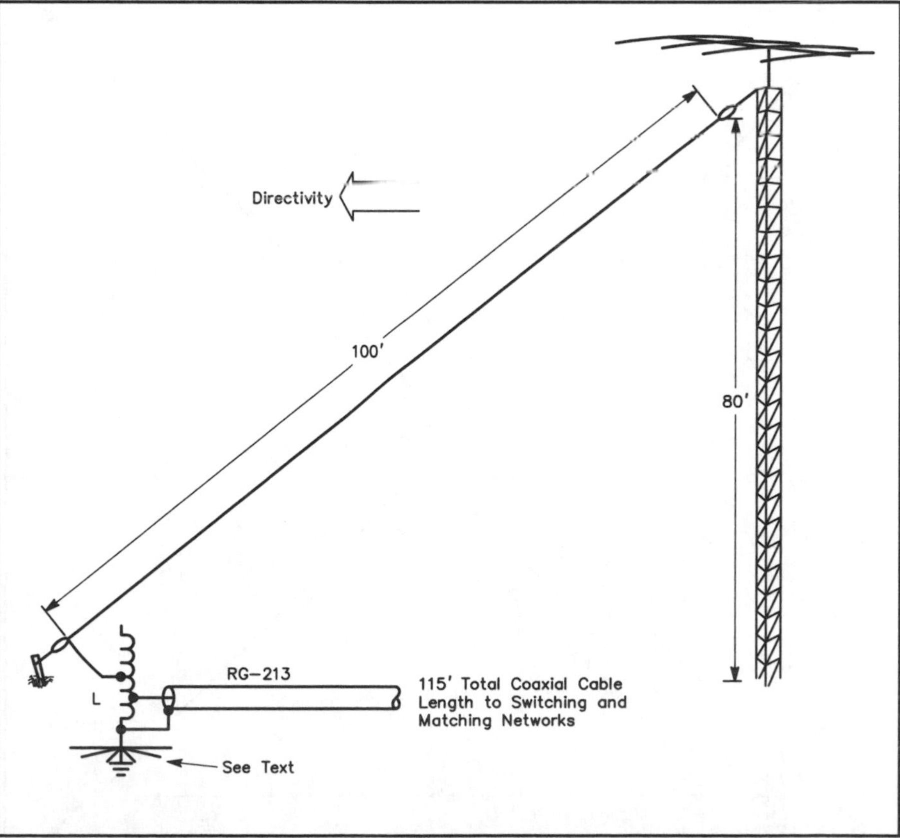

Figure 1—Basic tilted ground-plane vertical. L has 25 turns of heavy conductor (see text); length is 5½ inches, and diameter is 3 inches. Radiator is tapped 15½ turns above ground.

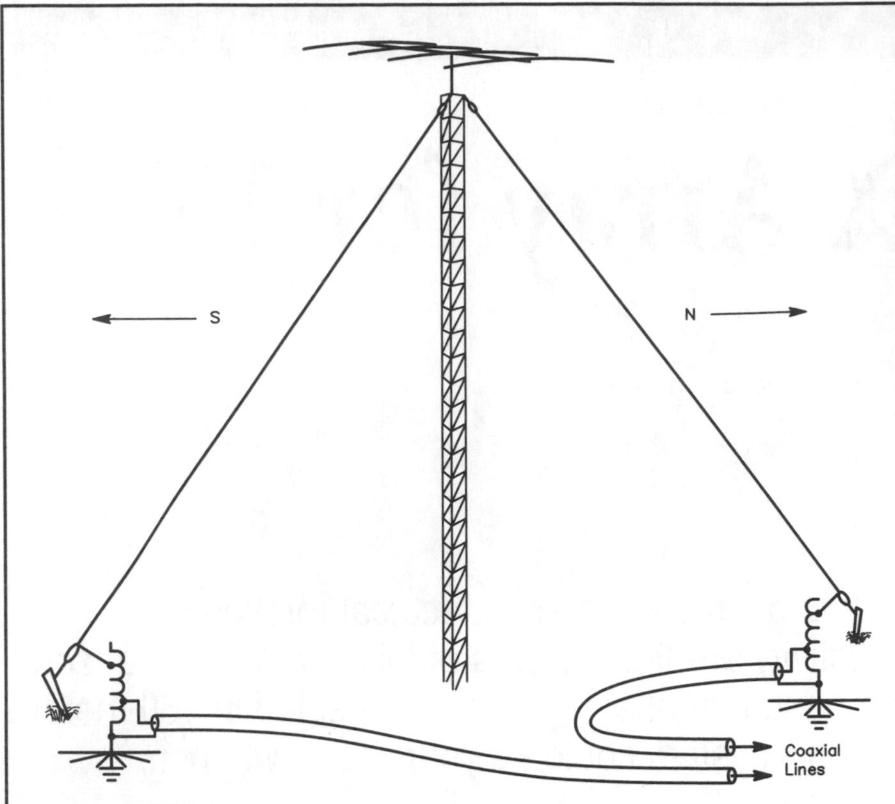

Figure 2—Two tilted ground-plane antennas that can be fed separately or in phase. The feed method is shown in Figure 1, with the method of Figure 7 used for feeding the antennas in phase.

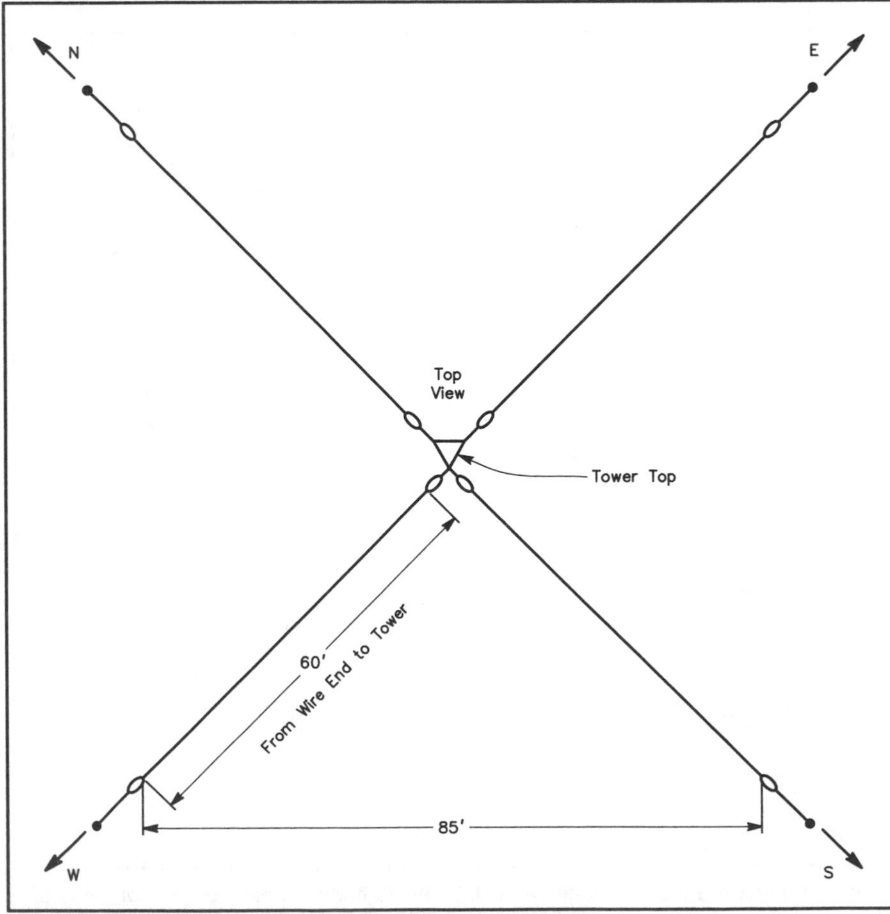

Figure 3—The 4X array as viewed from above the tower.

east. Some of you may want to explore the possibilities further. Figure 2 shows the details of the two-element phased system.

The 4X Configuration

Two more tilted ground-plane verticals were added, thereby providing east-west sloping radiators (Figure 3). A switching and phasing arrangement was added to my 4X array. It allows me to feed any of the slope wires separately, adjacent pairs in phase, or all four wires in phase. When using adjacent pairs in phase, maximum radiation is along a line that bisects the angle between the two antennas (NE, SE, SW or NW directions). When I feed all four wires in phase I note that the radiation is essentially omnidirectional. All of the unfed radiators are resonated to serve as reflectors. This concept is described in *The ARRL Antenna Book*.[2] My switching network is shown in Figure 4.

Phasing Networks

Most phasing methods call for long lengths of non-50-ohm coaxial cable.[3] I found this economically prohibitive. This negative factor inspired the approach I am using.

Each of my radiators is fed by means of 115 feet of RG-213 coaxial cable (formerly RG-8A/U 50-ohm line). The coil at the base of each wire is adjusted for the same resonance and SWR as the remaining three coils. The verticals to be fed in phase have their transmission lines connected in parallel through a suitable network for changing the reflected impedance back to 50 ohms. My networks are shown in Figures 7 and 8.

Tapped-Coil Matching

By using inductance and no intentional parallel capacitance for my matching coils, I am able to obtain greater effective antenna bandwidth because of reduced Q. Stray capacitance and antenna capacitance to the tower and ground are present, however. All electrical connections are soldered. A simple rain cover is used over each coil to protect it from moisture and dirt. The absence of switches, variable capacitors and rotary inductors enables construction of a highly reliable matching system without the need for weatherproof boxes.

This system is relatively easy to tune to obtain nearly identical performance from each antenna branch. This becomes a necessity when using "brute-force" parallel feed in the phased-pair and omnidirectional modes. Otherwise, the power distribution and phasing would be disturbed. This would distort the radiation pattern. Large-diameter, heavy-conductor, air-wound coils are best for this job. Two of my coils are made from silver-plated 1/4-inch-diameter copper tubing. The two remaining coils are made from large, flat conductor material of the kind found in some rotary inductors.

I used a dip meter to adjust the coils for resonance (coaxial cables disconnected). My coaxial cables were tapped initially one

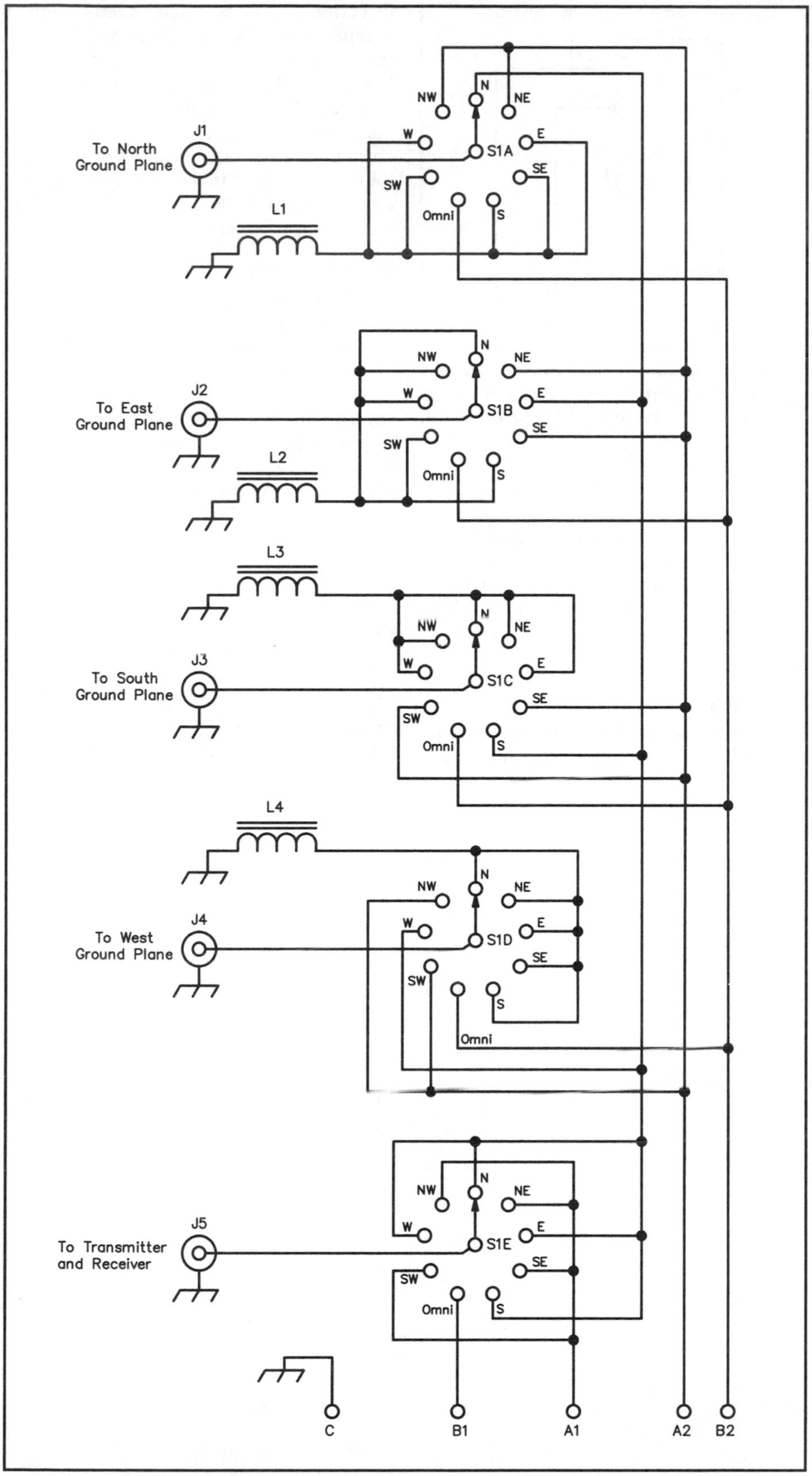

Figure 4—Switching system for the 4X array. L1 through L4, inclusive, are described in the text. A1, A2 and C are connected to the matching circuit shown in Figure 7. B1, B2 and C are connected to the matching circuit shown in Figure 8. S1 is a heavy-duty ceramic rotary switch, five poles, nine positions. J1-J5, inclusive, are coaxial connectors of the builder's choice.

Direction	SWR vs. Frequency				
	1800 kHz	1825 kHz	1850 kHz	1875 kHz	1900 kHz
N	1.4	1.0	1.2	1.8	2.6
NE	1.5	1.1	1.0	1.1	1.2
E	2.2	1.1	1.1	1.4	1.9
SE	1.2	1.0	1.0	1.0	1.1
S	1.4	1.0	1.1	1.4	2.3
SW	1.5	1.1	1.0	1.0	1.1
W	1.4	1.1	1.0	1.2	1.6
NW	1.2	1.0	1.0	1.0	1.2
Omni	1.4	1.1	1.0	1.0	1.0

Figure 5—Chart that shows SWR versus frequency in kilohertz.

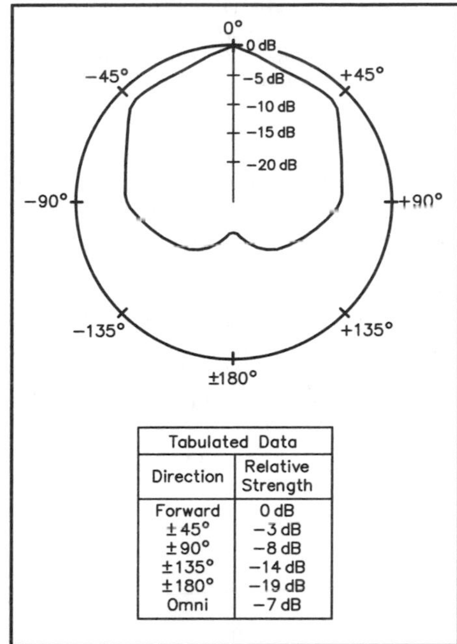

Tabulated Data	
Direction	Relative Strength
Forward	0 dB
±45°	−3 dB
±90°	−8 dB
±135°	−14 dB
±180°	−19 dB
Omni	−7 dB

Figure 6—Directivity pattern of the array. The pattern is a composite average of measurements made while receiving 10 different stations. There is no apparent difference between the two directive modes—single radiator or phased pair.

third of the way up from the ground ends of the coils. Final tap placement is made while feeding power to the antenna and observing an SWR meter. Alligator clips make this an easy matter to accomplish. When the SWR bottoms out at the same frequency for all four radiators, remove the alligator clips and solder the coil taps in place. Some interaction between the four antennas will occur, so make certain that all of the taps are where they belong before soldering them.

Reflector Tuning

The radiators not being fed are used as reflectors. This is done by switching small inductors in parallel with the ends of the coaxial feed lines. My inductors contain three or four turns of no. 16 wire wound around the center part of a 3/8-inch-diameter ferrite rod from a built-in AM broadcast receiver antenna. The coils are adjusted to give a resonance that is four percent lower than the resonant frequency of the radiators.

Ground Conditions

The efficiency and performance of the antenna depends on the quality of the ground system. Note 3 provides a good reference for ground systems, and a bibliography. Each of my radiators is worked against a counterpoise that contains two or three 1/4-wavelength wires, bent to fit in the boundaries of my property. In addition, I use a 10-foot rod in the ground. Water pipes

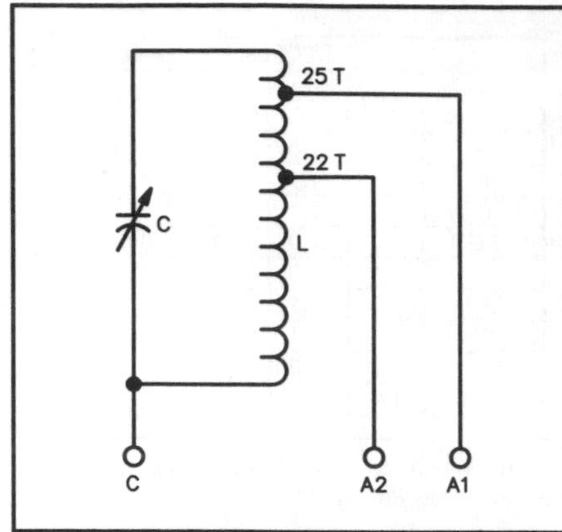

Figure 7—Phased-pair matching network. C is a 1000-pF variable, rated at 1000 V or greater. L is 30 airwound turns of heavy conductor (see text), 7 inches long and 3½ inches in diameter.

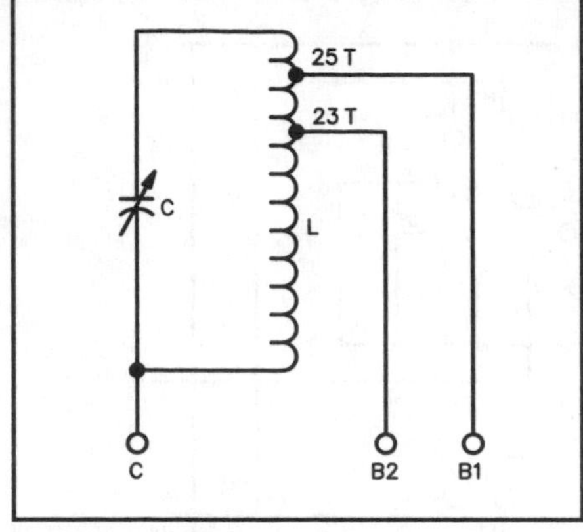

Figure 8—Network for omnidirectional use. C and L are the same as for Figure 7.

and all other available underground metal objects are tied to my ground system. You should try to extend your radials in the direction of the preferred radiation.

Insulators

The top ends of the radiators contain high RF voltage. I use 15-inch-long Plexiglas strips as insulators, after having problems with 8-inch-long commercial plastic insulators. Moisture and air pollutants caused these problems. The present insulators need to be cleaned periodically. High-quality glass insulators of the type used aboard ships should be excellent and, with luck, should not require periodic cleaning.

Receiving

The antenna directivity enhances reception by rejecting signals from unwanted directions. I was encouraged when I compared my 4X array to an 800-foot unterminated Beverage antenna that is bidirectional NW and SE. In the past, the Beverage wire showed an 8-10 dB S/N advantage over the tilted north vertical alone. During long-haul QSOs to North America, the 4X array, used in the phased-pair mode, comes within 3 dB of the S/N ratio provided by the Beverage.

In Conclusion

Many European stations tell me my signals are as strong as local ones. On occasion they remark that I have the loudest signal on the band.

It's a pleasure to have a directional array on top band. I simply turn a knob to rotate the pattern — much faster than a motor can rotate a typical beam antenna! The 4X array is compact and can be supported by the existing HF-antenna tower. I hope that some of you will try this antenna, and I look forward to hearing from you about your results.

Notes

[1]m = ft × 0.3048; mm = in × 25.4; km = mi × 1.609.

[2]G. Hall, ed., *The ARRL Antenna Book* (Newington: ARRL, 1984), p. 8-12.

[3]J. Devoldere, *80-Meter DXing* (Greenville, NH: Communications Technology, 1978).

By Rod Newkirk, W9BRD From *QST*, June 1990

The 'BRD Zapper: A Quick, Cheap and Easy "ZL Special" Antenna

Here's a multipurpose directional wire antenna that has its origins in Kraus, Windom and Newkirk *(who?–Ed.)*.

Working Europe from Chicago with low power on 21 MHz, using an indoor, bent-ends W8JK bidirectional wire beam (Figure 1)[1], is fairly straightforward fun. Voltage feed via a λ/4 stub, gamma-matched to coax, is hard to beat for spartan simplicity. No worry about balancing element currents, no nit-picking with element lengths. Just make the whole system symmetrical and dip it to your favorite frequency by adjusting the stub-shorting point.[2] The 8JK has a similar pattern and good gain all the way from the fundamental to the second harmonic, so it's really a multiband antenna. You can roll the whole thing up in three minutes when it's not in use.

One problem: The bidirectional characteristics of the 8JK cause my ears to be flattened regularly by undesired signals from the direction opposite Europe. Unless you're in the geographical center of a three-way QSO, the unused lobe of this antenna can be a nuisance. Question: Is it feasible to convert easy 180° bidirectionality to tricky 135° *uni*directionality without resorting to clumsy center-feeding, multi-wire-dipole elements, balanced-gammas, etc?[3]

Yes—because there's a *uni*directional ZL Special lurking in our little 8JK. The principal difference between the W8JK and the ZL Special lies in the phasing: In the 8JK, the elements are fed 180° out of phase; in the ZL Special, they're driven 135° out of phase. Therefore, instead of feeding the antenna near the stub's shorting point, we'll need to feed it λ/16 from the stub's shorted end. That's where the feed path to one element is λ/8 (45°) longer or shorter than the other, which, after the stub's 180° phase reversal, produces the 135° phase shift that were looking for between the elements.

You can find the proper feed point on the stub by "sniffing" signals of known origin along *one side* of the stub with the insulated center conductor of some coax hooked to a receiver. (Start looking for this point by measuring λ/16 up from the bottom of the stub.) It's refreshing to hear Europeans rolling in while most of the murderous rearward signals now barely budge the S meter! Directivity is reversed at the opposite stub point (Figure 2).

So he's in there all right, that ZL Special, but coaxing the wily rascal out for an honest day's work is a challenge. The ZL

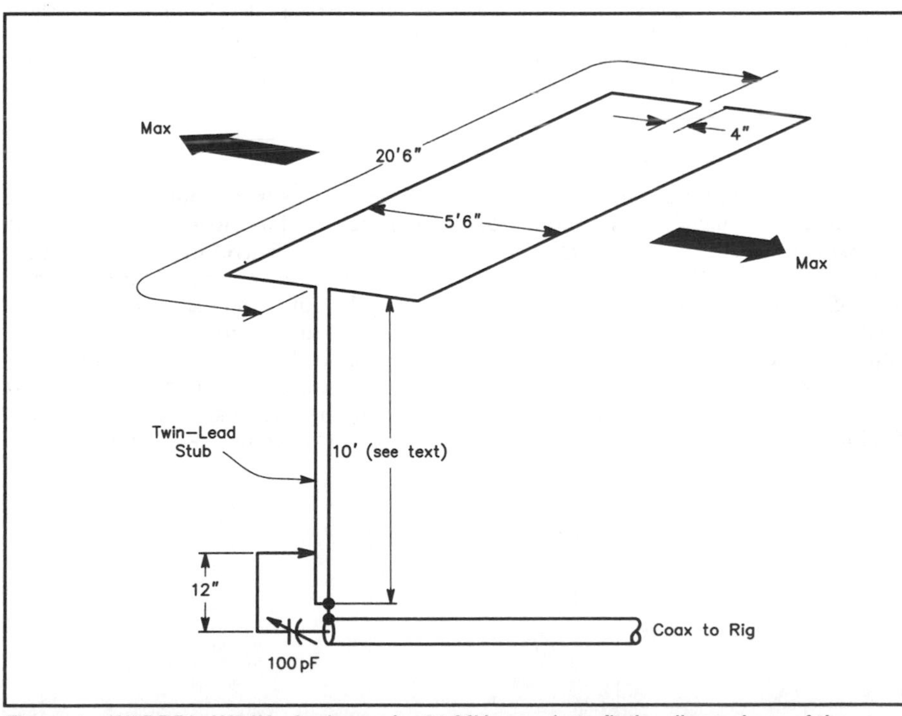

Figure 1—W9BRD's W8JK wire beam for 21 MHz, made to fit the dimensions of the townhouse bedroom that it occupies. Its dimensions aren't critical, although the element lengths should be close to λ/2 (total, each), the stub should be λ/4 or odd multiples, and the element spacing should be close to λ/8. At the bottom of the stub, connect both stub wires and the coax braid together, after resonating the system as outlined in Note 2. The 12-inch tap distance and 100-pF variable capacitor constitute a gamma match that brings the stub impedance to 50 Ω.

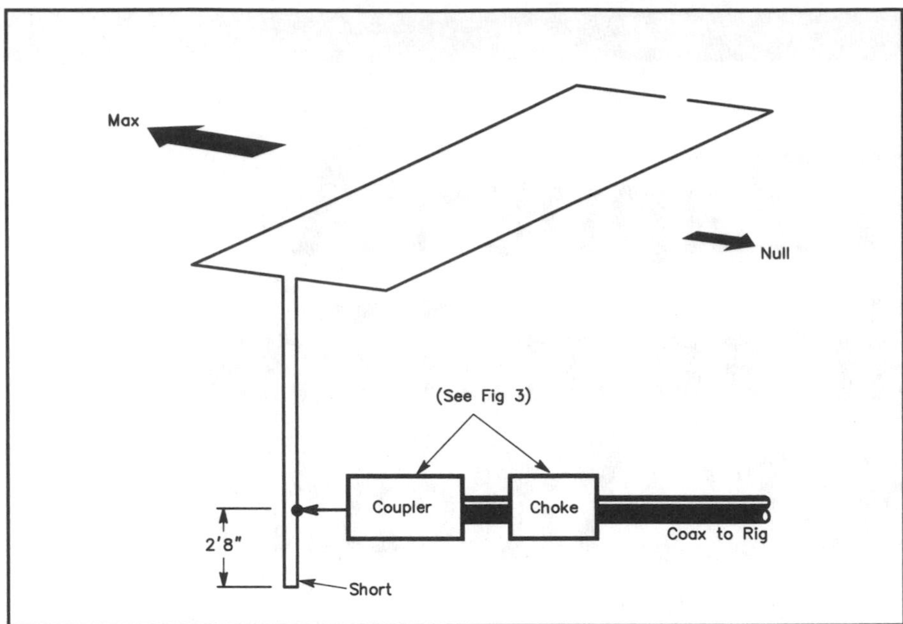

Figure 2—The antenna of Figure 1 configured for unidirectional operation with 135° element phasing. Resonate the system as described in Note 2 and connect the coupler via a very short wire to one leg of the stub about λ/16 above the stub-shorting point. The coaxial choke is needed to eliminate feed-line radiation, which can ruin the rearward null that this system provides.

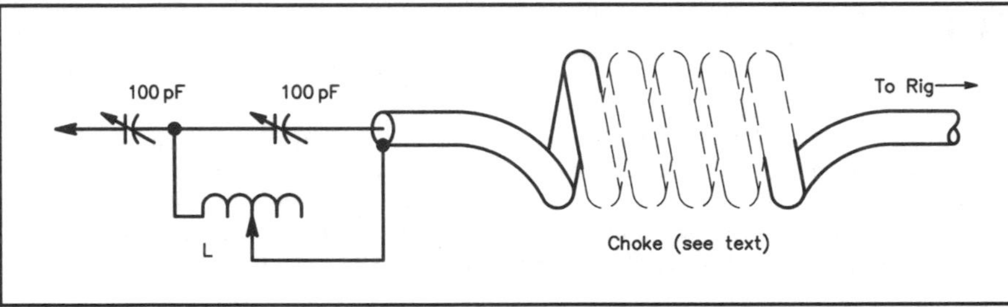

Figure 3—Matching-network and coaxial-choke details for the 'BRD Zapper. L consists of 10 turns of no. 16 wire, 1½ in. diameter, air core, space wound. For 100 W or less, broadcast-type variable capacitors are suitable. Adjust the inductor-tap point for lowest SWR. The coaxial choke is made of 30 turns of RG-58 wound on a ferrite rod.

Special has a very sharp null in one direction when the element phasing is right, but unbalancing the stub could cost us that null. Fortunately, when the system is resonant, the 135° tap point has a moderately high resistive impedance, making for easy matching of the antenna to 50-Ω coax via a T network (Figure 3).

Pattern distortion through incidental radiation and pickup must be minimized. Excessive leakage would put us right back at the mercy of those loud signals from the other direction. Therefore, the coupler must be built in the most compact form possible, mounted right at the stub, and isolated to keep the feed line from distorting the pattern. Such isolation is done at W9BRD via a home-brew coaxial choke made of 30 turns of the antenna's RG-58 feed line wound on a ferrite rod just before the matching network.

Because feeder radiation could upset the stub balance—and thus spoil the antenna's rearward null—the single-wire run from the matching network to the antenna had better be no more than an inch or two. Here, I'm borrowing on the single-wire-feed theme by Windom. Any circuitry above the choke will be hot with RF, so a bulky meters-bells-and-whistles matching unit will not do. It's best to use a compact, barebones, dedicated network. Run the stub all the way down from the flat top if possible. Incidentally, the tap point on the stub corresponding to the best rearward null is higher on the stub for element spacings wider than λ/8. System Q is also lower with wider spacings, but gain is maximum at λ/8.

As for the stub itself, TV twin lead is okay indoors, but close-spaced, end-fire elements mean high voltages and currents. If the stub is to be longer than a quarter wavelength (odd multiples only) or outside in the weather, low-loss, open-wire line is a must. You can run the stub of an outdoor version directly into the shack for handy directivity reversal.

Though we've been talking horizontal, this setup should be equally interesting as a vertical. By the way, when the elements are placed horizontally, one above the other, 135° phasing accentuates radiation and reception at high angles. Neat for local traffic nets, Sweepstakes and Field Day!

Notes

[1]Dr John D. Kraus, W8JK, first described this antenna in *QST* in "Directional Antennas with Closely Spaced Elements," *QST*, Jan 1938, pp 21-23, 37. Kraus also discussed this antenna in three articles in *Radio* magazine in 1937 and 1939, and in *Proceedings of the IRE*, Feb 1940.

[2]The easiest way to do this is with a dip meter. Jab a straight pin into each leg of the stub near the bottom, short the pins together and couple the dip meter Into the system by bringing the dip motor close to the shorted pins. Locate resonance using the dip meter and measure the dip-meter frequency on your receiver. Move the pins up or down the stub and repeat the process until the antenna is resonant at or near your frequency of Interest. When you've done this, cut the stub, strip and solder a short in the wires where the pins were located at resonance.

[3]Of course, this antenna can be center fed in either Its bidirectional or unidirectional modes. Such flat tops are described in J. Hall, Ed., *The ARRL Antenna Book*, 15th ed, (Newington: ARRL, 1988), Chapter 8, and J. Devoldere, *LowBand DXing* (Newington: ARRL, 1987), pp 2-101 through 2-102.

Combination Driven Arrays

Broadside, end-fire and collinear elements can readily be combined to increase gain and directivity, and this is in fact usually done when more than two elements are used in an array. Combinations of this type give more gain, in a given amount of space, than plain arrays of the types just described. The combinations that can be worked out are almost endless, but in this section are described only a few of the simpler types.

The accurate calculation of the power gain of a multi-element array requires a knowledge of the mutual impedances between all elements, as discussed in earlier sections. For approximate purposes it is sufficient to assume that each set (collinear, broadside, end-fire) will have the gains as given earlier, and then simply add up the gains for the combination. This neglects the effects of cross-coupling between sets of elements. However, the array configurations are such that the mutual impedances from cross-coupling should be relatively small, particularly when the spacings are 1/4 λ or more, so the estimated gain should be reasonably close to the actual gain.

FOUR-ELEMENT END-FIRE AND COLLINEAR ARRAY

The array shown in Figure 1 combines collinear in-phase elements with parallel out-of-phase elements to give both broadside and end-fire directivity. It is popularly known as a "two-section W8JK" or "two-section flat-top beam." The approximate gain calculated as described above is 6.2 dB with 1/8-λ spacing and 5.7 dB with 1/4-λ spacing. Directive patterns are given in Figures 2 and 3.

The impedance between elements at the point where the phasing line is connected is of the order of several thousand ohms. The SWR with an unmatched line consequently is quite high, and this system should be constructed with open-wire line (500 or 600 Ω) if the line is to be resonant. With 1/4-λ element spacing the SWR on a 600-Ω line is estimated to be in the vicinity of 3 or 4 to 1.

To use a matched line, a closed stub 3/16 λ long can be connected at the transmission-line junction shown in Figure 1, and the transmission line itself can then be tapped on this matching section at the point resulting in the lowest line SWR. This point can be determined by trial.

This type of antenna can be operated on two bands having a frequency ratio of 2 to 1, if a resonant feed line is used. For example, if designed for 28 MHz with 1/4-λ spacing between elements it can be operated on 14 MHz as a simple end-fire array having 1/8-λ spacing.

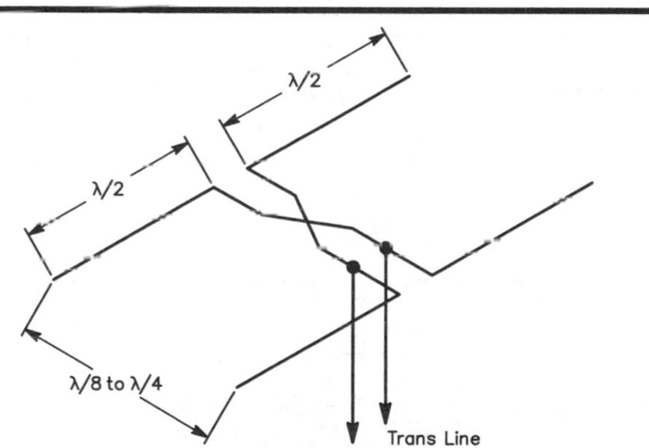

Figure 1—A four-element array combining collinear broadside elements and parallel end-fire elements, popularly known as the W8JK array.

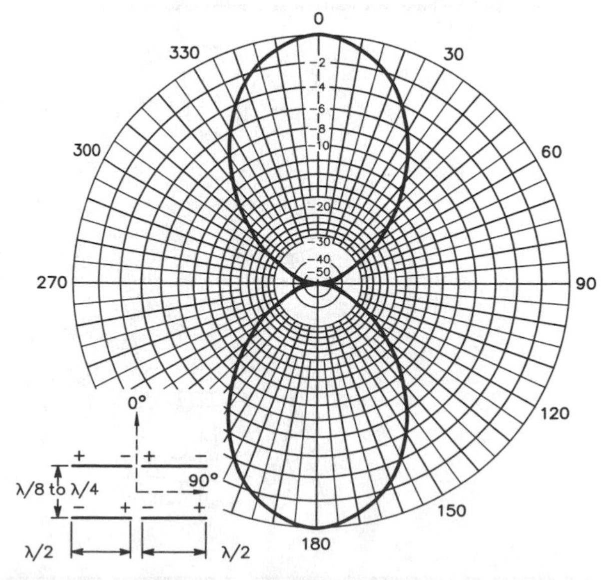

Figure 2—E-plane pattern for the antenna shown in Figure 1. The elements are parallel to the 90°-270° line in this diagram. Less than a 1° change in half-power beamwidth results when the spacing is changed from 1/8 to 1/4 λ.

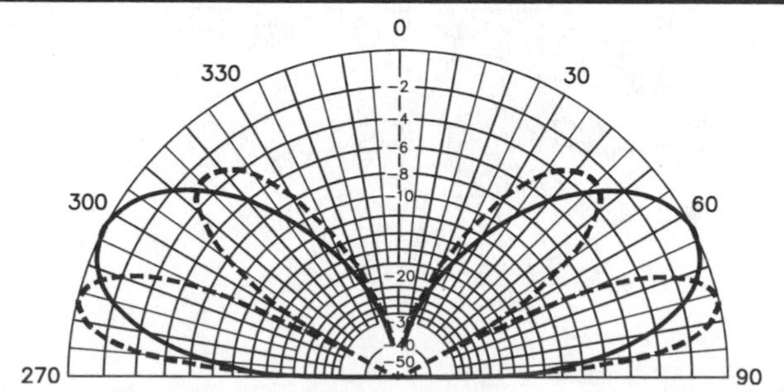

Figure 3—Vertical pattern for the four-element antenna of Figure 1 when mounted horizontally. Solid curve, height 1/2 λ; broken curve, height 1 λ above a perfect conductor. Figure 2 gives the horizontal pattern.

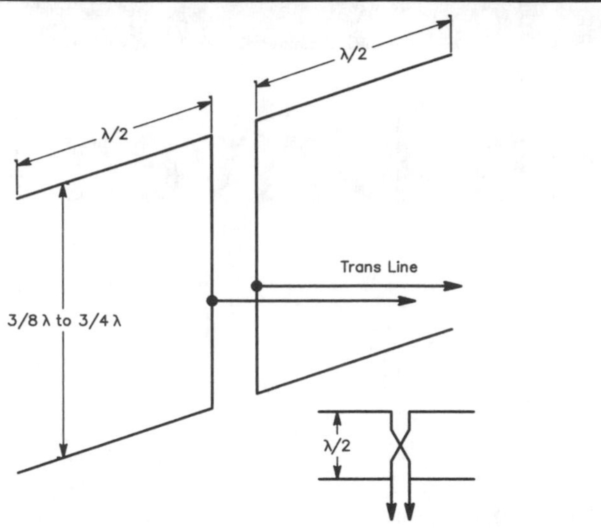

Figure 4—Four-element broadside array ("lazy H") using collinear and parallel elements.

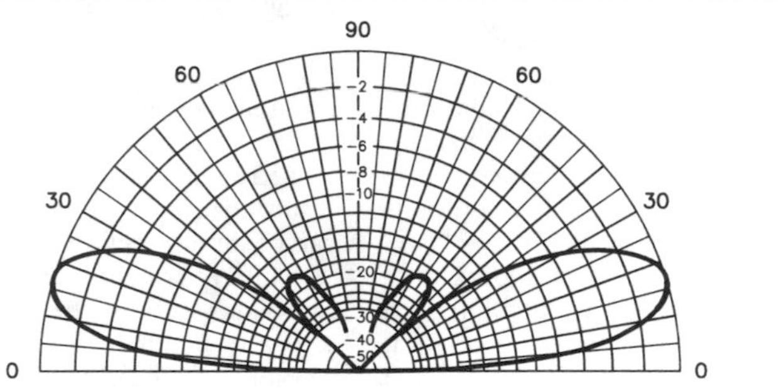

Figure 6—Vertical pattern of the four-element broadside antenna of Figure 3, when mounted with the elements horizontal and the lower set 1/2 λ above a perfect conductor. "Stacked" arrays of this type give best results when the lowest elements are at least 1/2 λ high. The gain is reduced and the wave angle raised if the lowest elements are close to ground.

FOUR-ELEMENT BROADSIDE ARRAY

The four-element array shown in Figure 4 is commonly known as the "lazy H." It consists of a set of two collinear elements and a set of two parallel elements, all operated in phase to give broadside directivity. The gain and directivity will depend on the spacing, as in the case of a simple parallel-element broadside array. The spacing may be chosen between the limits shown on the drawing, but spacings below 3/8 λ are not worthwhile because the gain is small. Estimated gains are as follows

3/8-λ spacing—4.4 dB
1/2-λ spacing—5.9 dB
5/8-λ spacing—6.7 dB
3/4-λ spacing—6.6 dB

Half-wave spacing is generally used. Directive patterns for this spacing are given in Figures 5 and 6.

With 1/2-λ spacing between parallel elements, the impedance at the junction of the phasing line and transmission line is resistive and is in the vicinity of 100 Ω. With larger or smaller spacing the impedance at this junction will be reactive as well as resistive. Matching stubs are recommended in cases where a nonresonant line is to be used. They may be calculated and adjusted as described in Chapter 26.

The system shown in Figure 3 may be used on two bands having a 2-to-1 frequency relationship. It should be designed for the higher of the two frequencies, using 3/4-λ spacing between parallel elements. It will then operate on the lower frequency as a simple broadside array with 3/8-λ spacing.

An alternative method of feeding is shown in the small diagram in Figure 3. In this case the elements and the phasing line must be adjusted exactly to an electrical half wavelength. The impedance at the feed point will be resistive and of the order of 2 kΩ.

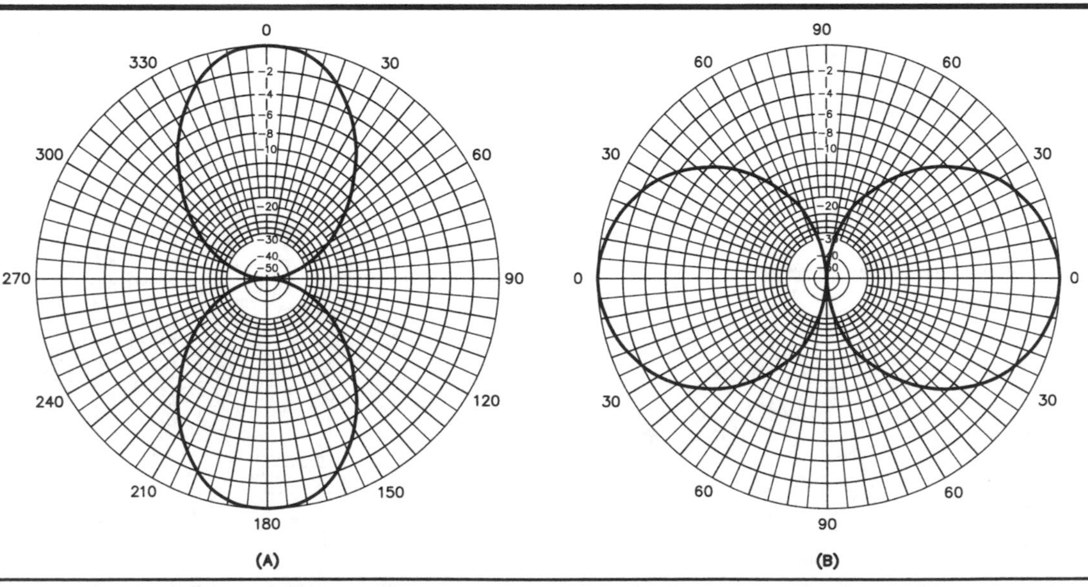

Figure 5—Free-space directive diagrams of the four-element antenna shown in Figure 4. At A is the E-plane pattern, the horizontal directive pattern at low wave angles when the antenna is mounted with the elements horizontal. The axis of the elements lies along the 90°-270° line. At B is the free-space H-plane pattern, viewed as if one set of elements is above the other from the ends of the elements.

CHAPTER SEVEN

VERTICALLY POLARIZED

More on "Half Slopers"

The July 1979 *QST* article on quarter-wave slopers has prompted me to write you on the subject. I have had considerable experience using them on 160 and 80, and I have tried them on 40.

The first one I know who used this type antenna was W3AU. His success with the system on 160 caused him to recommend it to W7RM and me. The original configuration used more than one sloping element fed in parallel against the tower. It resembled a top-loaded vertical with a simulated ground system. All my results have supported the view that this is an omnidirectional, vertically polarized radiating system, even when only one sloping element is used.

All the antennas of this type I have tried have been at my father's station, W6UA, and have been on a single tower. The tower is 82 feet (25 m) high, approximately 1½ feet (0.45 m) in uniform triangular cross section, and has a single set of guys at the 70-foot (21.3-m) level. The guys are broken with insulators at about 20-foot intervals, and the top section of each guy is electrically connected to the tower. The base has whatever insulation the concrete provides and there are 10-, 15- and 20-meter Yagis on top of the tower. (This is an AB105 tower, the same as is used by W3AU and W7RM.)

The first antenna of the type in question was tried on 80 meters. It was attached at the 60-ft level, so that top-loading effects of the guys, as well as the rotaries, were probably effective. The attached wire came off at 30° to vertical and was trimmed for minimum VSWR. The resultant length was within 2 feet of the formula length, and the antenna then had a VSWR less than 1.5 from 3.5 to 4.0 MHz. It worked well on transmit but was extremely noisy on receive. The original wire sloped east, so I put up an identical system, attached to the same tower level, sloping west. Electrical characteristics were nearly identical, and no directivity difference between the two could be detected either on transmit or receive.

This is logical considering how little of the antenna current in the sloping wire has any horizontal component, other antenna systems used on 80 at W6UA have been dipoles, inverted Vs, and a delta loop. The quarter-wave sloper is better than any of the others for transmitting, but it is poor on receiving. I have taken down all but the two slopers. They are used for transmitting and receiving, with frequent use of a 40-meter collinear antenna on receiving weak signals.

I also tried this system on 40, with a wire attached to the tower at 30 feet, coming off at 30°. Perhaps because there was so much tower electrically above the attachment point – at least half a wavelength – this one was much harder to tune. The wire length, which was very critical, was some feet shorter than expected, and the VSWR behavior across the band less benign. Not much time was spent on this particular antenna because it seemed no better than a $\lambda/2$ vertical. Both worked well, but neither was the equal of the collinear types available.

The most recent quarter-wave sloper system I have used has been on 160 meters. My first attempt was a single sloping element from the 75-foot level. This places whatever contribution there is from the top guy sections below the point of attachment. I couldn't get this antenna to match by altering the length of the sloping wire, even while making rather severe adjustments. There did seem to be some promise for the system, so the next season I used two sloping elements of about the "correct" length and fed the arrangement with a matching network at the bottom of the tower, with 300-ohm twin-lead going up the tower (Figure 1). This was the best antenna I have used on 160 at W6UA. Previously I had used an inverted V, a center-fed 80-meter dipole at 90 feet and a $\lambda/2$ inverted L. Since the wires must come off the tower at about 45-50° to vertical, the rationale of using two wires instead of one was to attempt to cancel the horizontal current component. As a matter of curiosity I tried adding a third wire and could find no change in setting of the matching components or in the

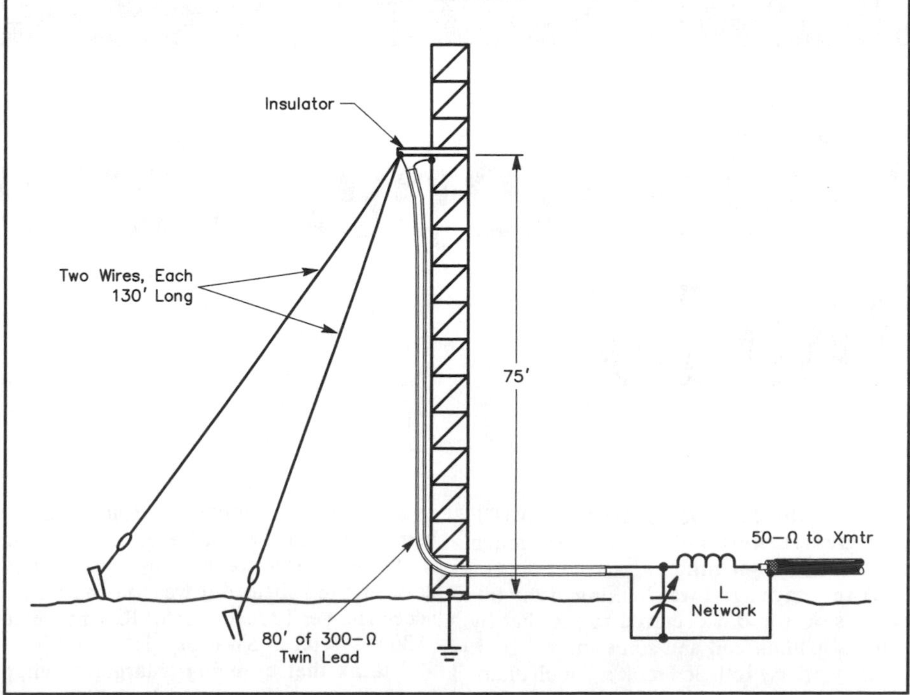

Figure 1—Illustration of the arrangement used at W6UM for a 160-meter half sloper.

overall effectiveness. Using some rough estimates of component values required for matching, length of feed line, etc., I calculate the feed impedance of the antenna itself to be approximately 100–*j*300 ohms.

My estimate of how well this antenna performs must be tempered by the realization that conditions during the past two years have been much worse than the period before. During the 160-meter contests, openings to W1, W2 and W3 have been marginal. I have worked nothing with this antenna I hadn't worked before, but the impression remains that the antenna is better than those used earlier. It has done a consistent job into the Caribbean, SA, Pacific and Japan, as well as across the USA. Receiving remains a problem because of the noise. I have taken down the 160-meter version of this antenna. If I put it up again I will definitely install some auxiliary receiving antennas.

The net result seems to be that this is an effective antenna. Getting a good match directly into coax appears easier if the sloping wire is more nearly vertical and if there is not too much loading above the attachment point. The system should be more efficient, up to a point, as there is more vertical structure above the feed point. There is no compelling reason to insist upon using a resonant configuration. The next attempt I make on 160 might well be with the existing 80-meter system, fed at 60 feet and with 80-meter sloping elements, but using a transmission line and matching network at the bottom of the tower as before.

Given that the antenna does work, the areas I feel need further investigation are better characterization and description of it for different parameters, e.g., measurement of the feed impedance as a function of frequency for different sloper attachment points, lengths, angles and conditions of top loading. I would also like to see the current flow on all of the structure mapped via a probe. If some of the results are forthcoming I hope to see them in an early issue of *QST*. — *Charles Weir Jr., W6UM*

From *QST*, December 1979 (Technical Correspondence)

Half-Sloper Variation For 160 M

I read the July *QST* article by W1CF with great interest. I enjoyed it very much and learned something.

A few days ago I was looking at my 80-meter sloper and it occurred to me that by using a loading coil and some more wire it would work on 160. So I took a 4-inch diameter coil of 10 turns of no. 10 wire out of the junk box, attached the coil to the far end of the half sloper, added about 30 feet of wire and ran the wire to the corner of my garage. The noise bridge indicated it was too short, so I added another 10 feet. Presto! Resonance at 1830 kHz with an SWR of 1.1:1.

I think that by using a larger (higher inductance) loading coil, the added wire could be shortened considerably without seriously affecting the 2-to-1 SWR points on the curve.

I have not made too many contacts on 160 yet, as there is very little activity this early in the season around here. I have had several contacts though, always getting at least as good a report as I have given. —*Philip True, W7AQB.*

By Gary E. Myers, K9CZB From *QST*, June 1980

A Two-Band Half-Sloper Antenna

When an off-the-wall empirical design like this works well, suspicion and skepticism are warranted. Maybe you'll agree that this sloper idea is an exception!

The popularity of the half-sloper antenna seems to be increasing, as evidenced by recent articles in *QST*.[1,2,3] This type of antenna has some worthwhile advantages, particularly for the lower frequency operator – low-angle radiation for antennas of modest height, compactness and simplicity of construction. On the minus side, narrow bandwidths and difficulties in resonating the system have been reported.[3]

The antenna system to be described here evolved from a simple-minded attempt to design a two-band half sloper for 80- and 40-meter operation. A trap-type of antenna was selected as the design basis because of previous experience with trap antennas and because the inductive loading of the trap on the lower-frequency band allows a somewhat shorter overall length. The same inductive loading, however, was also expected to increase the antenna Q and thereby further decrease the bandwidth. For this reason I was prepared to experiment.

It is well that I was so prepared because the final form of the antenna bears little resemblance to the initial concept. But most interesting — and exciting — is the impressive bandwidth on 80 meters.

The Design

A diagram of the two-band half-sloper antenna system is shown in Figure 1. It can be seen that there is no obvious quarterwave dimension in the entire system. In fact, the radiator itself is a nonresonant device.

Initial attempts to prune the wires to resonance resulted only in a mound of wire clippings and one frustrated amateur. After many hours of cut-and-try experimentation, accompanied by a growing, gut-level appreciation for what apparently was happening, the magic combination of wire lengths and trap component values was found. Impedances of 40 $-j80$ ohms at 7.2 MHz and 60 $-j40$ ohms at 3.6 MHz were measured with a noise bridge. It was then a simple matter to cancel out these capacitive reactances with an inductor.

For convenience, the inductor was placed in the transmission line, rather than at the feed point – final "tweaking" of the system is more easily performed on the ground than at 40 feet (12.2 m) in the air. A Smith Chart exercise shows that the SWR on the transmission line between the feed point and inductor L is 5:1 at 7.2 MHz and 2:1 at 3.6 MHz. When RG-8/U is used, the additional loss incurred because of these SWRs is less than 0.5 dB at 7.2 MHz and is almost nonexistent at 3.6 MHz.

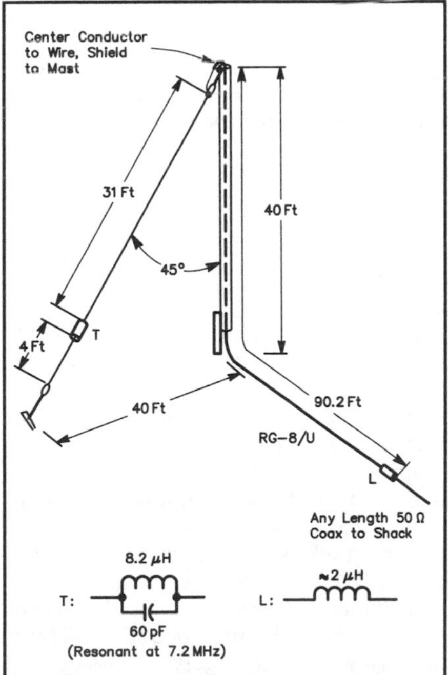

Figure 1—The two-band half sloper. To shield the transmission line from the antenna field, the line is routed up the inside of the support mast, which is grounded. Inductor L is needed to resonate the system.

Performance

This antenna performs very well. Operation at K9CZB is primarily 80-meter cw, with some 40-meter ssb. Power output is nominally 100 watts. Signal reports on 80 have been uniformly good, with comments such as U R LOUDEST 9 ON BAND AND VY FB SIG, VY STRONG. Voice operation on 40 has also resulted in good signal reports, although the praise has not been so lavish. This is my only 80/40-meter antenna, so direct comparisons were not possible. However, it appears to greatly outperform two previous antennas, a 160-foot end-fed wire and a trap dipole, both strung 30 feet above ground. All in all, it is about what I expected from a half sloper.

This kind of performance is nice, but nothing to write an article about, since that has already been done. The real performance story about this antenna can be summed up in one word: bandwidth. A glance at the SWR curves in Figure 2 will open the eyes of any 80-meter operator. As far as I know, this is unheard-of bandwidth for such a simple and compact antenna. It is a real treat to QSY 400 kHz And see the SWR meter needle barely move. This isn't a low-Q antenna—it's a *no*-Q antenna! The bandwidth on 40 is far less impressive and is, in fact, similar to what has been reported previously for half slopers.

Construction

During experiments with prototypes of this antenna, I noticed some sensitivity to feed-line placement and length. Therefore,

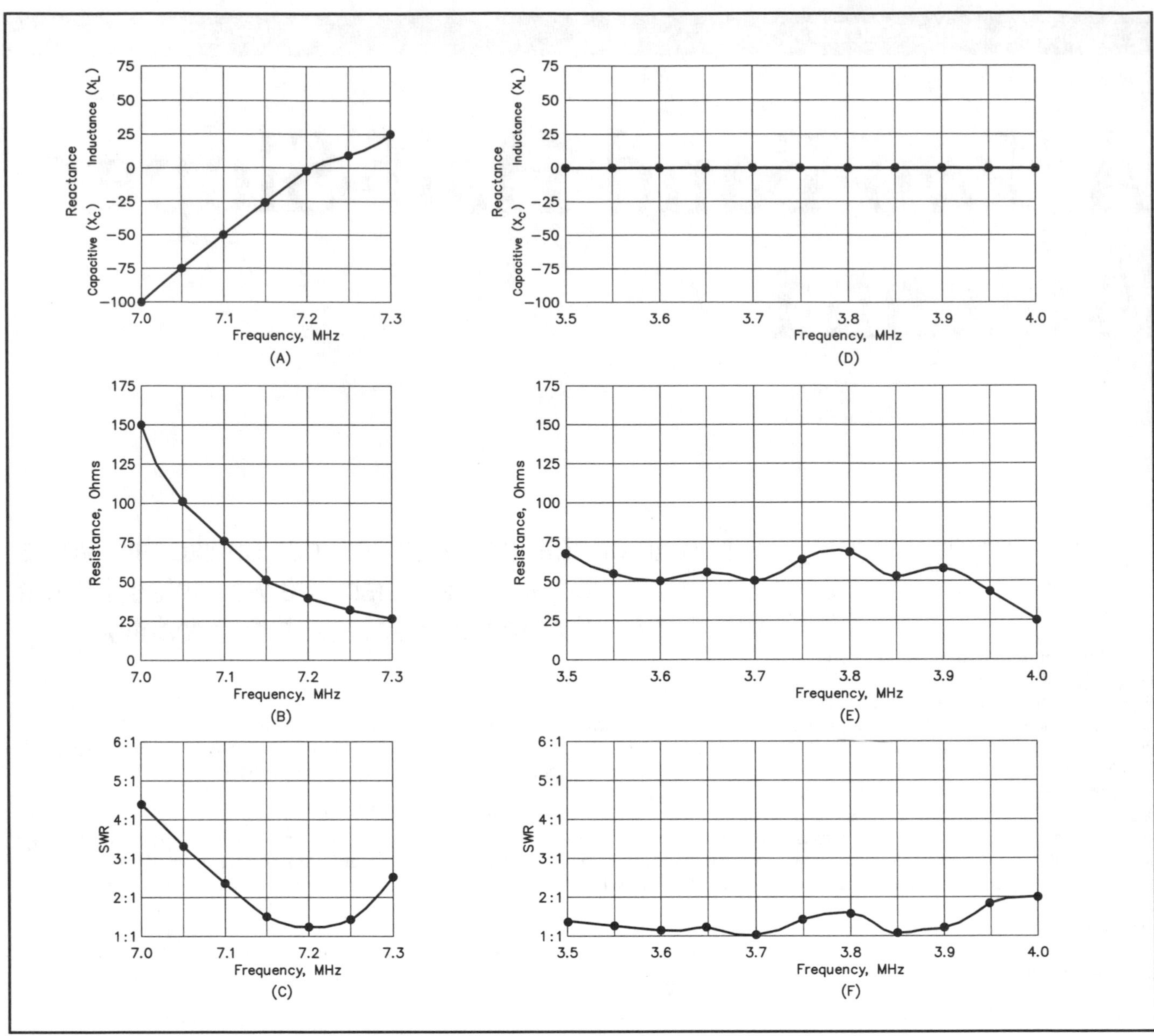

Figure 2—Loading characteristics of the antenna. The SWR curves were determined from a Smith Chart. Impedance values were measured with a noise bridge.

in later versions I ran the feed line up the *inside* of the support mast to shield it from the antenna field. This precaution seems to be effective, for no such sensitivity has been observed since. (I can't help but wonder if this might improve the behavior of any cantankerous half sloper.) If a nonmetallic support is used, or if there is no possibility of placing the feed line inside the support, double-shielded coaxial cable should serve equally well, provided the outer braid is connected to the inner braid at the top of the tower and a ground is connected to the outer braid at the bottom of the tower.

I used standard 10-foot TV mast sections for my support, simply because they were on hand. This results in a very flimsy and flexible mast in a 40-foot length, though. The first 20 feet must be doubled up with long U-bolts if the mast is to be walked up. The price of six sections of TV mast is about the same as a 36-foot telescoping push-up mast, but the latter is much sturdier and far easier to erect. If support materials are not already on hand, the telescoping mast is the better choice. I should mention that I took the trouble to bond all sections of mast together electrically to ensure good conductivity and to guard against TVI from rectification at joints after inevitable corrosion sets in. The base of the mast should be grounded. Effects from the guy wires can largely be avoided by breaking them into nonresonant lengths with strain insulators placed at the mast and every 19 feet thereafter.

The inductance and capacitance values shown in Figure 1 must be used for the trap. Construction techniques for the trap are covered in *The ARRL Antenna Book*.[4] A novel and inexpensive method of trap construction has been described by WB9OQM.[5] I built my trap using the method shown in Figure 3. Traps made in this fashion are much stronger than they appear. I've never had one break, even in high winds that caused property damage. In this antenna, however, the radiator also serves as one of the top guys. For that reason the trap was reinforced. Two 3/16-inch thick pieces of plastic were used with three layers of glass cloth and epoxy sandwiched between them. A rotary wire brush serves well to rough up the inner surfaces of the plastic to ensure good adhesion. However, one may use coarse sandpaper for that purpose. Glass cloth and epoxy are sold as a repair kit in many hardware stores.

Before the antenna wires are connected,

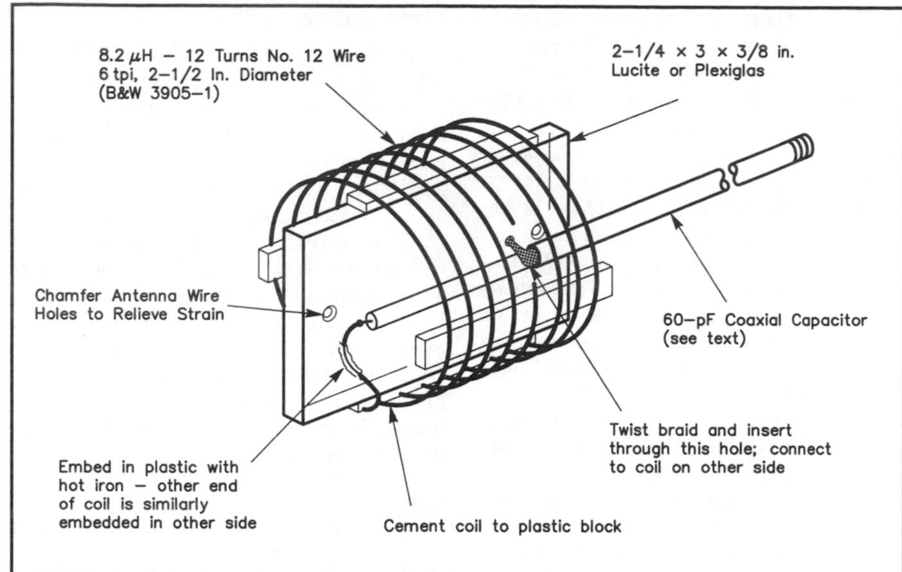

Figure 3—A simple, sturdy trap. The coaxial capacitor should be taped to the antenna wire after installation. It is not necessary to enclose the trap.

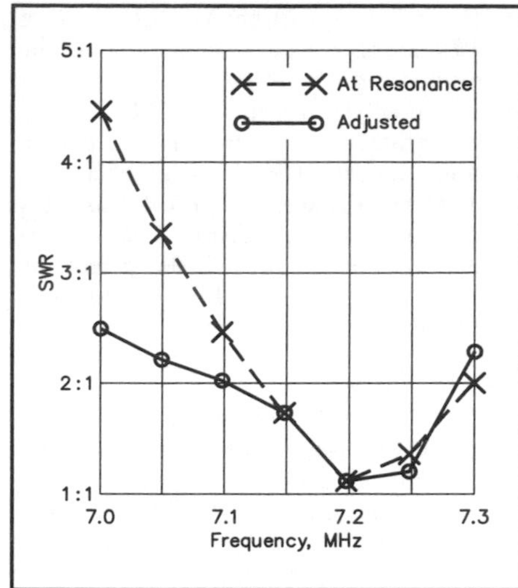

Figure 4—SWR curve for 40 meters, showing the effect of tapping L and adjusting feed-line length between L and the transmitter to obtain the best SWR curve. Such adjustments have little effect on 80-meter characteristics.

the trap must be tuned to resonance. A dip meter or noise bridge can be used to measure the resonant frequency. Start with about 30 inches of RG-8/U for the coaxial capacitor. After connecting it to the coil, as shown, 26 or 27 inches of braid will remain. At this point, the resonant frequency should be below 7.0 MHz. Trim the braid at the far end, a little at a time, snipping off the center conductor as you go. Recheck the resonant frequency each time. As 7.2 MHz is approached, continue trimming the braid, but stop cutting the center conductor. To increase the leakage path, the polyethylene dielectric should extend beyond the braid 1/8 to 3/16 inch (3.2 to 4.8 mm) when the trap is resonated at 7.2 MHz. Very close to 24 inches of braid should remain at completion. Tightly tape this end with several layers of plastic electrical tape.

The component values for this trap are exactly the same as those used in the W3DZZ trap dipole,[4] so there are several commercially made traps that may be suitable for this antenna. Traps made for a five-band, two-trap dipole, 108 feet long should have the proper values of capacitance and inductance.

In any antenna system, the radiator feed point impedance repeats itself every half wavelength along the transmission line. Inductor L must be inserted in the transmission line at a half-wave point in order to exactly cancel the capacitive reactance of this antenna system. It is, of course, advantageous to place L as close to the feed point as possible in order to minimize losses. A half wavelength at the lower frequency is as close as you can get without going to the feed point itself. The 90.2-foot length of RG-8/U shown in Figure 1 is an electrical half wavelength at 3.6 MHz for solid polyethylene dielectric coaxial cable *only*. If cable having a velocity factor other than 0.66 (e,g., foam-dielectric coaxial cable) is used, this length will have to be recalculated from the equation

$$\ell = 492\ V/3.6 \text{ feet}$$

(Feet × 0.305 = m)

In this equation, V is the velocity factor of the cable to be used. Only RG-8/U or a similar type such as RG-213/U should be employed for this section of the transmission line in order to keep the losses low. If you don't mind a dB or so of loss on 40 meters, RG-58/U is acceptable. The loss on 80 meters will be negligible in any case.

The value of inductor L should be 1.75 µH, but I recommend that a coil having about 3-µH inductance be used to allow some latitude for final tune-up of the system. I mounted 12 turns of a no. 3018 Miniductor ($1^1/_4$ inch in diameter, 8 turns per inch) in a small Minibox with SO-239 coaxial connectors placed at each end. After tapping the coil for the best SWR curve on 40 meters, the entire assembly was sealed and waterproofed with bathtub caulk.

Tune-up

As seems to be characteristic of half slopers, this antenna can be very touchy to tune up. If the length of transmission line between the radiator and L is not an exact integral multiple of a half wavelength at 3.6 MHz, tune-up can be a real "can of worms." Since the oft-quoted value of 0.66 for the velocity factor of standard RG-8/U is only a nominal value and can vary appreciably from brand to brand (in cheap cable from lot to lot), this length should be determined with a noise bridge.

If a noise bridge is not available, the following procedure may be tried. Cut this section of cable about 6 feet shorter than the calculated length. Prepare a section of RG-58/U, 12 feet long, with solderless connectors on each end. Connect it to both the shortened feed line, using a PL-258 double female connector, and to L. Tap L to obtain the best combination of SWRs at 3.6 and 7.2 MHz. Record the SWR figures and tap position.

Now shorten the RG-58/U by 6 inches and repeat — and repeat — until you are certain you have passed through the point where the SWR values simultaneously bottom out at both frequencies. Prepare a length of RG-58/U (from the same lot) exactly as long as the best experimental length, using permanent coaxial connectors. Seal and waterproof all connections. RG-58/U is recommended for relative ease of pruning. If you don't mind unsoldering a PL-259 each time, RG-8 could be used. However, the additional loss from such a short section of RG-58/U will be infinitesimal at these frequencies.

This procedure is obviously tedious, but it is necessary to obtain good performance on 40 if a noise bridge is not available.[6] In fact, some adjustment of this section of transmission line may be necessary even if a noise bridge is used to measure the electrical length to obtain optimum two-band performance. The exact half wavelength should always be used as a starting point, in any case.

Strangely enough, the above procedures are necessary only to optimize 40-meter performance. My experience has been that merely cutting the half-wavelength section of transmission line to the calculated length, then tapping L to obtain the best SWR at 3.6 MHz, is sufficient to obtain a ratio of 2:1 or less over the entire 80/75-meter band. So tune for 40, and 80 should take care of itself.

Once the antenna system has been resonated, it may pay to experiment with the value of L and the length of transmission line between L and the transmitter. Changing these values will change the shape of the 40-meter SWR curve somewhat. By so experimenting, you may be able to tailor the shape of the 40-meter SWR curve to your operating preference. Don't expect miracles, though, for the range of adjustment seems to be small. Figure 4 illustrates the results of such an effort. These adjustments will have very little effect on 80-meter bandwidth within the range of acceptable SWR on 40 meters.

Further Thoughts

The first prototype was constructed close to my house, and I was therefore concerned that the performance might not be reproducible. The next prototype was erected in a far corner of my yard, over 100 feet from the house and even further from any other structures or conductors. The final version was similarly located. Except for final tune-up parameters, all three behaved almost identically, even though a number of physical changes was made each time. As a final test of the soundness of the design, I built scaled versions for 40/20 and 20/10 meters. They exhibited very similar characteristics, although all parameters of the system seem to become very critical as the design frequency is increased.

The first two 80/40-meter systems were built using a 30-foot mast, yet their behavior was not markedly different from the final version with a 40-foot mast. Since the mast is an electrical part of the system, and since proximity to ground must play some role in the performance of the antenna, other heights, and supports that have beams attached, may yield different results.

The reactance of the trap at 3.6 MHz is 245 ohms. Therefore, there is a possibility of constructing an 80-meter-only version of this antenna by using a 10.8-µH inductor in place of the trap. This has not been tried, however.

Scaling up to 160/80 meters is an attractive possibility. Conceivably, a mast height as low as 50 feet could be used. A starting point would be the doubling of all wire lengths, and using a 16.4-µH coil with a 120-pF capacitor for the trap to preserve the 245-ohms reactance at 1.8 MHz.

There also seems to be a possibility that a slight increase in trap capacitor value, to resonate the tap at 7.1 or 7.15 MHz, might allow better coverage of 40 cw, but probably at the expense of the phone portion of the band. Such a change might have little effect on 80-meter bandwidth, but another round of cut-and-try could prove necessary.

Conclusions

At this point, this is still an experimental design. Further development may eventually allow a cut-to-formula type of construction, but until that happens, be prepared to experiment. The dimensions given in Figure 1 will put you in the ball park and should yield immediate results on 80.

The convenience of Transmatchless operation overall of 80/75, plus a reasonable portion of 40, coupled with an excellent radiated signal, is ample repayment even for many hours of cutting and trying. Once tuned up, this antenna is very well behaved and enjoyable to use. I would like to hear from others who construct antennas based on this design.

Notes

[1]Hopps, "A 75-Meter DX Antenna," *QST*, March 1979, p. 44.

[2]Atchley, "Putting the Quarter-Wave Sloper to Work on 160," *QST*, July 1979, p. 19.

[3]DeMaw, "Additional Notes on the Half Sloper," *QST*, July 1979, p. 20.

[4]*The ARRL Antenna Book*, 13th edition, 1974.

[5]Mathison, "Inexpensive Traps for Wire Antennas," *QST*, February 1977.

[6][Editor's Note: A third alternative is to use a dip meter to determine an exact half wavelength of line. See Downs, "Measuring Transmission-Line Velocity Factor," *QST*, June 1979.)

By Deane J. Yungling, KI6O From *QST*, April 1986

The KI6O 160-Meter Linear-Loaded Sloper

No room for a top-band antenna? Try this one on for size!

After having good success using my linear-loaded, inverted-L antenna over the winter 1984-85 160-meter season, I decided to try linear loading on a different type of 160-meter wire antenna.[1] I have had considerable success with a quarter-wave sloper on 80 meters, but my city-sized lot isn't deep enough to accommodate a full-sized sloper for the 160-meter band. About 120 feet would be required for the sloper to be 10 feet off the ground at the low end; I have about half that distance to work with. Linear loading solved the problem.

This antenna provides an effective bandwidth of about 70 kHz with an SWR of 2:1 or less. At the design frequency of 1.840 MHz, SWR is 1.1:1 with 50-ohm feed line and no matching network or tuner.

Although the individual dimensions are not critical, the sloping wire and the ladder line must resonate at the desired frequency. If the sloping wire is less than 65 feet long, the ladder must be longer, and vice versa, If you use a different length for the sloping wire, you will need to experiment a bit to see how much to add or remove from the ladder length.

Construction

My sloper is hooked onto the tower at about the 55-ft level using a strain insulator (see Figure 1). The coaxial-cable feed fine is securely taped to one tower leg, and the shield is connected to the tower leg with a radiator hose clamp. The tower is grounded at the base with several ground rods. The center conductor of the coaxial cable is soldered to the sloper at the strain insulator and is taped for weather protection where the center conductor and shield separate.

The ladder portion of the antenna is made with the same type of wire as the sloping portion. The ladder spacers are made of 3/8-inch hardwood dowels that have small holes drilled 1 inch from each end to hold the wires. The dowels should be soaked or sprayed with a wood preservative, prior to assembly, for weather protection. Plastic spreaders could also be used, if desired. The wires are firmly tied to the dowels with waxed lacing twine or similar material where the wires pass through the holes. The dowel spacing can be adjusted, with some difficulty, after the dowels are tied.

One end of the ladder is tied to the tower with heavy monofilament fishing line, with the other end tied to a tree or whatever else is handy. The tie-off lines are fanned vertically to keep the ladder from twisting and should be tensioned enough to eliminate any sagging in the ladder. Monofilament fishing fine provides some cushioning if the supporting tree sways or if the wind is strong.

Adjustment

Initially, the ladder should be a few feet longer than shown, to allow for adjustment. The resonant frequency should be checked, and a few inches removed at a time from the tower end of the ladder until the desired frequency is achieved. At this point, the wires should be soldered together and then taped to the end dowel. It appears that there must be an BF beam, or something similar, on the tower for any quarter-wave sloper to work properly. This sloper is no different in this respect. Finally, the antenna, particularly the sloping section, should be kept as far as possible from surrounding guy wires and other objects.

This antenna requires no ground system, balun or matching network over its operational bandwidth. The linear-loaded sloper is simple to construct and easy to adjust, and its performance is superior to my linear-loaded, inverted L.

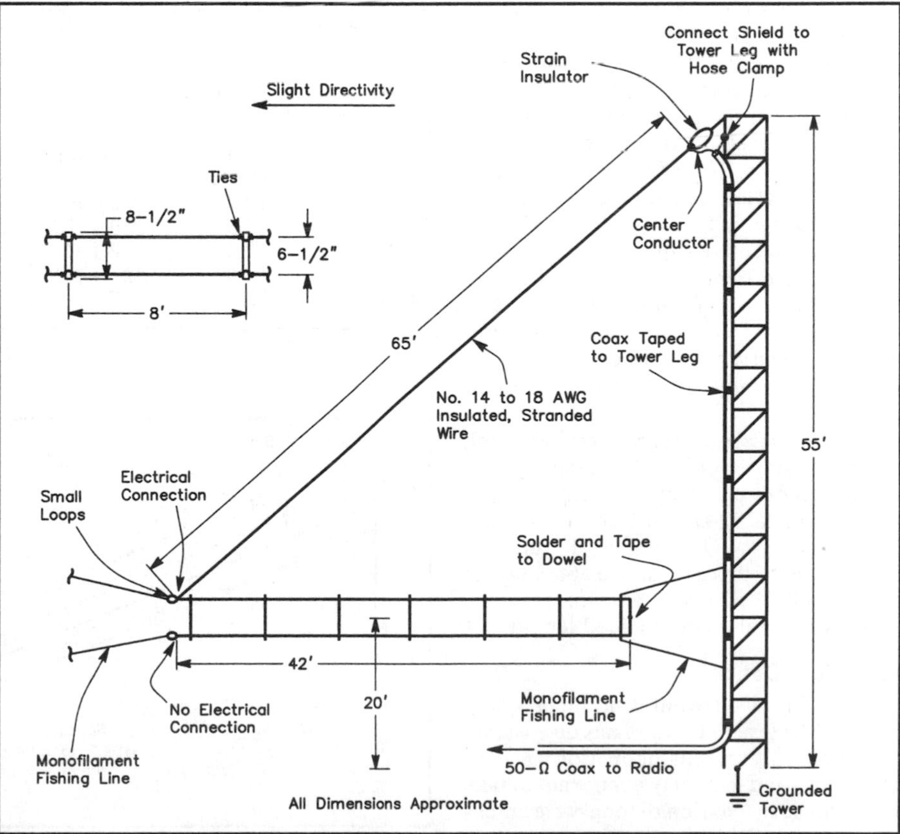

Figure 1—Construction details for the KI6O 160-meter linear-loaded sloper.

[1]D.J. Yungling, "The KI6O Top-Linear-Loaded 160 Meter Inverted 'L' Antenna," *CQ*, Apr 1985, pp 38-39.

The Super Sloper

Dramatically improve the front-to-back of a sloper.

You can imagine my surprise when *ELNEC*[1] revealed that a parasitic element can be combined with a long wire to form a whole family of directional antennas. There on my screen, I was looking at useful combinations not found in that bible of antennas, *The ARRL Antenna Book.* I had been exploring combinations of full-wave parasitic elements with wire arrays. How far off center can I slide a parasitic element, I wondered? "A lot," was the answer coming from *ELNEC* [and confirmed by *EZNEC—Ed.*]. As I offset a closely spaced (0.015 to 0.046 λ) parasitic element by 1/8 to 5/8 λ, a whole family of antennas appeared (see Figure 1).

I call the configuration a Super Sloper because the pattern resembles that of the well-known sloper, but it is greatly enhanced (see Figure 2). Super Slopers provide gain in the direction from the tall to the short pole. The amount of gain depends on antenna length, or more exactly, on the number of half wavelengths in each element.

Like slopers, Super Slopers require two supports, one tall and another shorter. At my station, and in this article, I considered only nonconductive supports. Other builders can explore the possibilities of Super Slopers suspended from metal towers. Super Slopers are very inexpensive to build (if you already have suitable supports). They require only wire and a few feet of PVC pipe.

Unlike slopers, Super Slopers have high feedpoint impedances. A matching network is required when feeding a Super Sloper with a 50-Ω line. The result is a broadband, low-Q antenna (Figure 3). Don't think of this as just one antenna, but rather a whole family of antennas, one for each half wavelength of added length.

Technical Concepts

If you just want to build an antenna, skip ahead to the Practical Antennas discussion. Those with a technical inclination can continue here, and they may even want to take a look at the discussion of long-wire antennas in *The ARRL Antenna Book.*

An antenna is called a long wire if it is one wavelength or longer. The radiation pattern can be described as the surfaces of two opposed cones that are coaxial with the wire and have their apexes meeting at the feed-point. The apex angle becomes smaller and the lobes grow stronger as antenna length increases. An azimuth plot of radiation pattern for single long-wire antennas over ground shows four principal lobes at low radiation angles. There are two principal lobes, *in one direction*, when the antenna is terminated in a matched, resistive load.

ELNEC shows that a parasitic element added to a single long-wire antenna creates a *unidirectional* radiation pattern similar to that created by a resistive termination. Energy eliminated from the back goes into useful forward gain. The parasitic element can be tuned as either a director or a reflector.[2]

Experience with Yagis (and many other applications of parasitic elements) led me to believe that all parasitic elements must be nearly a half wavelength long and located within the span of the driven element. That concept is completely *incorrect*. Parasitic elements can be *any* resonant length and offset from the driven element, as long as there is adequate coupling between the elements.

Here's how Super Slopers produce front-to-back ratio (F/B) and gain. The best

Table 1
Spacing and Phase Angle for Various Antenna Lengths

At the spacings listed, the current in the resonant parasitic element is equal to driven-element current ±10%. (Derived from *ELNEC*, under free-space conditions, 0.25-λ offset, at 7.1 MHz.)

Antenna Length (λ)	*Spacing (λ)*	*Phase Angle (degrees)*
0.5	0.046	–156
1.0	0.030	–147
1.5	0.023	–144
2.0	0.018	–142
2.5	0.015	–142
3.0	0.015	–144

Table 2
Impedance and SWR versus Frequency (Figure 5)

Frequency (MHz)	*SWR (200-Ω)*	*Impedance*
14.0	1.54	293 + *j*48
14.1	1.25	248 + *j*13
14.2	1.16	172 + *j*0
14.3	1.75	127 + *j*54
14.4	2.61	107 + *j*113

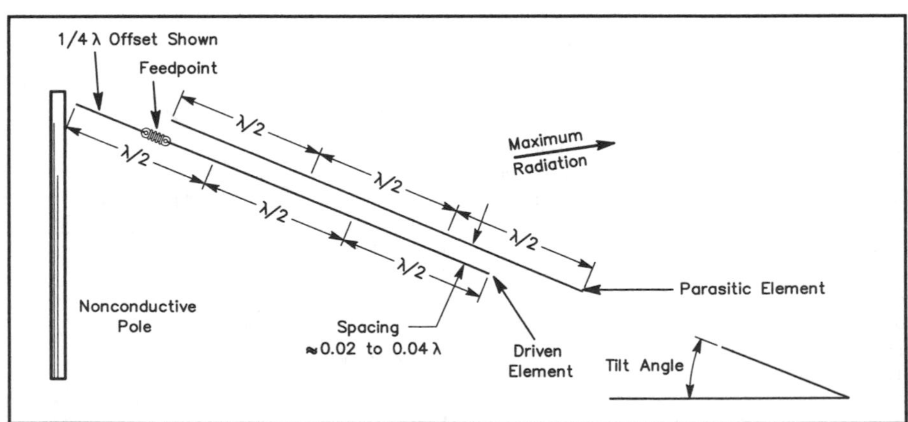

Figure 1—Drawing of a Super-Sloper antenna showing nomenclature. This is a 3-λ/2 model. See Figure 7 for construction details.

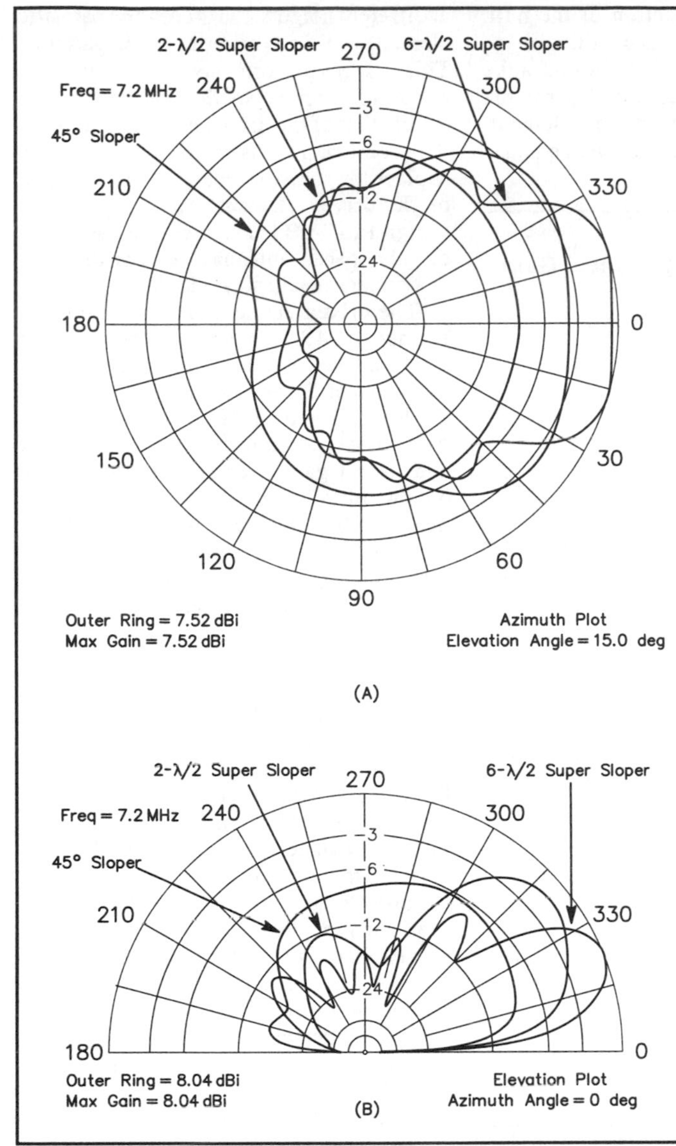

Figure 2—Azimuth (A) and elevation (B) plots for a 45° sloper, 2-λ/2 (0.25-λ offset, spaced 0.04 λ, low end 0.119 λ above ground) and 6λ/2 (0.25-λ offset, spaced 0.015 λ, low end at 0.066 λ above ground) Super Slopers. The high end of each is λ/2 high.

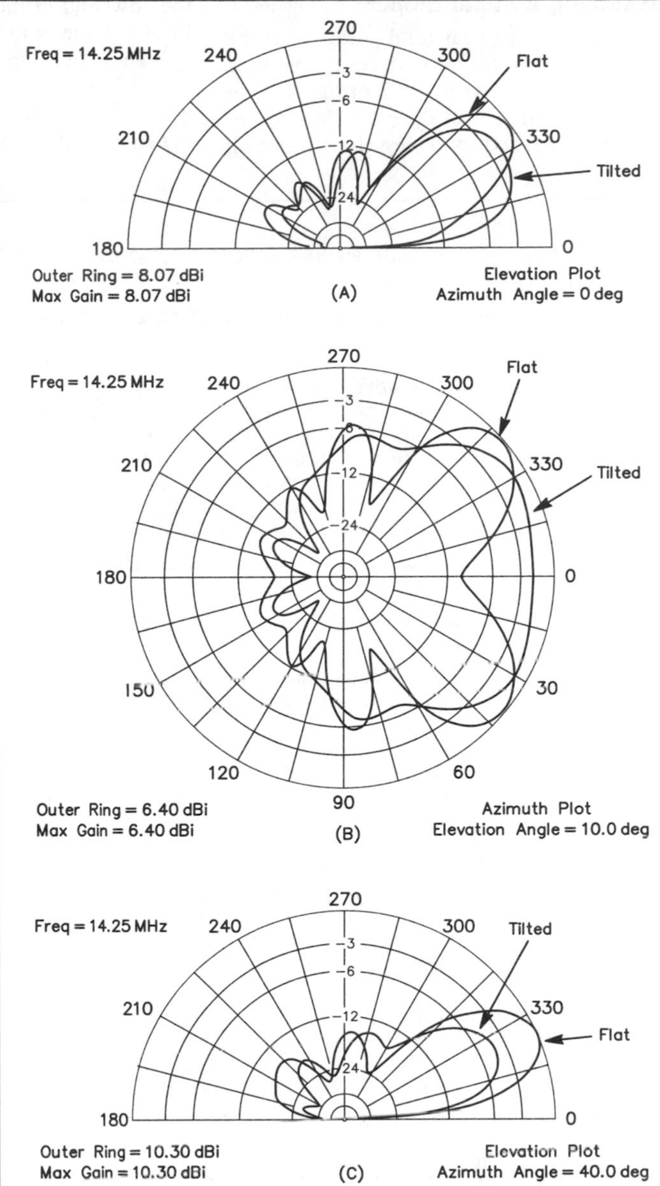

Figure 4—Patterns of a 3-λ/2 antenna with 0.36-λ offset and spaced 0.04-λ oriented flat at 1/2 λ vs sloping from 1/2 λ to 0.045 λ. A shows the elevation patterns at 0°; B shows the azimuth patterns; C shows elevation plots at azimuth=40°.

F/B results from a 180° phase shift between the currents in the driven and parasitic wires, as seen from a distant location where a pattern null is desired. A phase difference of 180° means that the signals completely cancel each other, if each element delivers equal signal strength. A phase shift of 90° is possible with a 1/4-λ offset, and the remaining 90° phase shift can come from the tuning (length) of the parasitic element. If the fields cancel in one direction, they will reinforce in some other direction, to produce gain.

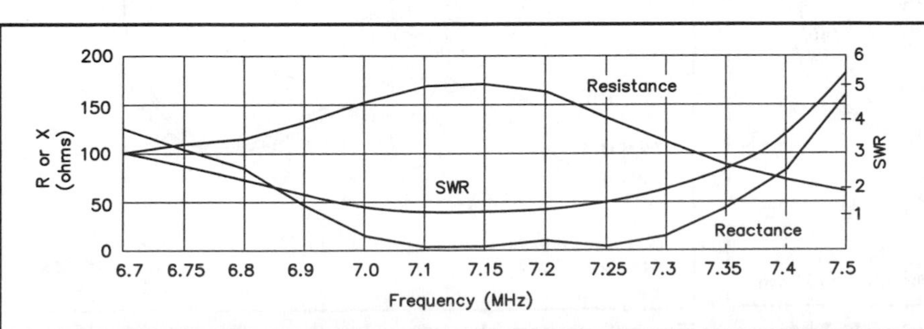

Figure 3—A plot of impedance and SWR versus frequency for the 2-λ/2 antenna described in Table 6, placed as described in Figure 6. The data source is *NEC/WIRES 1.5* (see Note 3).

By increasing the phase angle of the current in the parasitic element, the angle of maximum cancellation can be moved. This is particularly useful in Super Slopers because the principal lobes of long wires lie at some small angle to the direction of the wire. From Table 1, notice that the current in a resonant parasitic element has about –144° to –156° phasing, which is about right for correct reverse-lobe canceling.

If we want each element to deliver equal signal strengths to a distant location, nearly equal currents must flow in both the driven and parasitic elements: The coupling between elements must be very close. Table 1 suggests the approximate spacings for currents to be equal within ±10%. (The antenna also has gain at closer spacings because gain is affected less by unequal currents than is F/B.)

Designing a Super Sloper

When placed near (and parallel to) the ground, Super Slopers show twin-lobed azimuth patterns with a null, often 10 dB or greater, on the axis of the wire. This center null can be filled by tilting the wire, with the low end in the direction of the null. Figure 4 shows the effects of tilting.

When the antenna is tilted, the end of the parasitic element can become closer to the ground than that of the driven element. When considered as two separate antennas at different heights, the driven and parasitic elements will not have the same patterns. This reduces performance, but locating the parasitic element so that its end height equals that of the driven element can solve the problem. That is, increase the spacing to place the parasitic element end above that of the driven element. Models give this design 1 to 2-dB gain advantage over more closely spaced antennas, but construction is a greater mechanical challenge.

The driven element is fed at a current antinode, 1/4 λ from an end. The antennas can be made to exhibit a wide range of feed resistances: 20 to 300 Ω, or more (see Tables 2, 3 and 4). High resistances and wide offsets combine to produce low SWR and useful gain over an unusually great bandwidth (Table 2). I've used 1/4-λ matching lines (using RG-62, a 93-Ω line), 4:1 baluns and ladder lines to successfully match the antenna at high impedances.

You can vary the antenna length (gain increases with length), height, offset, phasing and spacing as needed. These many variables would be difficult to work with if each were a critical adjustment, but fortunately, they are not critical. Tables 1, 3, 4 and 5 show the tolerant design features of these antennas. From these tables, we can observe several trends:

- Current balance greatly affects F/B, with less effect on gain.
- Phase angles change very slowly with antenna length.
- Gain changes very little with changes of current and phasing.
- Gain varies by a little over 1 dB with a spacing increase from 0.01 to 0.04 λ.
- Gain is nearly constant with offsets moving from 0.1 to 0.4 λ.

Good F/B is easy to achieve, even

Table 3

Element Spacing versus Phase Angle, Relative Current, Impedance, Gain and Lobe Angle

Derived from *ELNEC* free-space model, 14.2 MHz, 2-λ/2 elements, offset 0.25 λ

Spacing (l)	*Phase Angle (degrees)*	*Relative Current*	*Impedance of Driven Element*	*Gain (dBi)*	*Lobe Angle (degrees)*
0.01	−136	1.62	205 − *j*230	4.98	48
0.02	−141	1.29	128 − *j*138	5.54	48
0.03	−144	1.11	103 − *j*94	5.78	48
0.04	−149	0.97	80 − *j*59	6.1	48

Table 4

Offset versus Phase Angle, Relative Current, Impedance, Gain and Lobe Angle

Derived from *ELNEC*, free-space model, 14.2 MHz, 2-λ/2 elements, spaced 0.02 λ. Notice how little gain changes with different offsets.

Offset (λ)	*Phase Angle (degrees)*	*Relative Current*	*Impedance of Driven Element*	*Gain (dBi)*	*Lobe Angle (degrees)*
0.1	−167	1.08	19 −*j*54	5.7	48
0.15	−157	1.18	48 − *j*101	5.67	48
0.2	−148	1.26	89 − *j*135	5.6	48
0.25	−141	1.29	128 − *j*138	5.57	48
0.3	−139	1.16	134 − *j*111	5.6	48
0.35	−138	0.963	123 − *j*70	5.58	48
0.4	−134	0.71	108 − *j*14	5.4	48
Pattern Reversal Begins					
0.5	−26	0.25	83 + *j*46	3.0	54
0.6	39	1.31	126 − *j*57	5.2	61
0.7	43	1.71	192 − *j*180	4.0	64

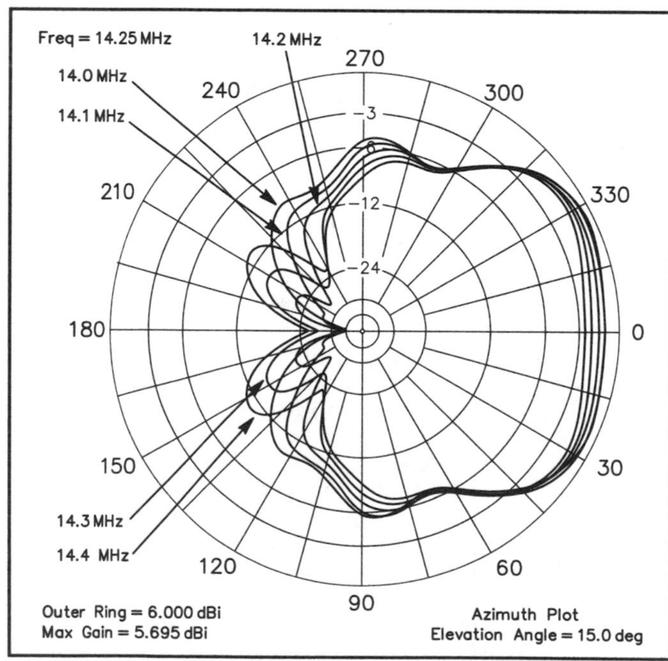

Figure 5—A radiation pattern over a range from 14.0 to 14.4 MHz. The pattern is that of a 3-λ/2 antenna with 0.36 λ offset, with the ends at 1/2 λ and 0.045 λ. See Table 2 for the SWR tabulation.

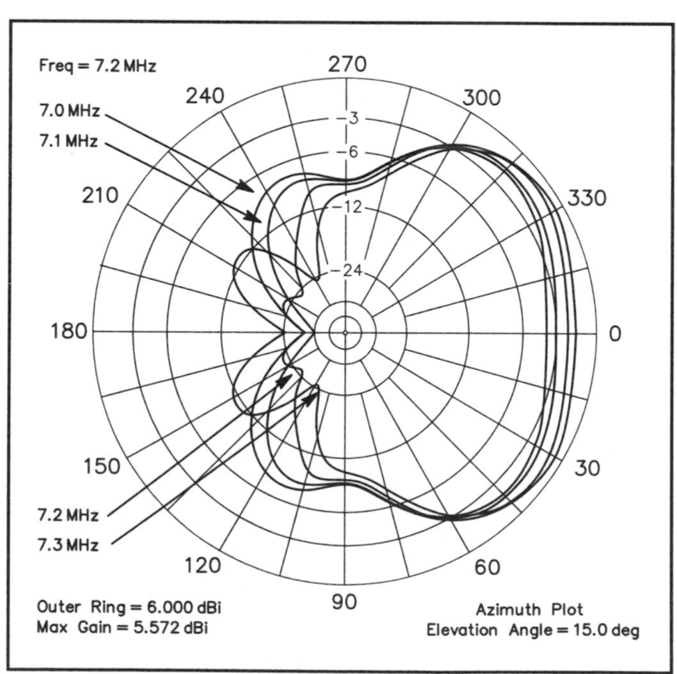

Figure 6—Pattern plots over a frequency range from 7.0 to 7.3 MHz. The antenna is 2-λ/2, with 0.25-λ offset and 0.04-λ spacing. The driven element is 1.02-λ long; the director is 0.98-λ long. See Figure 3 for an SWR curve.

though the antenna is designed for best gain or easy construction. Improved F/B ratios and reverse-oriented patterns are possible from designs optimized at specific frequencies. Computer modeling is the best way to optimize reverse patterns. Figures 5 and 6 show some possible patterns and how the patterns vary with frequency. Table 2 and Figure 3 show predicted SWR.

Practical Antennas

Table 6 describes Super Slopers constructed at my station (refer to Figure I for nomenclature). The principal difference between the two 3-λ/2, 20-meter antennas in Table 6 is the increased offset from 17½ feet to 22 feet. This changes the phasing, resulting in a slightly improved F/B ratio and slightly improved gain. The trade-off is a narrower pattern and a longer antenna.

A 60-foot mast supported the 3-λ/2 and 2-λ/2 20-meter versions. They were mounted back-to-back and tilted to fill the center null. Another 3-λ/2, 20-meter unit was mounted from a 38-foot mast. The 40-meter version sloped from 60 feet down to 6 feet. All antennas performed as predicted. Additional height at either end definitely increases low-angle radiation, but it reduces high-angle radiation.

Figure 7 shows how I assembled and spread the wires. The departure from straight lines (as depicted in Figure 1) has no practical effect on the antenna. Be sure to place the director over the driven element when erecting the antenna. A director placed at the side will skew the pattern, favoring the side of the director. Keep the Super-Sloper support lines tight to minimize sagging. Severe sagging leads to improper phasing and degraded results. I put a support under my 40-meter Super Sloper at midspan, but I just pull the 20-meter antenna support lines tight. The two elements have a tendency to twist and wrap together in the wind. You can prevent this by using two support ropes at the low end or by using spacers as on transmission lines. Both methods work well. Super Sloper performance suffers from excessive ground losses when the low end is at ground level. I strive for an antenna slope (tilt angle) of 10° to 20° and a minimum height of 6 feet. Higher is better.

A 200-Ω feedpoint impedance is easily transformed to 50 Ω with a 4:1 balun. Ladder line and a tuner are another option. Refer to *The ARRL Antenna Book* for other methods of transforming high feed impedances to values acceptable for modern transceivers. Do not feed the Super Sloper directly with coax unless the coax is part of an impedance-matching section.

Table 5
Parasitic Element Length versus Phase Angle, Relative Current, Gain and Lobe Angle

Derived from *ELNEC*, under free-space conditions with 14.2 MHz, 2-λ/2 elements, spaced 0.02 λ apart.

Percent Short (λ)	*Phase Angle (degrees)*	*Relative Current*	*Gain (dBi)*	*Lobe Angle (degrees)*
0	–140	1.30	5.54	48
1	–125	1.47	5.33	49
2	–107	1.53	5.02	51
3	–89	1.43	4.70	52
4	–75	1.23	4.44	53
5	–65	1.05	4.19	53

Table 6
Working Antenna Dimensions

Use a 4:1 step-down transformation to match 50-Ω line.

Antenna Number	*1*	*2*	*3*	*4*	*5*	*6*
Band	40 m	20 m	20 m	20 m	15 m	10m
Length*	2	3	3	2	2	2
Driven (ft)	139.3	106.5	105.6	70	46	34.5
Director (ft)	134.3	104.0	102.2	68	45	32.8
Offset (ft)†‡	34.3	17.5	22	17.5	11.5	8.6
Spacing (ft)	6	1.5	1.5	1.5	1.0	1.0
Feedpoint (ft)‡	34.3	17.5	17.25	17.5	11.5	8.6

*Length expressed as a multiple of λ/2.
†Offset is λ/4, except for antenna 3, where the offset is 0.31 λ.
‡Feedpoint and offset are both measured from the high end of the driven element.

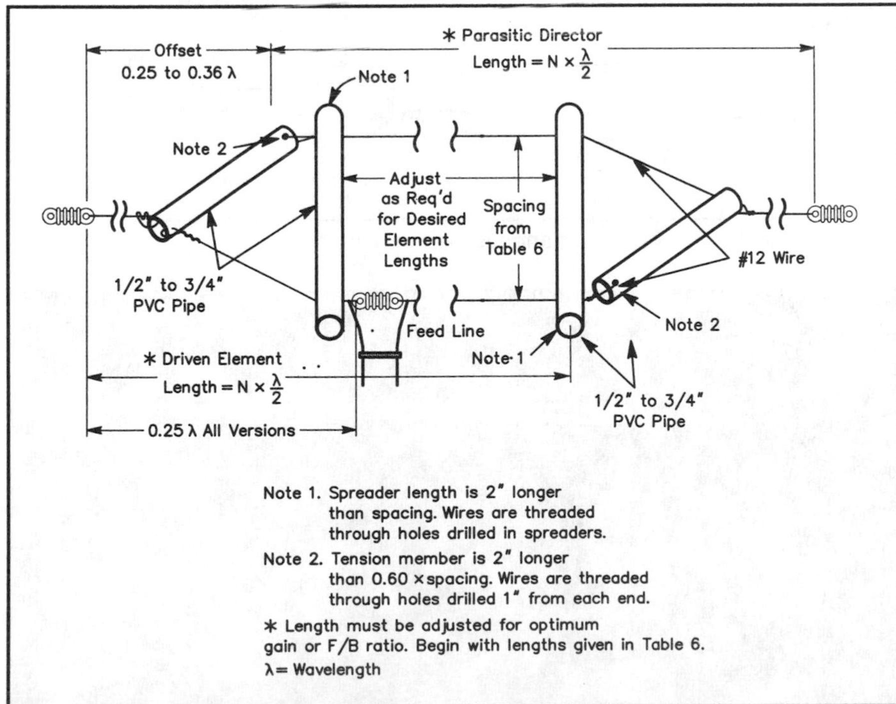

Figure 7—A method of constructing the long-wire parasitic antenna. Additional center spacers reduce the tendency of the wires to twist in the wind.

Results and Conclusions

I've built and used six Super Slopers very successfully. The F/B is often dramatic. A difference of six S units is not uncommon when switching between two antennas built to favor opposite directions.

This family of antennas has not yet been well researched and studied. Use an antenna modeling program before building designs that are substantially different (eg, longer or made from tubing) from those described in Table 6.

Notes

[1]*ELNEC*, and its successor *EZNEC*, are computer antenna modeling programs available from Roy Lewallen, W7EL. The description files for the antenna plots shown in this article are available as a self-extracting archive file named SUPSLOPE.EXE. This file can be found on the Internet (FTP to OAK.OAKLAND.EDU, directory pub/hamradio/arrl/qst-binaries).

[2]Parasitic elements offset more than 1/2 λ from the driven element, become reflectors.

[3]*NECIWIRES 1.5* is a computer antenna modeling program available from Brian Beezley, K6STI.

By John F. Lindholm, W1XX From *QST*, January 1983

The Inverted L Revisited

City dwellers, don't despair. Here is a *good* 160-meter antenna that should fit on your lot!

I'll never know what inspired me to make a few contacts in the ARRL 160Meter Contest. My antenna was made by tying together the open-wire feeders of my 80-meter dipole. The performance was not fantastic, but it was the first step in getting me "hooked" on the "gentlemen's band." Working three dozen European stations from a friend's house got my interest up. My friend has a good 160-meter antenna system – and *lots* of property to fit it on!

Returning to my 60 × 150-foot lot made me feel depressed.[1] I suffered all winter while listening to the others working VKs, ZLs and even JAs at daybreak. What could I do to improve *my* signal?

Many hours the following summer were spent trying to figure out how to cram a 160-meter antenna within the confines of my small lot. Space restrictions dictated that my wire be no longer than a standard 80-meter dipole. I began to consider alternative antennas.

Shunt feeding my 50-foot tower was investigated but dismissed for various reasons. This arrangement would require disconnecting the shunt feed when cranking down the tower, and all guy wires would have to be broken up into nonresonant lengths with insulators. Additionally, my cables would have to be rerouted to ground level. I then considered a full-length dipole originating in a neighbor's yard three houses to the east, crossing my property and finally terminating in the yard two houses to the west! This idea was rejected. The legal negotiation fees would have run into six figures! After much head scratching, I settled on the inverted L, an antenna made popular by the grand master of 160-meters, Stew Perry, W1BB. I credit Stew for coming to my rescue!

The Inverted L

The inverted L was selected because it requires no more space than an 80-meter dipole and I could utilize my 50-foot crank-up tower for attachment. The vertical part is 50 feet long, and the horizontal part measures 130 feet, for an overall length of 180 feet (Figure 1). This makes the antenna approximately 3/8 λ, with the horizontal part providing top-loading. W1BB advises making the vertical section as long as possible (depending on tower height), and that an overall length of 160 to 180 feet works well.

Construction

With a bow and arrow, I successfully attached the far horizontal end of the antenna to the top of a 60-foot fir tree. From there, I ran the wire back to the top of the tower, where I bracketed a 30-inch-long two-by-four with an insulator screwed in the end (Figure 2). No. 14 Copperweld wire is used for the horizontal section, and no. 10 copper wire is used for the vertical section, which is spaced about 2 feet from the tower. These two wires are soldered to-

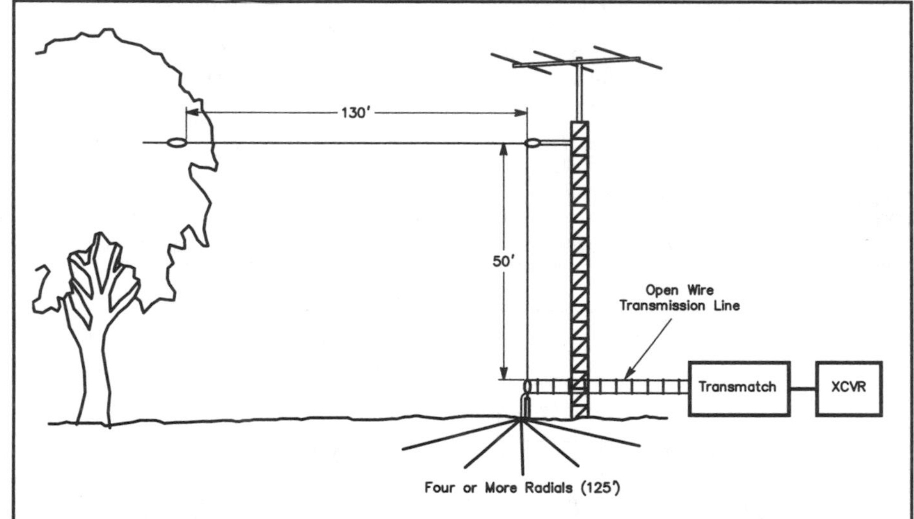

Figure 1—The W1XX inverted L is arranged in this manner.

Figure 2—Detail of how the antenna is mounted to the top of the tower. TV-mast clamps are used to secure the wooden insulator.

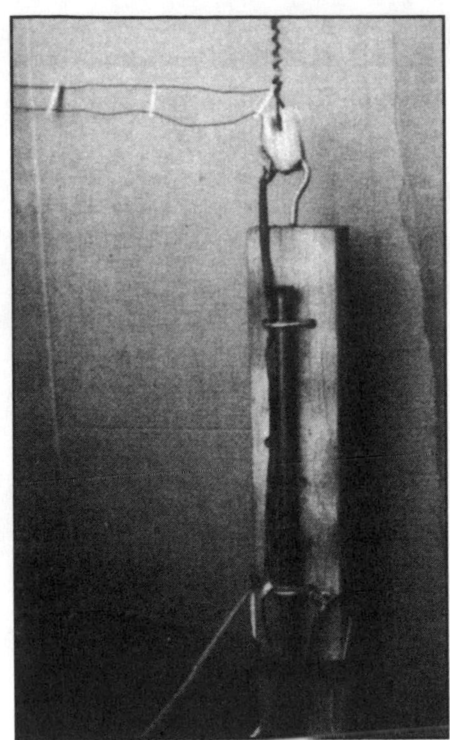

Figure 3—A ground rod at the tower base serves as a physical support for the insulator, and as an rf ground. Radial wires are connected to the rod by means of a ground bus, as described in the text.

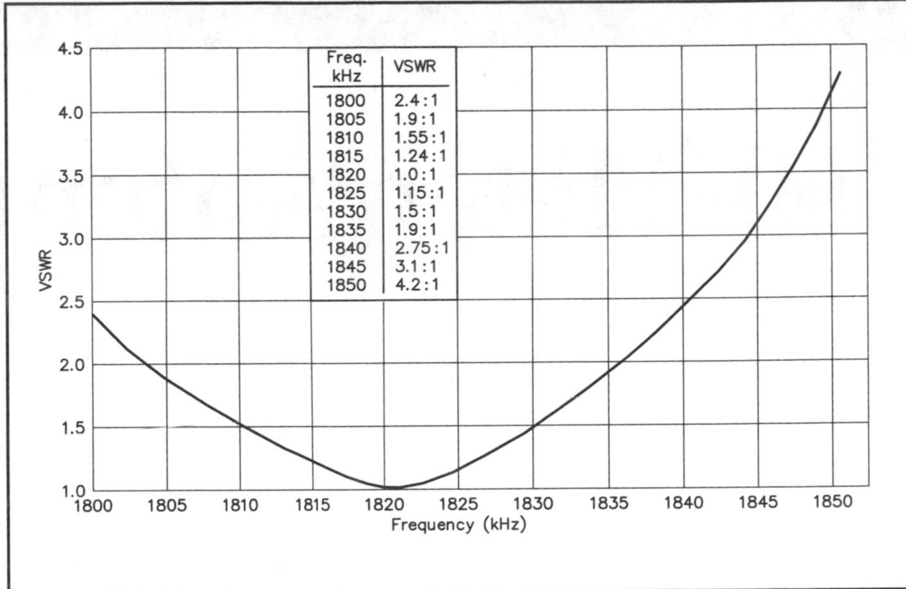

Figure 4—SWR curve for the inverted L, with the Transmatch adjusted for an SWR of 1 at 1820 kHz.

gether at the insulator.

A plumb line was used to locate a point on the ground directly below the 90° bend in the "L." At this point, a 10-foot copper clad ground rod was pounded into the earth, leaving 18 inches sticking out. A two-by-four is clamped to the ground rod by means of TV-mast U bolts. The lower end of the vertical wire is attached to an insulator that is screwed into the top of the wooden block (Figure 3). Next, I stripped some coaxial cable (RG-8/U) of its outer braid and used this to make a ground bus around the antenna base. One side of the open-wire transmission line is soldered to the base of the vertical antenna element, and the other side to the bus, which is attached to the ground rod with a clamp. This bus also serves as a connection point for the radials. So far, less than an afternoon of work had been invested.

Radials — the More the Better

The next day, my objective was to install radial wires, which are necessary because the inverted L is essentially a top-loaded vertical radiator. Previous meditation convinced me that several radials would fit on my lot. I'd *make* them fit! All radials were cut to 1/4 λ (125 feet), using scrap wire. About 300 feet of surplus telephone ground wire provided a good start. Stripping some old coaxial cable with a single-edged razor blade produced two radials from one length of wire (outer and center conductors). My technique for radial installation consisted of creasing the earth with a spade and shoving the wire in; afterwards the turf was pressed back into place with my heel.

With space a problem, it may be impossible to place all your radials in a straight line; don't worry, because it is not necessary. My installation followed a zigzag path to avoid fixtures like the house and driveway. W1BB advises putting some radials under the horizontal part of the antenna. Unfortunately, the location of my garage prevented this. Initially, only four radials were planted, but more were added later. As with all vertical antennas, the more radials you can put in the ground, the better the performance!

Matching System

Voltage-fed antennas approximately ½-λ long, such as the inverted L, will have a fairly high feed-point impedance. They will also exhibit inductive or capacitive reactance, depending on whether the antenna is slightly longer or shorter than ½ λ. Since the inverted L does not have a 50-ohm impedance, a matching system is needed. To make tuning adjustments easier, I opted for locating the Transmatch *in* the shack, using open-wire line to the antenna. Being able to adjust the antenna match conveniently is recommended, since the SWR climbs rapidly as you shift frequency. For example, adjusting the Transmatch for a 1:1 SWR at 1820 kHz produces an SWR of 2.4:1 at 1800 kHz (see Figure 4). With the Transmatch located in the shack, you can easily adjust for a 1:1 SWR no matter where you operate in the band. My Transmatch consists of a plug-in, link coupled coil and variable capacitors – all scrounged at flea markets. Any of the configurations found in the *ARRL Antenna Book*[2] should work well.

Performance

Does the antenna work? Having no comparison antenna, my conclusions are subjective. But I've been on the air enough to know when, as they say, "it plays." With only four radials in place, my first night of operation yielded plenty of U.S. contacts, plus a Caribbean DXpedition on the first call. Subsequently, many European stations have been worked from my northeast location with good signal reports. Contest activity has yielded some respectable scores, including many QSOs with the Caribbean, and South and Central America — even Antarctica!

By adjusting my Transmatch, I made limited tests with the L on 75-meter ssb. Comparisons were made to a 75-meter dipole at 50 feet. For signals close in, the L was down by some 3 to 5 dB, but equal or superior to the dipole for signals from eastern Europe. Apparently, the 160-meter inverted L also provides a low angle of radiation on this band too. On bands higher than 75 meters, the radiation angle will be tilted *upwards*, rendering the antenna inefficient for DX work. This phenomenon is explained in *The ARRL Antenna Book.*[3]

Giving up the 160-meter band for lack of sufficient real estate is unwarranted. With the inverted L, you can work Top Band from your urban lot. Installation is a breeze, and the performance is admirable. *Now* what's your excuse for missing out on the excitement of 160 meters?

Notes

[1]m = ft × 0.3048.

[2]*The ARRL Antenna Book*, 14th ed. (Newington: ARRL, Inc., 1982), pp. 4-1 through 4-8.

[3]*Ibid.*, pp. 2-23 through 2-24.

160-Meter Antennas

A recent QST article[1] described two 160-meter inverted-L antennas. The total length of one of the antennas (the vertical element and the horizontal arm) is approximately 1/4 λ; the other antenna is 1/2 λ long. The current-driven 1/4-λ antenna requires a fairly extensive ground screen to realize good radiation efficiency. On the other hand, the voltage-driven 1/2-λ antenna requires only a ground stake-or no ground at all. Here's some additional information to aid you in selecting between these alternatives.

Because the current distribution of these two antennas is quite different, the radiation characteristics are also different. That explains why some signals are ". . . very loud or very weak," depending on station distance, propagation conditions and which antenna is being used.

In the captions for the figures illustrating the two antennas, it's stated that the support poles could be metal or wood. Because the vertical element of the inverted L runs parallel—and rather close to—the support pole, one may wonder whether, if the pole is metal, this proximity will influence antenna performance.

The three antenna types I'll discuss are illustrated in Figure 1; they are (A) the 1/4-λ inverted L; (B) the 1/2-λ inverted L; and (C) the T antenna. Figures 2 and 3 show the calculated[2] radiation patterns for the two inverted-L antennas, assuming average ground ($\sigma = 3$ mS/m, $\varepsilon = 13$). The support poles are trees (or other wooden supports) 50 feet tall. Note that the horizontal arm of the antenna is close to the ground, in terms of wavelength (0.1 λ), and that, in this case, *ELNEC* overestimates the gain of horizontally polarized antennas at a height of 0.1 λ by about 3 dB.[3]

The reason the patterns are so different is because the vertical antenna element carries a heavy current when the antenna length is 1/4 λ, whereas the vertical antenna element carries only a small current when the antenna length is 1/2 λ. Hence, there is a great difference between the vertically and horizontally polarized components of the radiation field for these two inverted Ls. The former antenna has a monopole-like pattern, whereas the latter antenna has a dipole-like pattern.

The lengths given in Figure 1 are those for resonance at 1.9 MHz. DeMaw's dimensions give wire lengths that are a bit too short for resonance (according to *ELNEC*). (In practice, one would trim the antenna length for resonance.)

If you want to work DX, it is advantageous to deploy an antenna that has a deep overhead null in its radiation pattern. This minimizes high-angle sky-wave signals, noise and interference, and improves the signal-to-noise ratio of distant weak signals. The T antenna exhibits this feature. Figure 1C provides dimensions for a T antenna with a resonant frequency of 1.9 MHz; the radiation patterns are given in Figure 4. This antenna's radiation field is almost entirely vertically polarized.

For these plots, the antenna was in the X-Z plane (Z is the vertical axis and 0° azimuth is in the + X direction). The 0° azimuth for the inverted Ls is the direction that the horizontal arm points (away from the feed). There is a slight azimuthal pattern asymmetry for these antennas because the antenna structure is not symmetrical

Figure 1—Sketches for simple wire antennas for 160 meters (1.9 MHz); A, 1/4-λ inverted L; B, 1/2-λ inverted L; and C, 1/4-λ T antenna.

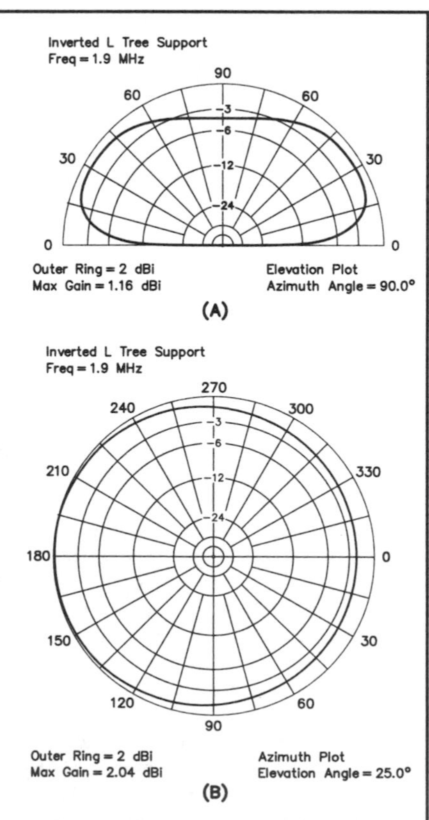

Figure 2—Radiation patterns for Figure 1A's antenna over average ground. The radiated field is dominantly vertically polarized—a monopole-like pattern. The absence of an overhead null in the vertical-plane pattern (compare Figure 2A with Figure 4A) is due to the horizontally polarized component of the field radiated by current on the antenna's horizontal arm.

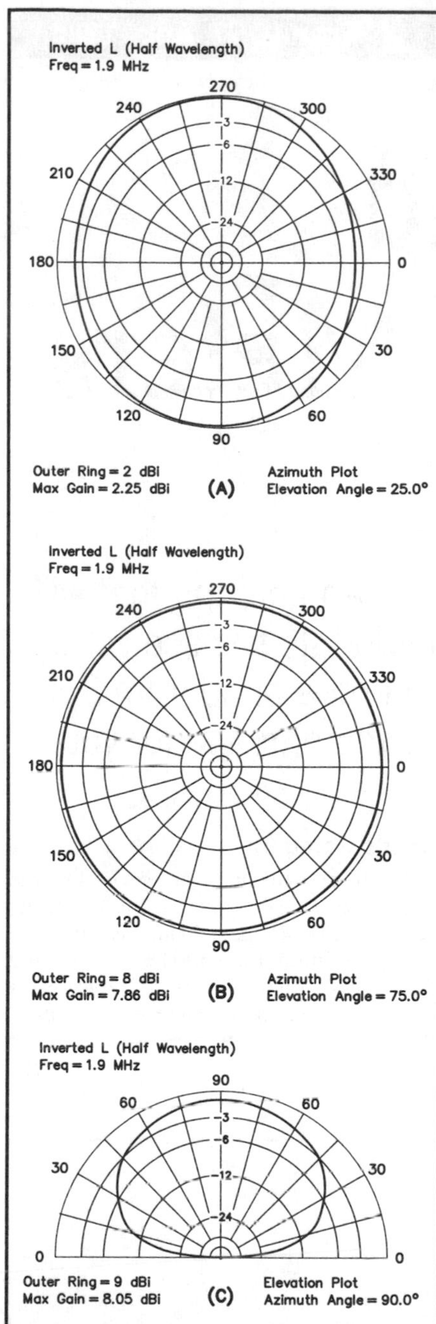

Figure 3—Radiation patterns for Figure 1B's antenna over average ground. The radiated field is dominantly horizontally polarized in the plane broadside to the antenna, and vertically polarized in the plane of the antenna—a dipole-like pattern. Hence the almost circular pattern at the high elevation angle, 75° (B), and the broadside directivity at the lower elevation angle, 25° (A).

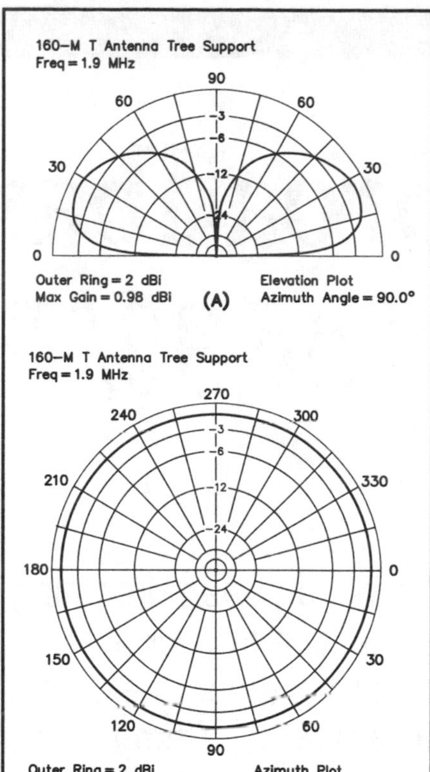

Figure 4—Radiation patterns for Figure 1C's antenna over average ground. The radiated field is almost entirely vertically polarized. The horizontally polarized component of the field radiated by current on one horizontal arm of the T is canceled by that radiated by the other arm.

with respect to the feed.

The effect of a well-grounded metal support tower was also studied. Surprisingly (to me), the effect is rather small, particularly for the 1/4 λ inverted L and 1/4 λ T antennas. The effect on the impedance of the 1/2-λ inverted L looked at first to be significant, but this is only because a small change in the resonant frequency of the antenna (the metal tower couples reactance into the antenna system and so changes the system's resonant frequency) results in a significant impedance change at frequencies near the anti-resonant frequency. Only a slight retuning of the matching circuit is necessary; the radiation patterns are affected relatively little. A 6½-foot nonconductive rope connecting the top ends of the horizontal arm to the support towers is assumed. For 50-foot support poles, it matters little (for 160-meter antennas) whether the poles are metal or wood.

But this is not the case if the support tower is also the mount for a 20-meter Yagi, because the tower and Yagi are much more nearly resonant than the tower alone. In this case, the tower carries a heavy current (0.8 A compared with the 1-A base current in the 1/4-λ inverted L). Because the phase of the tower current is +145° with respect to the phase of the current on the vertical element of the inverted L, the antenna's vertically polarized field is almost canceled. Also, the inverted L is no longer resonant: In fact, it is far from resonant; its impedance is 1547 –*j*2635 Ω, compared with the inverted L's impedance of 14 Ω with tree support.—*John S. Belrose, VE2CV*

Notes

[1] D. DeMaw, "The 160-Meter Antenna Dilemma," *QST*, Nov 1990, pp 30-32.

[2] *ELNEC* was used for all calculations. This program is available from Roy Lewallen, W7EL.

[3] *ELNEC* is a version of *MININEC*. *MININEC* assumes that the antenna is over a perfect earth for the purpose of calculating the antenna's impedance and the current distribution. The result of this approximation is that for horizontally polarized antennas, *MININEC* overestimates the gain of the antenna by an amount that increases as the height of the dipole decreases below a height of about 0.2 λ. The effect of real ground on the shape of the radiation pattern is correctly predicted, however, since for the calculation of the far-field pattern the program employs Fresnel reflection coefficients for the specified earth type.

By Brian L. Wermager, KØEOU From *QST*, April 1986

A Truly Broadband Antenna for 80/75 Meters

Have you been dreaming about an antenna that will do justice to your no-tune, solid-state transceiver by letting you operate across the entire band from 3.5 to 4.0 MHz? Then this may be the antenna for you!

With declining sunspots and poor conditions on the higher HF bands, 80 meters has suddenly become very popular. But, unfortunately, many hams are not able to use this band to its full potential. It offers every kind of ham activity from CW to phone, from nets and ragchewing to great DXing, but many hams are too limited by the frequency range of their antennas to enjoy this band completely.

Antenna-matching networks are one answer, but they spoil the advantage of the no-tune feature of modern transceivers. Matching networks also are often less effective than many hams think; they introduce losses. The losses can be significant at some settings which provide a match. With these things in mind, I decided to try some ideas that might give me a more broadbanded antenna. The prime requirement was that it be fed with common 50-ohm coaxial cable, with no traps, coils or capacitors.

First, I tried a quarter-wave sloper. This antenna worked very well, with a bandwidth of 300 kHz between the 2:1 SWR points. It still, however, limited me from operating CW DX at the bottom of the band and the phone nets at the top of the band. There had to be a better antenna.

Antennas can be broadbanded by using large-diameter elements. With this in mind, I began experimenting with two-wire slopers, attached to a common feed point, but with the wire ends fanned out from each other. (See Figure 1.) This seemed to help, but not as much as I had hoped. It did, however, shorten the length required for the sloper. For those with a short tower, this idea could make an 80 or 160-meter sloper possible when a single-wire sloper would be too long.

The Fickle Finger of Fate Strikes!

While I was trying one of these two-wire antennas at a low height on my tower, the SWR was less than 2:1 from 3.5 to 4.0 MHz! After several attempts to get it to work the same way at the top of the tower, I discovered that these results could be attained only when my old quarter-wave sloper was at the top of the tower *and grounded* to the tower. (See Figure 2.) The two antenna elements were obviously interacting with each other, broadening the bandwidth tremendously. Further pruning of the lengths of both the sloper and the two-wire element resulted in the amazing SWR curves shown in Figure 3.

What's Going On?

I will leave the question of *why* it works to the experts. (See the sidebar to this article.—Ed.) Like a true ham, I subscribe to the old saying, "If it works, leave it up and don't mess with it." My guess, however, is that it is something like one-half of a two-element log periodic seeking its mirror

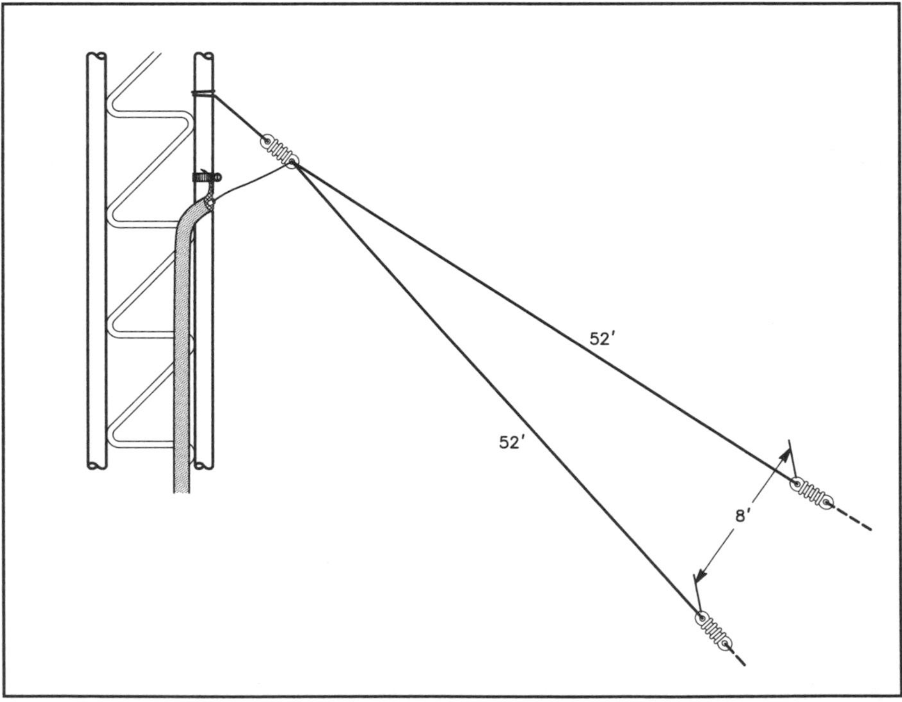

Figure 1—Arrangement of the original KØEOU experiment. The coaxial cable shield is connected to the tower, with both wires of the two-wire element connected to the center conductor. The wires are spread approximately 8 feet at the ends and are each approximately 52 feet long.

The MININEC Analysis of the KØEOU Three-Element Half Sloper

The Mini-Numerical Electronics Code (MININEC) analyzes thin-wire antennas, solving an integral equation representation of the electric fields using a method-of-moments technique. MININEC solves for the currents, impedance and patterns for antennas composed of wires in arbitrary orientations in free space and over perfectly conducting ground.

The impedance at the feed point calculates to be 79.4 + *j*1859.8 ohms. This assumes a perfect ground beneath the structure and simply a 3-foot extension of the tower above the connection of the upper wire. In practice, the impedance will be affected by both the ground conductivity and the top-loading effect of a beam antenna atop the tower.

A fair amount of current flows in the top section above the upper wire connection point. Top loading will affect the phase of this current, which will be reflected as a change in impedance at the feed point. In other words, the calculated data is not absolute. Use it as an approximation only.

A relatively high current flows at the base of the tower to ground—more than in any other part of the system. This indicates that a good earth connection, and even a radial system, would offer highest efficiency.

The antenna patterns, Figures A, B and C, are also approximations. Polarization is predominantly vertical–at low angles it may be considered to be almost completely vertical. Broadside to the direction of the wires, the polarization becomes horizontal at high radiation angles, ie, above 75 degrees. At 60-degrees elevation, the vertical component is almost 16 dB greater than the horizontal. The vertical component increases significantly at lower elevation angles, being in excess of 30 dB above the horizontal component at a 5-degree elevation. These figures all apply in a direction broadside to the wires. In the direction of the wires, both "front" and "back," the radiation is entirely vertically polarized.–*Gerald L. Hall, K1TD*

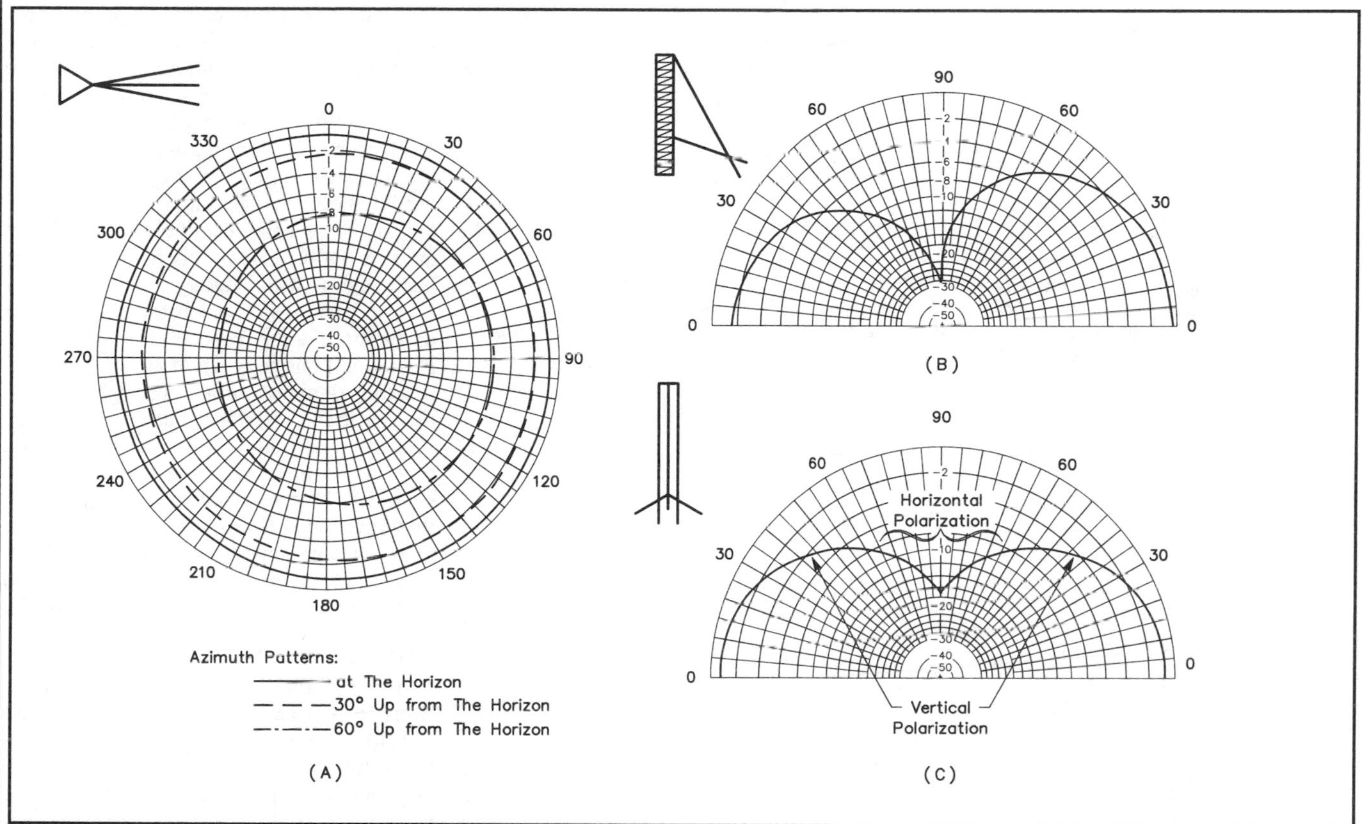

Figure A–Antenna azimuth radiation pattern for the KØEOU three-element half sloper antenna. Values are in dBi. Add 6.0 dB to the values shown.

Figure B—Antenna elevation radiation pattern, in the direction of the wires, for the KØEOU antenna. Values are in dBi. Add 6.0 dB to the values shown.

Figure C—Antenna elevation radiation pattern, in a direction broadside to the wires, for the KØEOU antenna. Values are in dBi. Add 6.0 dB to the values shown.

image in the grounded tower. The top element is tuned for the lower portion of the band and the two-wire element for the upper portion. In fact, there is a little SWR "bump" in the middle of the band that seems to give further evidence of this.

How Well Does It Work?

Although I have no way of scientifically plotting the antenna pattern, it does seem to be vertically polarized. Good DX performance from the antenna seems to verify this. Because many contacts have been made in all directions, the antenna probably has a fairly omnidirectional pattern. On-the-air comparisons with a quarter-wave sloper across town show that the antenna performs at least as well as the sloper. It also seems to have a little less noise on receive than the sloper.

Getting One Up for Yourself

If you have a tower over 40 feet high, you should be in business. The element dimensions will vary according to the height of your tower. My friend Kelly Davis, KD7XY, constructed one of these antennas on his 50-foot tower so we could see how the dimensions would change.

Measurements of his antenna are shown in Figure 4.

It is interesting to note that the height of the feed point on the tower does not appear to be critical at all. The angle of the wires in the two-wire element does not seem to be critical either. The sloper element, however, should come down *between* the two wires of the lower element. The sloper angle is about 45 degrees. As you attempt to get the

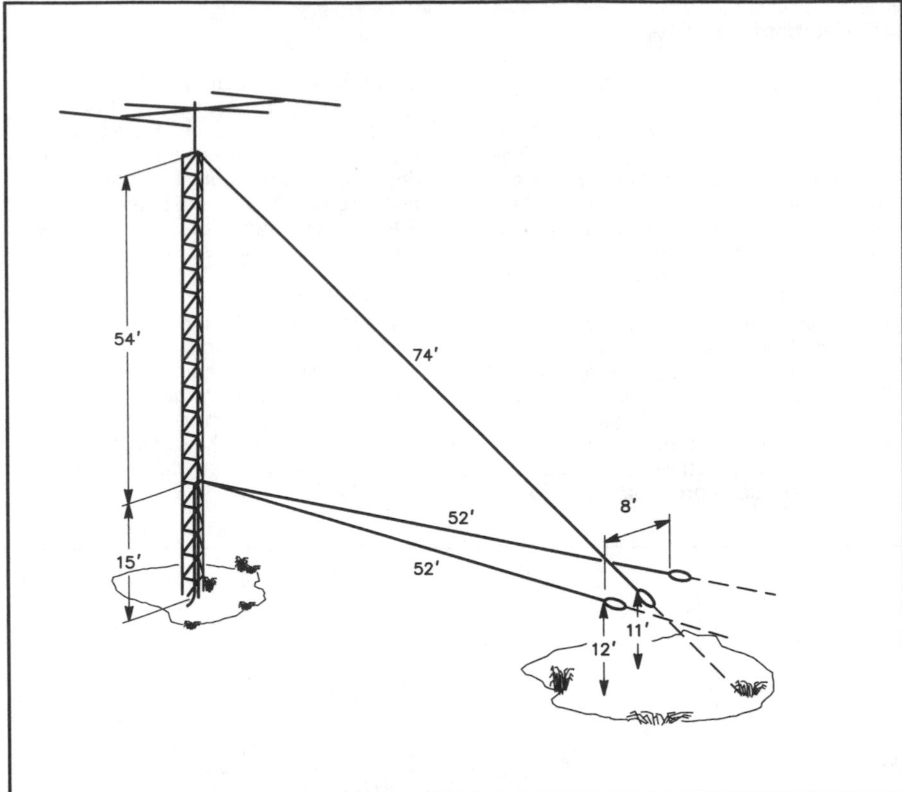

Figure 2—The antenna as constructed at KØEOU. The tower is 70 feet high. The feed point is as described in Figure 1 and is 15 feet above ground level on the tower. The sloper element is 74 feet long, is connected to the tower at the top end and slopes to a point 11 feet above ground level at the end.

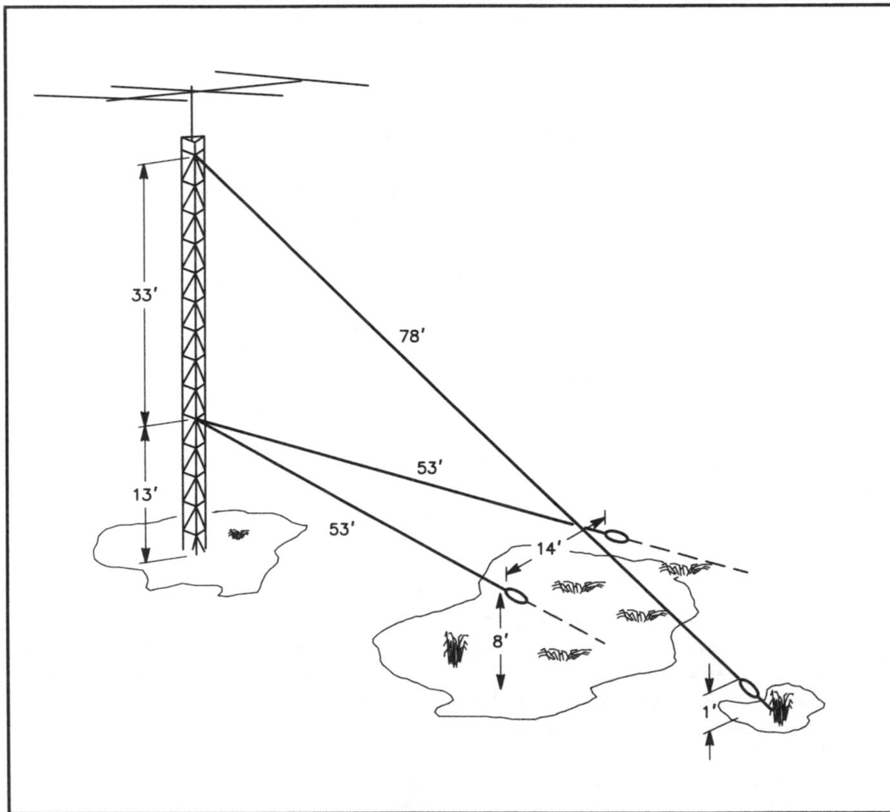

Figure 4—The antenna installation at KD7XY. The tower is 50 feet high. The sloper element is attached at 46 feet, is 78 feet long, and is only 1 foot above ground level at the end. The two-wire element is 53 feet long and is attached to the tower at 13 feet above ground level. Like KØEOU's antenna, it is virtually "flat" across the band; highest measured SWR is 1.4:1.

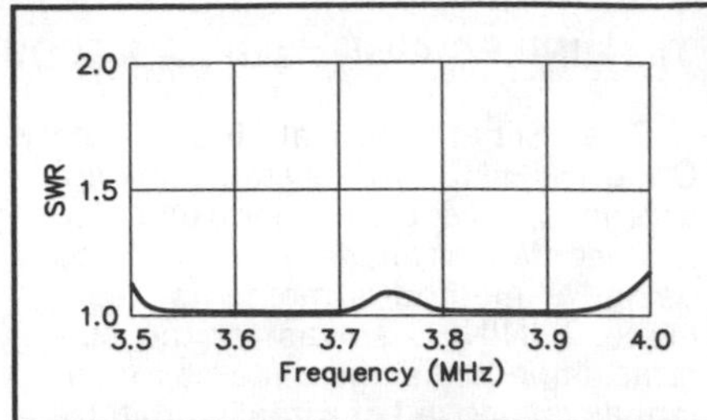

Figure 3—SWR measurements for the antenna at KØEOU. The highest SWR measurement between 3.5 and 4.0 MHz is 1.2:1.

lowest possible SWR from the antenna, remember that the angle of the sloper to the tower and its distance from the ground at the end will have an effect on the bandwidth. Because these antennas can be pruned from the ground, the trial-and-error method is easy. When pruning the antenna, remember that the two elements are cut for different frequencies. Changing the length of the top element changes the performance at the lower part of the band. Changing the lengths of the two-wire elements changes the performance at the top part of the band. Don't give up until your antenna SWR is, at the very most, 1.5:1 from band edge to band edge.

Use good insulators at the ends of each of the three wires. I used nylon fishing line at first, but one foggy, wet night the wire in the top element burned through in three places. It surprised me that there could be such high currents in a part of the antenna not even connected to the feed line, but I should have known better. Remember also that other objects around the antenna, such as other antennas or guy lines, could adversely affect antenna performance. Keep the antenna as much in the clear as possible.

Some Untried Ideas

I hope others will try some modifications to this antenna. For example, there should be no reason why a single wire for the bottom element won't work. My small city lot doesn't give me room to try a single wire, as it would surely need to be longer than the two-wire element. I would also like to see someone try cutting the top element for the high end of the band, and the bottom element for the low end. This could shorten the sloper element for someone with a shorter tower and may even give the antenna some gain in the direction of the two-wire element. It might also be possible to construct an antenna for another band inside of this one (40 meters, for example). The same feed point could be used, but with another sloper element for the second band. Another possibility is to construct the antenna with two dipoles. It is my guess (and only a guess), that it is the merging of the ends of the elements that causes the 50-ohm impedance of the antenna.

Winter gives us the best conditions on the 80/75-meter band. You can be ready to use the whole band with this simple-to-construct and very broadband antenna.

Improved Broadband Antenna Efficiency

When Brian Wermager's article. "A Truly Broadband Antenna for 80/75 Meters," arrived at HQ, I wanted to see if the antenna worked as well as claimed before publishing the information in *QST*.[1] I stopped tuning the antenna when the SWR was less than 1.6 across the entire band.

For working DX the antenna seemed to work as well as, or better than, a dipole at 50 feet. Casual contacts were not difficult to make, but in contests, it was a different matter. The results were satisfactory, but could "satisfactory" be changed to "outstanding"?

The key to improving the efficiency of this antenna is found in the sidebar accompanying the article. The MiniNumerical Electronics Code (MININEC) computer analysis done by Gerald Hall, K1TD, shows that: "A relatively high current flows at the base of the tower to ground—more than in any other part of the system. This indicates that a good earth connection, and even a radial system, would offer highest efficiency."

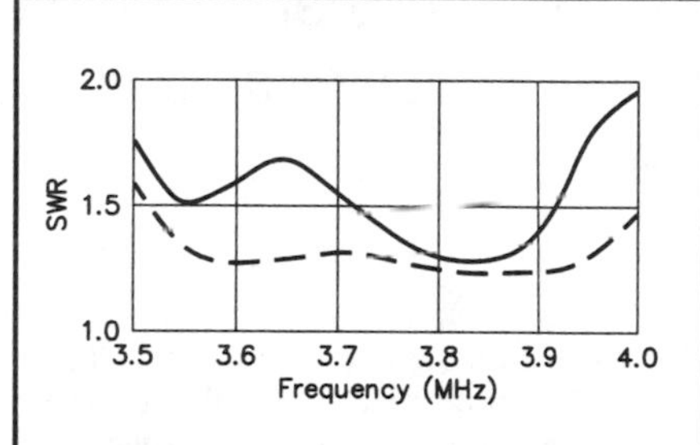

Figure 1—Graph showing SWR vs frequency for the KØEOU 80/75-meter antenna. The lower curve depicts results before adding ground radials; the upper curve was plotted after adding radials and retuning the antenna.

A system of radial wires improves antenna efficiency by cutting ground resistance losses. The decreased resistance in ground losses will also narrow the antenna bandwidth. The question now becomes: Is the narrowing of the bandwidth too much? I decided to investigate that question next.

With help from my brother-in-law, I installed 42 radials around, and bonded to, the tower base. As expected, SWR increased across the entire band. After I retuned the antenna, the SWR was below 2:1 across the entire band (see Figure 1). As for results, DX stations are typically 2 to 6 S units stronger on the KØEOU antenna than on the dipole at 50 feet. Contest QSOs are now a reality. I am not the loudest East Coast station on the band, but I am very pleased with the performance of the broadband antenna. (The dipole has been removed and stored!)—*Chuck Hutchinson, K8CH*

[1]B. Wermager. "A Truly Broadband Antenna for 80/75 Meters," *QST*, Apr 1988, pp 23-25.

Simple, Effective, Elevated Ground-Plane Antennas

Here's an easier and better way to use your grounded tower as a vertical antenna on 160 or 80 meters.

This article describes a simple and effective means of using a grounded tower, with or without top-mounted antennas, as an elevated ground-plane antenna for 80 and 160 meters.

Grounded towers have been used as shunt-fed verticals on the low-frequency amateur bands for many years. Generally. they required a gamma- or omega-type matching network and an extensive radial system for efficient operation. Recent computer studies reveal that simple elevated radial systems consisting of only four wires can produce results equivalent to 120 buried radials. Typically, these antennas are modeled as isolated monopoles. Presumably, grounded towers could be used with an appropriate shunt-fed matching network.

I've found an even easier method!

From Sloper to Vertical

Recall the quarter-wave length sloper, also known as the half-sloper. It consists of an isolated quarter wavelength of wire, sloping from an elevated feedpoint on a grounded tower. Best results were usually obtained when the feedpoint was somewhere below a top-mounted Yagi antenna. You feed a sloper by attaching the center conductor of a coaxial cable to the wire and the braid of the cable to the tower leg. Now, imagine four (or more) slopers, but instead of feeding each individually, connect them together to the center conductor of a single feed line. *Voila*! Instant elevated ground plane.

Now, all you need to do is determine how to tune the antenna to resonance. With no antennas on the top of the tower, the tower can be thought of as a fat conductor and should be approximately 4% shorter than a quarter wavelength in free space. Calculate this length and attach four insulated quarter-wavelength radials at this distance from the top of the tower. For 80 meters, a feedpoint 65 feet below the top of an unloaded tower is called for, The tower guys must be broken up with insulators for all such installations. For 160 meters, 130 feet of tower above the feedpoint is needed.

That's a lot of tower to dedicate to a single-band antenna, especially for someone with limited real estate. What can be done with a typical grounded-tower-and Yagi installation?

A top-mounted Yagi acts as a large capacitance hat, top loading the tower. Fortunately, top loading is the most efficient means of loading a vertical antenna. The amount of loading can be approximated by using an empirical formula developed by John Devoldere, ON4UN.[1]

Devoldere found that the electrical height of a top-loaded tower can be approximated by:

$$L = 0.38F(H + \sqrt{2S - H/500}) \qquad \text{(Eq 1)}$$

where

L is the approximate electrical length in degrees

F is the frequency in MHz

H is the height of the tower under the Yagi in feet

S is the area of the Yugi in square feet

[1]J. Devoldere, *Antennas and Techniques for Low Band DXing* (Newington: ARRL, 1994).

Table 1
Effective Loading of Common Yagi Antennas

Antenna	*Boom Length (feet)*	*S (area, ft²)*	*Equivalent Loading (feet)*
3L 20	24	768	39
51 15	26	624	35
4L 15	20	480	31
31 15	16	384	28
5L 10	24	384	28
4L 10	18	288	24
3L 10	12	192	20
TH7	24	—	40 (estimated)
TH3	14	—	27 (estimated)

To check Eq 1, consider the case of no antenna on top, where S = 0. Then L = 0.38 × 3.6 × 65 = 88.9°, which is very close to the desired 90° quarter wavelength.

The effective loading of a Yagi is the portion of the equation under the radical. The examples in Table 1 should give us an idea of how much top loading might be expected from typical amateur antennas. The term H/500 is ignored as insignificant compared with 2S.

The values listed in the Equivalent Loading column of Table 1 tell us the approximate vertical height replaced by the antennas listed in a top-loaded vertical antenna. To arrive at the remaining amount of tower needed for resonance, subtract these numbers from the nonloaded tower height needed for resonance. Note that for all but the 10-meter antennas, the equivalent loading equals or exceeds a quarter wavelength on 40 meters. For typical HF Yagis, this method is best used only on 80 and 160 meters.

Construction Examples

Consider this example: A TH7 Yagi mounted on a 40-foot tower. The TH7 has approximately the same overall dimensions as a full-sized 3-element 20-meter beam, but has more interlaced elements. I estimate its equivalent loading to be 40 feet. At 3.6 MHz, 65 feet of tower is needed without loading. Subtracting 40 feet of equivalent loading, the feedpoint should be 25 feet below the TH7 antenna.

I ran 10 quarter-wavelength (65-foot) radials from a nylon rope tied between tower legs at the 15-foot level, to various

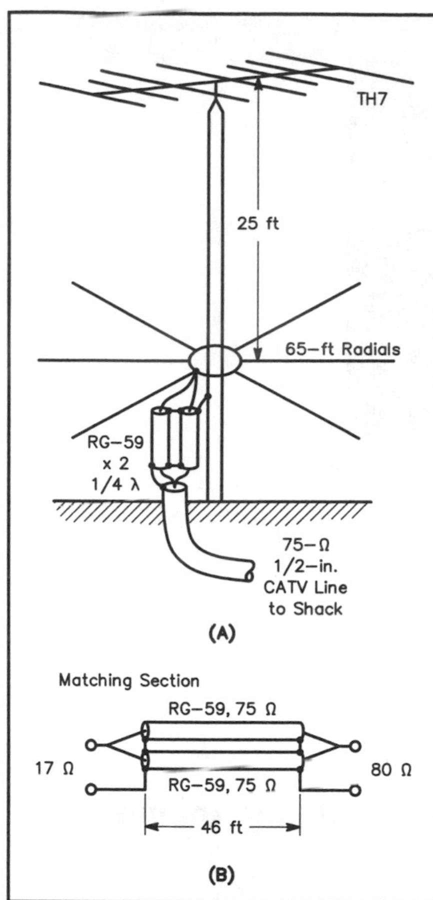

Figure 1—At A, an 80-meter top-loaded, reverse-fed elevated ground plane, using a 40-foot tower carrying a TH7 triband Yagi antenna. At B, dimensions of the 3.6 MHz matching network, made from RG-59.

supports 10 feet high. I tied nylon cord to the insulated, stranded, 18-gauge wire, without using insulators. The radials are all connected together and to the center of an exact half wavelength (at 3.6 MHz) of RG-213 coax, which will repeat the antenna feed impedance at the other end. Figure 1 is a drawing of the installation. I used a Hewlett-Packard low-frequency impedance analyzer to measure the input impedance across the 80-meter band.

An exact resonance (zero reactance) was seen at 3.6 MHz just as predicted. The radiation resistance was found to be 17 Ω. The next question is how to feed and match the antenna.

My approach to 80-meter antennas is to tune them to the low end of the band, use a low-loss transmission line, and switch an antenna tuner in line for operation in the higher portions of the band. With a 50-Ω line, the 17-Ω radiation resistance represents a 3:1 SWR, meaning that an antenna tuner should be in-line for all frequencies. For short runs, it would be permissible to use RG-8 or RG-213 directly to the tuner. Since I have a plentiful supply of low-loss 75-Ω CATV rigid coax, I took another approach.

I made a quarter-wave (70 feet × 0.66 velocity factor = 46 foot) 37-Ω matching line by paralleling two pieces of RG-59 and connecting them between the feedpoint and a run of the rigid coax to the transmitter. The magic of quarter-wave matching transformers is that the input impedance (R_i) and output impedance (R_o) are related by

$$Z_o^2 = R_i \times R_o \qquad \text{(Eq 2)}$$

For $R_i = 17\ \Omega$ and $Z_o = 37\ \Omega$, $R_o = 80\ \Omega$, an almost perfect match for the 75-Ω CATV coax. The resulting 1.6:1 SWR at the transmitter is good enough for CW operation without a tuner.

The Proof is In the Log

How effective is this antenna? Well, I used to install a 60-foot aluminum tower and 100 radials, 100 feet long in a clear one-acre field every winter, and remove it every spring for mowing. The top-loaded reverse-fed elevated ground-plane antenna has replaced that antenna with no regrets. My only other 80-meter antenna is a dipole at 110 feet broadside to Europe and the South Pacific,

I use the elevated ground plane for South Africa, South America. the Caribbean, parts of the Pacific and Asia, both long and short path. With it I have worked everything I can hear, with 1200 W output, including HL (rare in Alabama); HS; UA9; UAØ; UI; UJ; UL; most of the VK9s; XV; ZS1; ZS8MI; ZS9; 3Y5X; 8Q; 9M2; and 9V1. While running 5 W output. I have even worked two JAs with this antenna. which may say more for its effectiveness than anything else.

Will It Work on 160 Meters?

You bet it will, but it takes another tower. For the 160-meter band, a resonant quarter-wavelength requires 130 feet of tower above the radials. That's a pretty tall order. Subtracting 40 feet of top loading for a 3-element 20-meter or TH7 antenna brings us to a more reasonable 90 feet above the radials. Additional top loading in the form of more antennas will reduce that even more.

Recently, a friend moved to the country, and he needed a 160-meter antenna in a hurry for an upcoming contest. He had stacked TH6s on a 75-foot tower. I suggested he try four elevated radials at 10 feet above ground, with a tuner if necessary. He connected four radials about 120 feet long and a piece of RG-58. The SWR measured under 2:1 and he worked everything he heard in the contest. Figure 2 is a drawing of this installation.

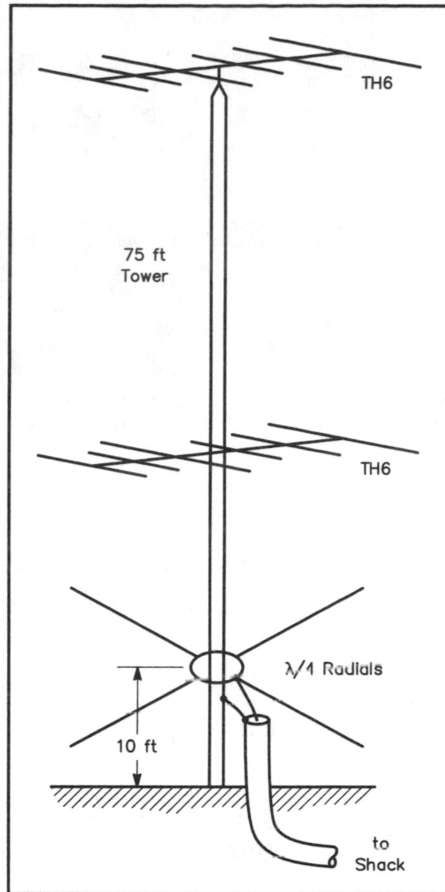

Figure 2—A 160-meter antenna using a 75-foot tower carrying stacked triband Yagis.

Another friend had a 120-foot tower with no antennas on it. He ran four elevated radials at 10 foet and obtained an SWR below 1.5:1 with a 50-Ω feed line. During the contest, he even beat out some big guns.

Elevated ground-plane antennas work! This simple, reverse-feed system makes it possible to feed grounded towers easily and efficiently.

References

A. Christman, "Elevated Vertical Antenna Systems," *QST*, Aug 1988, pp 35-42 (Feedback, *QST*, Oct 88, p 44).

A. Christman, "More on Elevated Radials," *QST*, Mar 1993, p 72 (Technical Correspondence).

CHAPTER EIGHT

RECEIVING ANTENNAS

Ungrounded Beverage Antennas

So you've heard how Beverages can help you hear better on the lower bands, but you're still not convinced they're worth the bother. K6STI analyzes the Beverage, especially an easy-to-construct "ungrounded" version.

Beverages are amazing receiving antennas. A good Beverage can turn a signal barely audible on your transmit antenna into solid, Q5 copy. A Beverage antenna (named after 1920s inventor H. H. Beverage) is a long, low wire used for receiving. They're most popular on the 160, 80, and 40-meter bands. Although occasionally pressed into service, Beverages are much too lossy for regular use as transmitting antennas.

Conventional Beverages are fed against ground. Although very long wires may develop a few dB of front-to-back ratio, unterminated Beverages are essentially bidirectional. Terminating a Beverage through a resistance to ground at the far end creates a highly unidirectional pattern. Beverages typically are 1 to 15 feet high and 1 to 4 wavelengths long.

Beverages are broadband, traveling-wave antennas. Power traveling toward the feedpoint is absorbed by the receiver, while that traveling in the opposite direction is absorbed by the termination resistance. Standing waves can't develop. Because waves traveling toward the feedpoint accumulate substantially in phase, while those arriving from other directions tend to cancel, the azimuth pattern of a long Beverage is highly directive. Beverages are nonresonant antennas and can be used over a wide frequency range.

A very useful Beverage variation uses two parallel wires, with one grounded at the far end. Waves traveling toward the far end reflect and return to the feedpoint differentially as transmission-line currents. With suitable termination and switching circuitry at the feedpoint, you can receive signals from either direction. A two-wire Beverage thus can replace two single-wire antennas.

For simplicity, I'll consider just single-wire Beverages here.

Eliminating Ground Connections

Beverages normally connect to ground at the feedpoint and far end. See Figure 1A. However, obtaining a good, reliable ground can be difficult. A short ground rod may exhibit thousands of ohms resistance to ground when installed in dry soil or when corroded. Since Beverages require a termination resistance of just several hundred ohms, a poor far-end ground can grossly misterminate an antenna and destroy its pattern. A poor feedpoint ground can reduce signal levels and encourage coax-shield pickup. You can use multiple ground rods, a ground screen, or a ground-radial system to lower ground resistance, but this greatly increases the work required to install a Beverage.

A ground connection isn't essential for Beverage operation. You can place the termination resistance about one-quarter wavelength from the far end of the wire. Since the impedance of a quarter-wave wire is low and resistive, the remainder of the wire just sees the termination resistor in series with a low resistance. Similarly, you can feed a Beverage about one-quarter wavelength from the near end. Both ends of the wire are left unconnected. See Figure 1B.

Ungrounded Beverages are simple to install and have stable properties. Their radiation patterns often are superior to those of their grounded counterparts because they don't suffer unwanted response from vertical wires. (This response can be reduced by lowering wire height. One way to do this is to slant the feed and termination wires over a long distance.)

The only real disadvantage of ungrounded Beverages is that they're frequency-dependent. For example, the low-impedance, quarter-wave wire sections become high-impedance, half-wave sections at twice the design frequency. I'll show you how to multiband an ungrounded Beverage later.

Antenna Modeling

While you can use a *MININEC*-based antenna- analysis program to model Beverage antennas, you'll have to settle for approximate results at best. The *MININEC* algorithm uses perfect-conductivity ground during current calculations. It takes lossy ground into account only later when calculating radiation patterns. Because Beverages are so close to ground, their current distribution is strongly affected by ground characteristics. *NEC*'s Sommerfeld-Norton ground model takes ground dielectric constant and conductivity into account when calculating wire currents. This lets you accurately analyze antennas in close proximity to earth. I used the *NEC/Wires 1.0* program to obtain all results presented here. I modeled antennas over average-quality earth (dielectric constant= 13, conductivity = 5 mS/m).

Wire Height

Figures 2A and 2B shows the azimuth and elevation patterns for a 1000-foot long, 10-foot high, ungrounded Beverage on 160 meters. The antenna uses # 18 wire and a 650-Ω termination resistance. It's fed and terminated 120 feet from each end. The remarkable pattern is better than that of many HF Yagis.

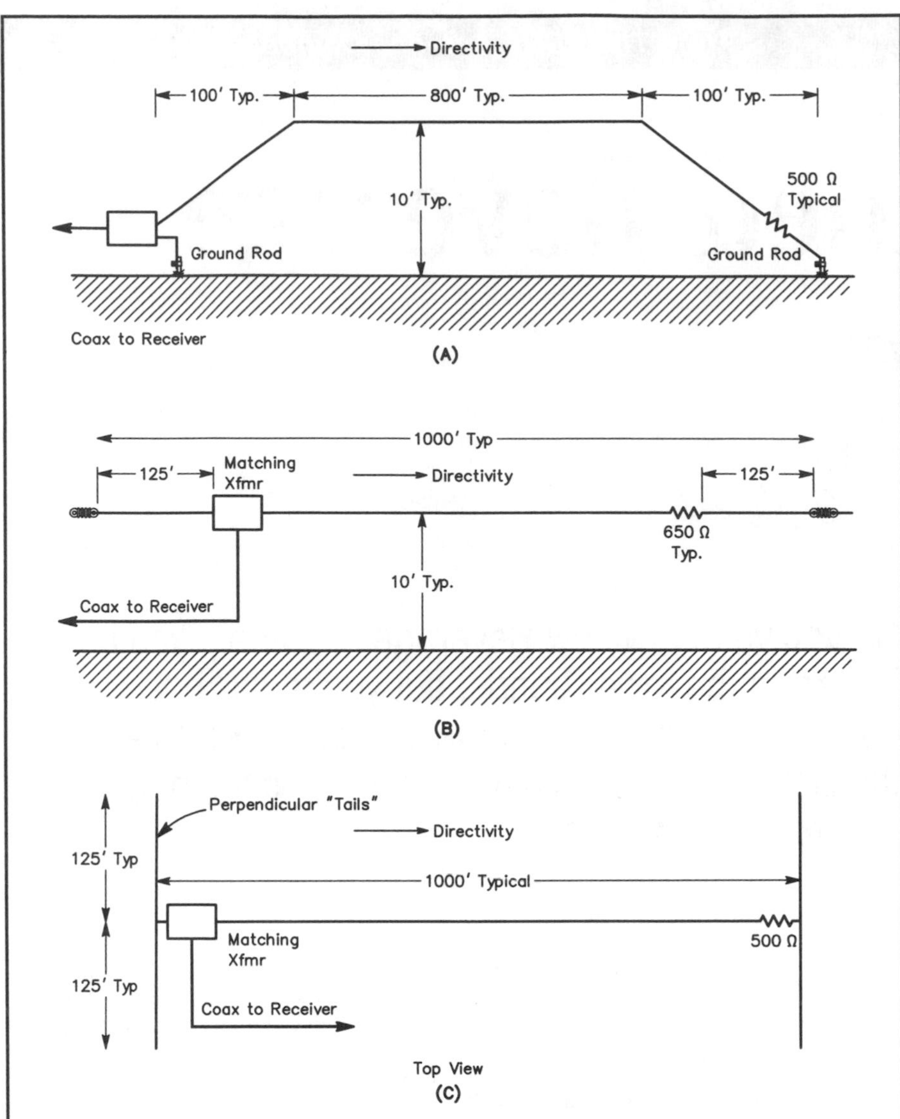

Figure 1—Single-wire Beverage antenna feeding and termination configurations. At A, conventional method is shown using slanted-wire terminations. At B, ungrounded Beverage is shown with matching transformer and terminating resistor inserted λ/4 in from either end of wire. At C, ungrounded Beverage using full span of wire by employing perpendicular "tails" at both ends. The tails function as elevated counterpoise radials at 160 meters.

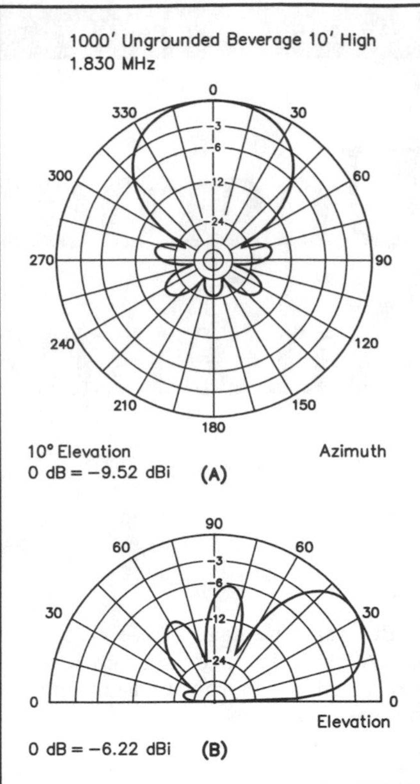

Figure 2—Azimuth and elevation patterns at 1.83 MHz for 1000-foot long ungrounded Beverage mounted 10 feet off flat ground with average conductivity and dielectric constant. The worst-case rearward lobes are better than 18 dB down compared to peak response. Note that the gain is only –9.52 dBi at 10° elevation, peaking at –6.22 dBi at about 30° elevation. If the transmission line to the receiver is very long, a preamplifier may be needed.

Figure 3 shows the azimuth pattern of a similar Beverage 1 foot above ground. The termination resistance for this antenna is 750 Ω and it's fed and terminated 125 feet from the wire ends. The pattern is even sharper than that of Figure 2. Output is about 1 dB lower.

If you're tempted as I was to just lay a Beverage on the ground, take a close look at Figure 4. This is the Beverage of Figure 3 but just 1 inch off ground (*NEC* provides accurate results for wire heights down to within several wire radii of ground). While still useful, the pattern has degraded considerably. Signal output is quite a bit lower. (Nevertheless, some users report excellent results from on-the-ground Beverages. Don't hesitate to try one if it's your only option.)

In the West you can just lay an ungrounded Beverage in the chaparral a few feet off ground. In the East you can unroll one on top of a foot or two of snow (snow has very low dielectric constant and conductivity and should be transparent to RF at low frequencies). Beverages like these are very easy to install (no support poles or ground rods). You can roll one out just before a contest.

Occasionally a Beverage may have to clear a height-sensitive area. Figure 5 shows what can happen when wire height varies. The first 200 feet of this antenna is 3 feet high (chaparral height). Next, a 50-foot section slopes from 3 to 15 feet. This is followed by a 200-foot span 15 feet high, a 50-foot downslope, and finally the remainder of the wire at 3 feet. This model is similar to a Beverage used by K6TQ which must clear an agricultural area with produce trucks. While the rear part of the pattern is relatively unaffected, the height variation causes a large sidelobe bulge. It doesn't matter much if your Beverage is

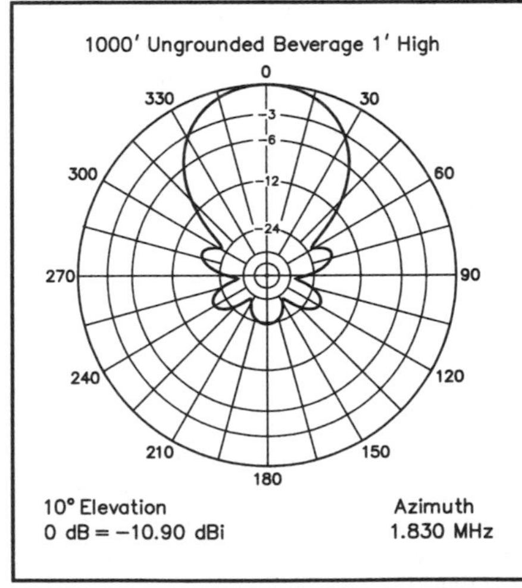

Figure 3—Azimuth response for 10° elevation angle at 1.83 MHz for 1000-foot long ungrounded Beverage mounted only 1 foot off flat ground. Compared to the higher antenna in Figure 2, this low Beverage has about 1 dB less gain, but the worst-case backlobes are suppressed even better.

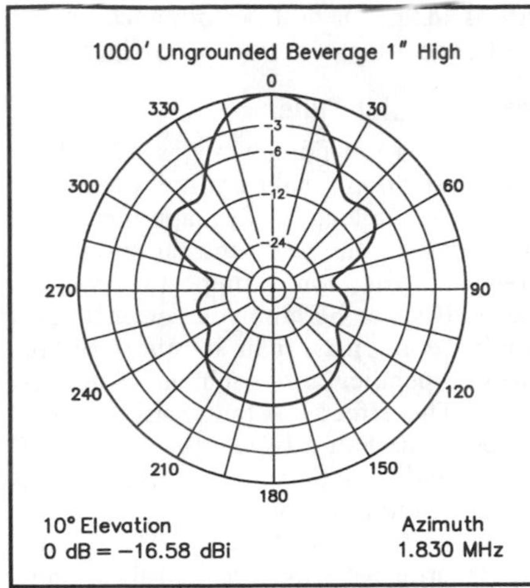

Figure 4—Azimuth response for same antenna as in Figure 3, except that the height is now only 1 inch off ground. The rearward pattern suffers and the gain fails about 5.4 dB compared to the Beverages in Figures 2 and 3.

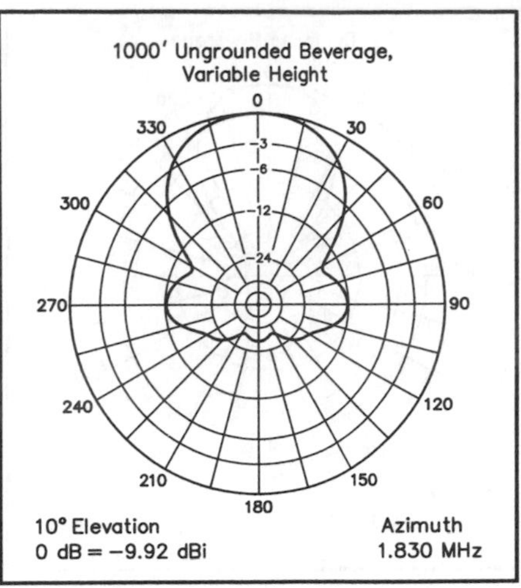

Figure 5—Azimuth response at 1.83 MHz for 1000-foot long, variable height ungrounded Beverage. The sidelobes at 90° and 270° are larger than the antenna in Figure 2, but the pattern and gain are still quite acceptable.

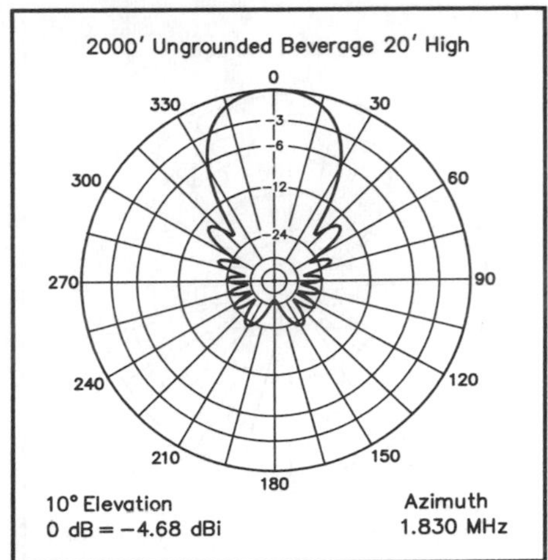

Figure 6—Azimuth response at 1.83 MHz for 2000-foot long ungrounded Beverage mounted 10 feet over flat ground. The gain is almost 5 dB higher than that of Figure 2, and the backlobes are all suppressed in excess of 23 dB. This is a great 160-meter receiving antenna, but must be oriented carefully at desired receiving bearing because of narrow frontal lobe.

high or low, but try to keep the height as constant as possible.

Wire Length

Figure 6 shows the azimuth pattern of a 2000-foot ungrounded Beverage. The termination resistance is 550 Ω. The pattern of a very long Beverage like this may be too narrow for complete coverage of a general direction.

Figure 7 shows the pattern of a 500-foot ungrounded Beverage. The termination resistance is 1100 Ω. While much broader than the patterns of longer antennas, it's still quite directive and can provide a substantial receive improvement.

Feeding Ungrounded Beverages

Conventional Beverages usually are fed with an autotransformer having a 3:1 turns ratio. This matches a 450-Ω impedance to 50 Ω. Ungrounded Beverages typically have an input resistance of 500 to 1000 Ω (roughly equal to the termination resistance) and an input reactance of up to 100 Ω. An exact impedance match isn't important; a better match just delivers a higher signal. Although usually not required to improve signal-to-noise ratio, a receive preamplifier can equalize Beverage and transmit-antenna signals for easier direct comparisons.

Both grounded and ungrounded Beverages can be sensitive to unwanted coupling at the feedpoint. If you connect coax through an autotransformer to a Beverage, the outside of the coax shield can become part of the antenna system if it exhibits an impedance similar to or less than whatever connects to it. It's the same thing that can happen with a coax-fed dipole. To prevent this, use a matching transformer with separate primary and secondary windings and low interwinding capacitance. This will isolate the outside of the coax shield from the antenna. Alternatively, you can use an autotransformer followed by a current-type balun. (Because the feed tail of an ungrounded Beverage is likely to provide a lower-impedance termination than a typical ground rod, these Beverages should be less sensitive to coax-shield pickup.)

Terminating Ungrounded Beverages

Optimal termination resistance depends on wire length, height, diameter, and ground constants. The termination resistance normally just affects the rear part of the pattern. It's easy to find the optimal value by modeling the exact geometry of your antenna with *NEC*. If you use long spans of wire with considerable sag, model the sag using multiple wires.

You can determine optimal termination resistance experimentally using a signal source to the rear of the antenna. Simply adjust the termination resistance for minimum signal.

(Actually, minimum response directly to the rear isn't quite optimal for most designs. This condition usually causes the pattern to degrade slightly over a broad region elsewhere. Nevertheless, adjusting for minimum response is close enough for all but the most fanatical Beverage enthusiast.)

Multibanding Ungrounded Beverages

If you provide a termination resistance one-quarter wavelength from the far end for each band of interest, you can use a Beverage on several bands. At the near end use a separate quarter-wave tail for each band and connect them in parallel at the feedpoint.

Figure 8 and Figure 9 show 160-meter and 80-meter patterns of a 1000-foot-long, 10-foot-high, dualband Beverage. This antenna uses a 700-Ω resistor 63 feet from the far end and a 350-Ω resistor 125 feet out. The two termination resistances provide near-optimum performance on 160. The backlobe of the 80-meter pattern is suboptimal but the sidelobes are still well down. I didn't experiment much with termination-resistance values or positions. I'll bet both can be improved.

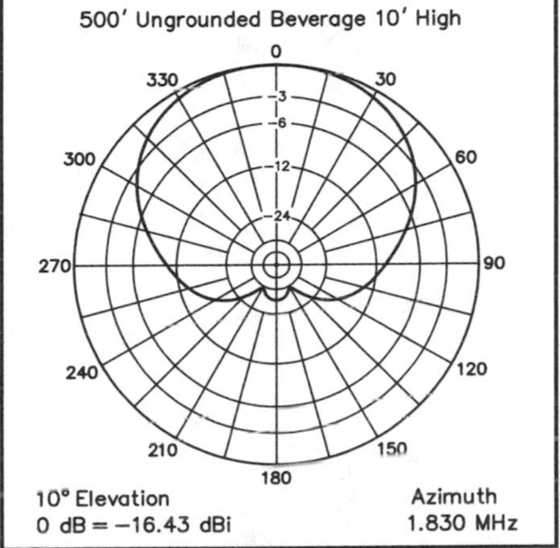

Figure 7—Azimuth response at 1.83 MHz for 500-foot long ungrounded Beverage mounted 10 feet over average ground. The gain is almost 7 dB less than Figure 2 at 10° elevation. The side lobes are large, even though the rear lobes are well suppressed.

Another alternative is to use a single termination resistance with multiple quarter-wave tails. The resistor value will be optimal on just one band since antenna length isn't constant in wavelengths. But the value won't be far off on any band if the

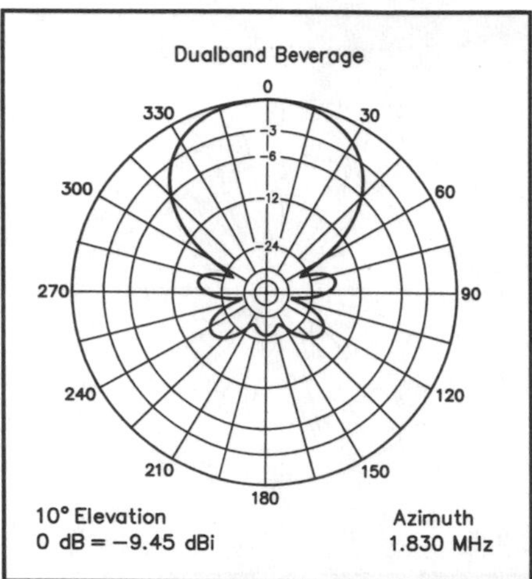

Figure 8—Dualband Beverage on 1.83 MHz, using a 700-Ω resistor 63 feet from the far end, and a 350-Ω resistor 125 feet out.

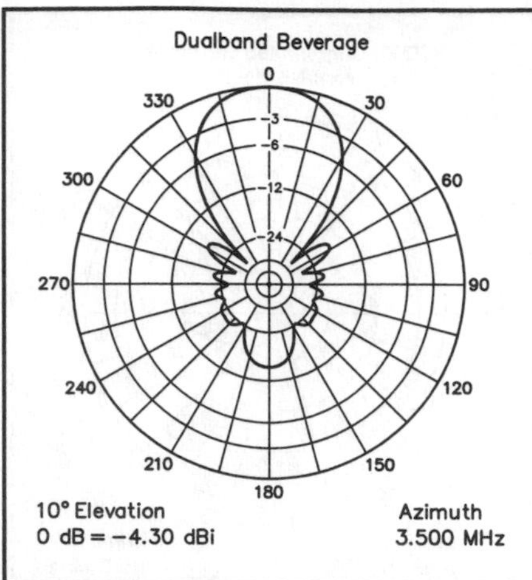

Figure 9—Dualband Beverage on 3.5 MHz, using same terminations as in Figure 7. The termination is not optimal for this band, but still yields desirable directivity.

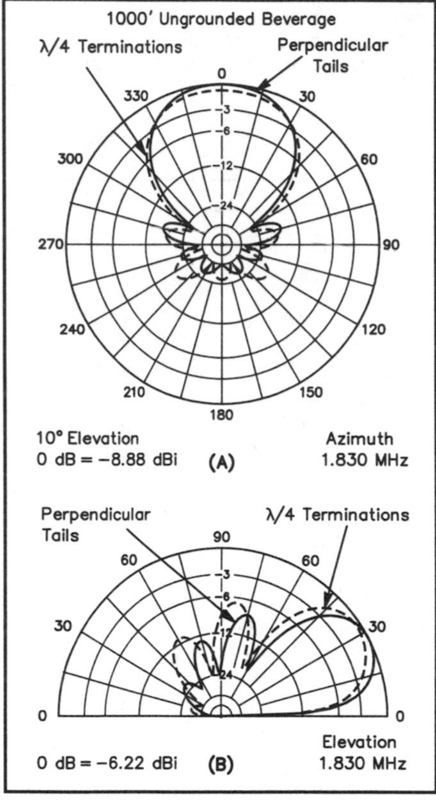

Figure 10—Ungrounded Beverage, 1000 feet long with two λ/4 perpen-dicular termination "tails." Pattern from ungrounded Beverage in Figure 2 is overlaid for comparison. At A, the azimuthal patterns at 10° elevation are shown; at B, the elevation patterns are shown.

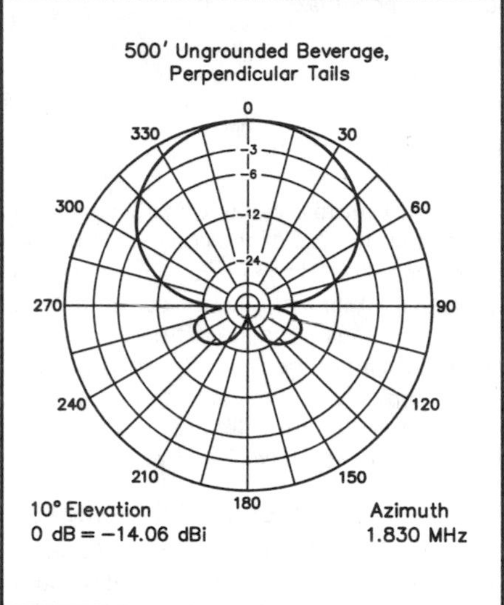

Figure 11—Azimuthal pattern for ungrounded Beverage, 500 feet long with two λ/4 perpendicular termination tails.

wire is long. For best performance on all bands, use a separate resistor for each tail.

Perpendicular Tails

The ungrounded Beverages described so far really don't make optimal use of the entire antenna length. Although they contribute power, the near- and far-end quarter-wave tails don't support traveling waves. If space permits you can feed and terminate against a pair of quarterwave wires perpendicular to each end. See Figure 1C. This effectively lengthens the Beverage by one-half wavelength. Essentially the perpendicular tails form two, two-wire, elevated radial systems.

Figures 10A and 10B shows the azimuth and elevation patterns for a 1000-foot-long, 10-foot-high Beverage with perpendicular tails, with the patterns from Figure 2 overlaid for comparison. This antenna uses a 500-Ω termination. The pattern is somewhat narrower than that of Figure 2 and the backlobes are even smaller. The elevation plot shows that overhead response is about 10 dB less with perpendicular tails and that the forward lobe cants at a lower angle.

Since the quarter-wave tails of a 500-foot Beverage comprise half its length, you might expect a more dramatic improvement for perpendicular tails on this antenna. Figure 11 shows the pattern of a 500-foot-long, 10-foot-high Beverage with perpendicular tails. This antenna also uses a 500-Ω termination. Side response is down quite a bit from Figure 7 and forward response is 2 dB greater.

While they provide better performance, designs with perpendicular tails are more complex and occupy additional space. You can make better use of the additional space by phasing two Beverages with in-line tails to obtain a much better azimuth pattern . . . but that's another story.

By Brian Beezley, K6STI From *QST*, September 1995

A Receiving Antenna that Rejects Local Noise

Simplicity and performance combine to give birth to a compact antenna you'll want to have!

Noise can make a ham's life miserable on any amateur band. As we approach the minimum of the sunspot cycle, many hams are discovering that noise can be particularly frustrating on the low bands. In summer, static crashes caused by thunderstorm lightning can totally mask weak signals on the 160, 80, and 40-meter bands. During other seasons, power-line noise, noise from household appliances, and incidental radiation from home electronic products often limits reception.

A recent *QST* article by Floyd Koontz, WA2WVL,[1] describes a small receiving antenna for the low bands that provides a cardioid directional pattern. This pattern can reduce noise and QRM from the rear. As I marveled at the elegance and simplicity of Floyd's design, I realized that the antenna did have one shortcoming: Because it is vertically polarized, the antenna responds strongly to local noise propagated by ground waves. I wondered whether it was possible to devise a receiving antenna to better reject local noise.

The Ground Wave

Most hams who operate HF are familiar with the sky wave (or space wave) that's responsible for long-distance ionospheric propagation. See Figure 1A. The space wave has two components: The *direct wave* propagates along a straight line from the transmit antenna toward the ionosphere. The *ground-reflected wave* bounces off the earth's surface and heads in the same direction.

The space wave also exists for local propagation, as shown in Figure 1B. The direct wave travels in a straight line between the transmit and receive antennas, while the ground-reflected wave takes a midpoint bounce. But when the antennas are close to ground, the direct and reflected waves nearly cancel, leaving a very small residual space wave. When the antennas are right at the earth's surface, the waves cancel completely. So what makes local communication possible? Answer: A third wave, called the *surface wave*, that exists for antennas close to ground. This wave diminishes in intensity as you increase antenna height. The surface wave exists *only near the surface of the earth*. The combination of the direct wave, the ground-reflected wave, and the surface wave is called the *ground wave*.[2]

Surface-wave intensity varies with frequency and ground conductivity. It's stronger at low frequencies and for highly conductive ground. But the most important property of the surface wave is its polarization sensitivity. The surface wave is much weaker for horizontal fields. For example, on 80 meters the broadside ground-wave response of a short piece of wire is about 34 dB lower when oriented horizontally. The difference is about 42 dB at 160 meters.[3]

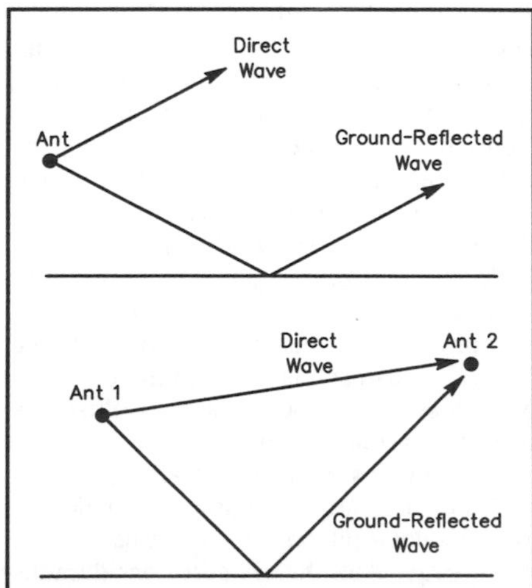

Figure 1—At A, the direct and ground-reflected components of the space wave for ionospheric propagation. At B, the space-wave components for local propagation.

The surface wave makes local AM radio broadcasting possible. Because of the poor propagation of horizontally polarized ground-wave signals, AM broadcasters universally use vertical polarization. The same phenomenon causes vertical antennas to pick up much more local noise than horizontal antennas do. Even if a noise source has a stronger horizontal component, by the time the field reaches the receive antenna, the vertical component almost always dominates.

These facts suggest that the first requirement of a receiving antenna with low response to local noise is insensitivity to vertically polarized radiation, the dominant component of the ground wave. Surprisingly, simply avoiding the use of vertical wires isn't enough. An antenna composed only of horizontal wires can still respond to vertical fields.

A Low Dipole

Figure 2 shows the ground-wave response of an 80-meter dipole 10 feet high. An easy-to-install, inconspicuous dipole like this is sometimes used for receiving when the transmit antenna is vertically polarized. The pattern shows the electric field strength 10 meters above ground at a distance of 1000 meters for an input power of 1 kW (the dipole exhibits the same pattern on receive). This geometry might be representative of that for a noisy power pole. Although the pattern may look similar to that of a free-space dipole, I think

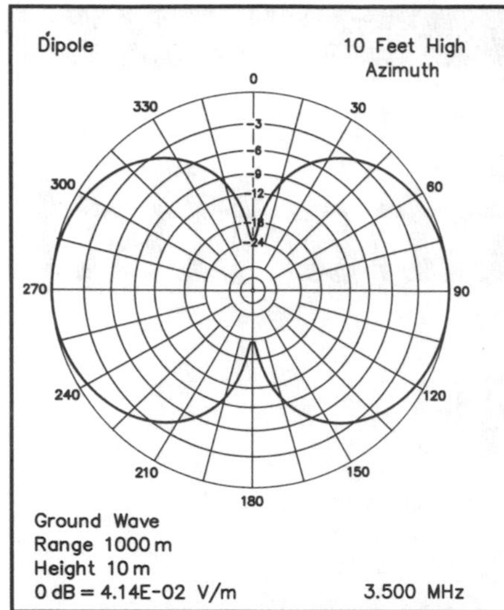

Figure 2—Azimuthal plot of the ground-wave response of a 10-foot-high 80-meter dipole. The input power to the antenna is 1 kW. The peak electric field is shown.

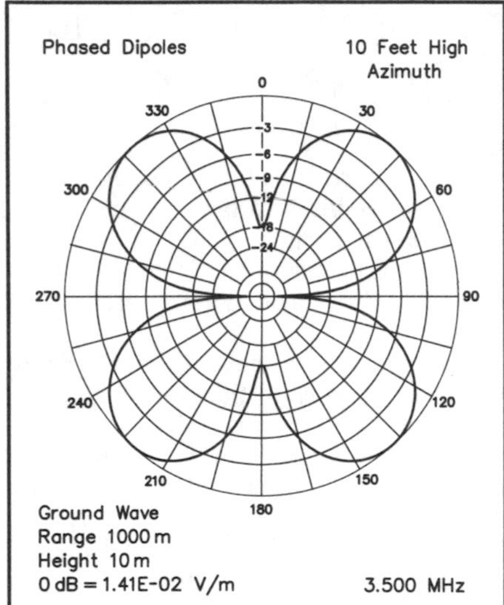

Figure 3—Azimuthal plot of the ground-wave response of two parallel dipoles 10 feet apart, 10 feet high and fed out of phase. The input power to the antenna is 1 kW. The peak electric field is shown.

you'll be surprised to know that the wire is oriented broadside to the pattern null. A low dipole actually responds to ground-wave fields best off its ends!

Here's an explanation for this peculiar behavior: The dipole has no response to the vertical component of a broadside ground wave because the electric field is perpendicular to the wire. The antenna responds only to the weak horizontal component. The vertical component also is perpendicular to the mirror image of the antenna formed by the ground-reflected wave. But away from broadside, the dipole, its image, or both, have a nonvanishing projection in the vertical plane. This enables the dipole to respond to the vertical component of a ground wave. In addition, lossy earth causes the surface wave to develop a radial component in the direction of propagation. Away from broadside, the radial component also projects onto the wire and induces current. The vertical and radial components induce maximum current when the surface wave arrives in line with the wire.

When compared to a quarter-wave vertical with four radials elevated 10 feet above ground, the low dipole has a 12.5 dB lower ground-wave response in its most sensitive direction in line with the wire. For sky-wave signals arriving at 20° elevation from their weakest direction (also off the ends), the dipole has 10 dB lower response than the vertical. Therefore, the low dipole has a signal-to-noise ratio advantage of 2.5 dB for signals and noise arriving from their worst-possible directions. In the most favorable directions broadside to the wire, the S/N advantage peaks sharply at 26.1 dB.

Raising the dipole broadens the broadside S/N peak and improves S/N off the ends. For example, for signals arriving at 20° elevation, a dipole at 50 feet has an S/N advantage over the reference vertical of 6.3 dB in line with the wire and 25.6 dB broadside. If you have just a single noise source and you can rotate a high dipole, you should be able to come within a few decibels of the latter figure most of the time. But when multiple noise sources in different directions arise (typical for power-line noise in times of low humidity), rotating the antenna won't help much. The S/N advantage of the high dipole then is likely to be near the worst-case figure.

These numbers illustrate the advantage of horizontal receiving antennas and substantiate the notion long held by amateurs that "verticals are noisy." But you can reduce ground-wave noise much more effectively if you don't rely on a simple horizontal wire.

Canceling Ground-Wave Components

Although a horizontal wire responds only weakly to a broadside ground wave, its response is substantial off the ends. If you could somehow eliminate the end response, you'd be left with the low broadside response and whatever residual response developed at intermediate angles.

Figure 3 shows the ground-wave response of two parallel dipoles 10 feet apart and 10 feet high fed out of phase. The phasing cancels everything arriving off the ends of the wires. The end nulls combine with the low broadside response to create a residual cloverleaf ground-wave pattern. The peak of the cloverleaf is 9.3 dB down from the peak end-response of a single dipole.[4] (Wire losses are ignored here to illustrate the cancellation principle.)

This antenna is just a very-close-spaced W8JK endfire array. Although it makes a good receiving antenna for the low bands, it's pretty large. And there's still considerable ground-wave pickup in the cloverleaf peaks. If the wires somehow could remain parallel for all directions, it might be possible to achieve complete cancellation of the vertical and radial components of the ground wave.

In some sense, the sides of a circular loop are parallel everywhere. Current amplitude and phase vary little in small loops of regular shape. Therefore, the currents in opposite sides of such loops are nearly equal and out of phase. Unlike a W8JK array, a small horizontal loop does not have a null anywhere along the ground. But its ground-wave response is uniformly low in all directions because the antenna responds only to the weak horizontal component. Everything else cancels out (or nearly so).

A small loop usually is defined as one with a total conductor length of less than 0.1 λ. But unless you use large-diameter conductors to minimize RF resistance, loops this small are inefficient. A preamplifier may be needed to overcome receiver noise. You can increase the output of a loop by increasing its size, but the larger you make it, the less constant the current becomes. This reduces groundwave cancellation.

The antenna of Figure 4 overcomes this difficulty by using two feedpoints, each on opposite sides of the loop to force current balance. One of the phasing lines is twisted to maintain proper phase.[5] This loop can be made quite large and still exhibit very low response to ground-wave noise.

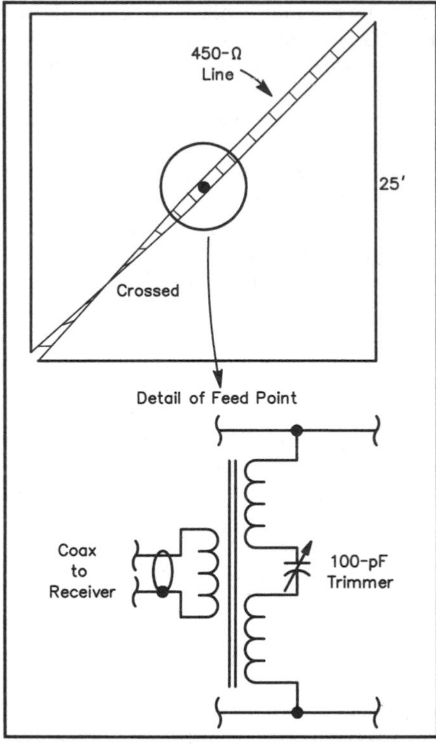

Figure 4—Basic diagram of the 80-meter low-noise loop antenna showing detail of the feedpoint arrangement. The antenna measures 25 feet on a side, is 10 feet high, and made of #14 wire. It's fed at opposite corners with phasing lines made of #14 wire spaced 1.5 inches. A small ferrite transformer at the junction of the phasing lines matches the antenna to 50 Ω coaxial feed line and also functions as a balun. The trimmer capacitor (a capacitance of about 40 pF is required) in series with the antenna-side winding resonates the loop at 3.5 MHz.

A Practical Design

The 80-meter loop of Figure 4 has a perimeter of 0.36 λ. It's 25 feet on a side, 10 feet high, and made of #14 wire. It's fed at opposite corners with phasing lines made of #14 wire spaced 1.5 inches. A small ferrite transformer at the junction of the phasing lines matches the antenna to 50 Ω and also functions as a balun. A trimmer capacitor (about 40 pF is needed) in series with the antenna-side winding resonates the loop at 3.5 MHz.

The ground-wave response of this particular loop is shown in Figure 5 and its sky-wave response in Figure 6. For signals arriving at 20°elevation, the worst-case sky-wave response is 20.5 dB below that of the reference vertical. This signal level is quite usable on 80 meters without a preamp. Because the loop reduces ground wave noise at least 45.1dB, S/N improvement is 24.6 dB for the worst-case combination of signal and noise directions. if you use a preamp and adjust for equal signal levels, ground-wave noise will be four S units lower on the loop *no matter what direction it comes from!* For signals arriving at higher angles, S/N enhancement approaches 30 dB.

Both the ground wave noise pattern and sky-wave signal pattern are very uniform in azimuth. The antenna is essentially omnidirectional with an overhead null, just like a vertical. An overhead null is useful for reducing near-vertical-incidence skywave signals from nearby stations. (The loop rejects their ground-wave signals along with local noise.)

Although the S/N performance of this loop is not particularly sensitive to height, you can increase output level substantially by raising the antenna well above ground. For example, if you raise the loop to 50 feet, the output increases 15.2 dB for signals arriving at 20° elevation. At this height, the output level is only 5.3 dB below that of the reference vertical. You'll never need a preamp with a loop this high. The S/N advantage drops 0.1 dB. At a height of 20 feet, output increases 7.2 dB and S/N drops 0.6 dB.

You can shrink the loop to 10 feet on a side. The S/N advantage increases 0.5 dB, but the signal level drops 6.5 dB. You can raise the signal level 3.4 dB by doubling the loop side lengths to 50 feet, but the S/N advantage then drops 7.9 dB. You can thus trade-off loop size, S/N enhancement, and signal level. If you use smaller wire, output drops. For example, it's 1.7 dB lower for #22 wire.

Although this loop is a narrowband device and must be carefully resonated, the resonance is much broader than that of a typical small loop. You can get away with a single capacitor setting for both 3.5 and 3.8 MHz if you're willing to accept somewhat lower signal levels. The capacitor setting does not affect the patterns or skywave-to-ground-wave ratio—it simply alters output level.

The input resistance at the junction of the phasing lines varies over a wide range with loop size, height, and phasing-line characteristic impedance. The input resistance is about 40 Ω for the loop of Figure 4 at a height of 10 feet. A transformer using a type-77 ferrite core (such as an FT-82-77, FT-114-77 or FB-77-1024) with 9 turns of any size enameled wire on the coax side and 8 wire turns on the antenna side provides a good match to 50-Ω coax.[6] The input resistance drops to about 20 Ω when the loop is raised to 20 feet and to about 15 Ω at 50 feet. Use 13 turns of wire on the coax side for 20 feet and 15 for 50 feet.

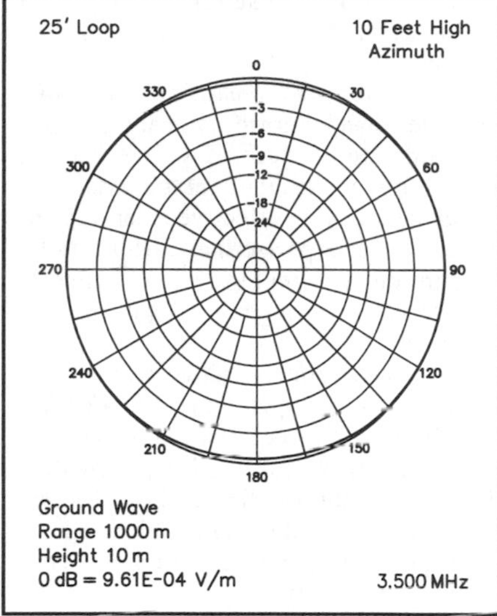

Figure 5—Azimuthal plot of the ground-wave component of a 25-foot-square, 80-meter loop at a height of 10 feet. The input power to the antenna is 1 kW. The peak electric field is shown.

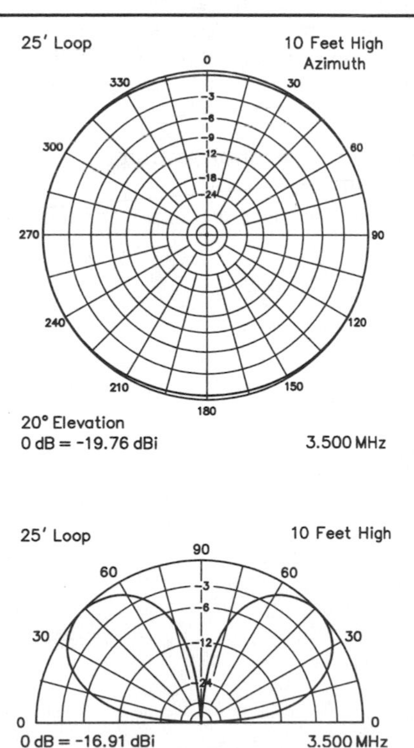

Figure 6—The sky-wave responses of the 10-foot-high, 25-foot-square, 80-meter loop. At A, the azimuthal plot; at B, the elevation plot.

Although the optimum transformer turns ratio varies with antenna height, little output is sacrificed if you use a fixed ratio.

For example, output is only 1 dB less than optimal when a loop designed for 10 feet is used at 50 feet. If you're determined to obtain the best possible match, use 16 turns of wire on the coax side and tap the winding for lowest SWR (alternatively, a switch can be used to select any of two or more taps). Use the lowest possible power when measuring SWR. It's easy to puncture the dielectric of a small trimmer capacitor with a momentary blast.

The antenna's resonant frequency shifts when the phasing lines get wet. If you use true open-wire line with plastic spacers, the frequency shift will be less than 100 kHz. But if you use 450-Ω line with segmented polyethylene dielectric, the resonant frequency decreases more than 200 kHz when the line becomes thoroughly damp. Although this won't affect signal-to-noise ratio, output drops in the desired frequency range. You may be tempted to try phasing lines of 300-Ω twinlead routed inside PVC tubing to avoid moisture effects, but line impedances this low work only for smaller loops.

This antenna should be constructed as symmetrically as possible to maximize cancellation of the vertical and radial components of the ground wave. Make the loop perfectly square and accurately align it in the horizontal plane. Cut the phasing lines to the same length. Although these loops perform well near houses, fences, and towers, try to install the antenna as far from other conductors as possible to maximize current balance. Use the shortest possible leads to interconnect the matching components. Although it's probably unnecessary, I like to split the antenna-side transformer winding and put the tuning capacitor in the center to promote equal currents in the phasing-line conductors.

To minimize the number of supports, a loop about 17 feet on a side can be constructed using a 20-meter quad spreader. You can mount the spreader well up on a tower to increase output. Alternatively, you may be able to eliminate supports altogether by stringing a loop in your attic or garage. However, the current balance of indoor loops may be degraded by electrical wiring, plumbing, heating ducts, or other nearby conductors.

On-the-Air Performance

Ed Andress, W6KUT, located in the San Diego suburb of Poway, constructed a loop 21 feet on a side and 10 feet high. Ed's location is subjected to strong, chronic power-line noise. On 80 meters, Ed uses a pair of phased quarter-wave verticals for transmitting. We used the verticals as a reference when evaluating the loop. A 20-dB preamp was available. With the preamp, signals near loop resonance were about equal to those from the verticals.

The loop performed as expected. It

dramatically enhanced the signal-to-noise ratio of most sky-wave signals. Sometimes the verticals did better during a momentary change in propagation; occasionally a particular noise arose that the loop didn't attenuate much. But overall, the loop was far superior. It made little difference on strong signals. It made listening to moderately strong signals much more pleasant. It let us copy weak signals that were buried in the noise and unreadable on the verticals.

During times of no detectable powerline noise, we often noticed a curious effect: The loop still enhanced signal-to-noise ratio by one or two S units, making copy of moderately weak signals more pleasant. On these occasions, we were unable to hear the telltale, raspy buzz of power-line noise (or any other noise signature) when we listened with the transceiver's AM detector. Unless noise happens to arrive at low angles and signals at high, there's no reason for the loop to enhance skywave S/N. We believe that the unidentified noise is local and propagates by the ground wave. We speculate that it may be the sum of hundreds of weak man-made noise sources in the densely populated suburb. (The superposition of a large number of noise sources tends to be characterless even when the individual sources aren't.)

Total Ground-Wave Cancellation

If you stack two of these loops vertically and bring both feed lines into the shack, you can form a deep null on the horizon for all azimuth angles by combining the signals with a fixed amplitude and phase offset. Except for azimuth-response irregularities caused by nearby conductors and small elevation-angle differences due to range, this system will cancel all ground-wave components. This noise canceler was to have been the original subject of this article. However, a single component loop worked so well in practice that I decided total ground-wave cancellation was overkill.

Comparison with Other Antennas

A conventional, single-feedpoint, small loop about two feet on a side—oriented horizontally—yields roughly the same S/N enhancement as the loop of Figure 4. However, output will be down about 46 dB from the reference vertical. You'll need a low-noise preamp with a loop this small unless your receiver has a very low noise figure. Still, even with a preamp, a conventional small loop makes an attractive, low-profile alternative. You must construct a low-output antenna like this carefully to avoid stray pickup. A single capacitor setting won't provide good output levels on both phone and CW.

The WA2WVL cardioid antenna attenuates thunderstorm static and ground-wave noise to the rear. If you're seldom troubled by omnidirectional local noise, it should make a more effective receiving antenna than the loop described in this article. The cardioid also requires fewer supports and less space. As a bonus, it's inherently broadband.

If you have room for a two-wavelength Beverage, it will outperform the WA2WVL cardioid on sky-wave noise and should reduce ground-wave noise arriving more than 45° off boresight by at least 15 dB. If local noise near boresight isn't a problem, a long Beverage can tremendously improve your receiving capability.

The easiest way to improve reception on 80 or 160 meters is to use the most sensitive horizontal antenna available at your antenna switch. Many hams with 80-meter verticals find that switching to a 40-meter dipole or beam improves copy of weak signals even though the antenna is nowhere near resonant on 80 meters. When just a single noise source is active, you should be able to null it by broadsiding a 40-meter rotary.

Scaling the Antenna to Other Frequencies

While I've used the 80-meter band for illustration in this article, it's easy to scale the design to other frequencies. Simply multiply lengths, heights, transformer turns and capacitor values by the number you get when you divide 3.5 by the target frequency in MHz.

That said, I don't recommend this antenna for use above 40 meters. If you're using a vertical antenna on the upper HF bands, do yourself a favor and replace it with the highest horizontally polarized antenna you can manage. Not only will your receive noise decrease, your transmit signal almost certainly will improve due to higher ground-reflection gain.[7] (Exception: If your vertical radiates over saltwater, keep it!) If you're already using a horizontal wire on upper HF, I think your next antenna project should be a rotary beam rather than a receiving loop.

Notes

1 Floyd Koontz, WA2WVL, "Is this Ewe for You?" *QST*, Feb 1995, pp 31-33.

2 Frederick Terman, *Radio Engineers' Handbook*, 1st ed (New York: McGraw-Hill, 1943), pp 674-709.

3 *MININEC*-based antenna-analysis programs do not compute the surface wave, nor can they accurately model antennas close to ground. I used *NEC/Wires 2.0* with its surface-wave and Sommerfeld-Norton ground options for all antenna models in this article. All models assumed average ground characteristics (dielectric constant 13, conductivity 5 mS/m).

4 A small cloverleaf peak requires very close wire spacing. For example, the peak is down only 5.4 dB for a spacing of 50 feet.

5 The feed arrangement is like that of an Alford Loop. In fact, the low-noise receiving antenna really is just a variant of the VHF/UHF loop devised by Andrew Alford for a different purpose in 1940. See Terman, pp 814-815.

6 Don't be tempted to eliminate the transformer and connect 50-Ω coax directly to a loop. The transformer functions as a current balun. It keeps noise current induced on the coax shield from entering the receiver. Just a little pickup by the shield can pollute the low-noise output of a loop.

7 For example, on 20 meters, a horizontal dipole only 35 feet high has 3 to 6 dB broadside gain over a vertical dipole at low angles over average ground.

By Gary Breed, K9AY From *QST*, September 1997

The K9AY Terminated Loop—A Compact, Directional Receiving Antenna

Wish you had enough room for an effective low-band receiving antenna? You do! This four-direction system fits in a 30-foot circle!

Low-band operators are always looking for ways to improve their hearing. As a low-band fan, I was impressed with the EWE antenna developed by Floyd Koontz, WA2WVL.[1,2] Koontz shows us how to build a compact, directional antenna—a design that quickly became very popular. But when I sat down at the computer to figure out the best way to install my own EWEs, a surprising new design emerged from my modeling experiments.

Ladies and Gentlemen...

Allow me to introduce you to the *terminated loop*, a concept that further shrinks the space required for a good receiving antenna *without sacrificing performance*! Figure 1 compares the real estate requirements of my four-direction loop system to that of an equivalent EWE array. The new system is not only smaller, it's easier to install, needing only one support instead of five.

The terminated loop (see Figure 2) is physically and electrically quite simple. It consists of a wire loop of any convenient shape (diamond, delta, etc), hung from a single support and with a ground rod at the bottom. A 9:1 impedance-matching transformer connects from one end of the loop to ground; a terminating resistor connects the other end of the loop to ground. This antenna is directional, favoring signals arriving from the feed point end, rejecting by several S units any signals arriving from the end connected to the terminating resistor, R_{TERM}.

Here's the real news: A very small terminated loop maintains its directional pattern. As the loop's size gets smaller, however, the desired signal strength is also reduced. The antenna described here is small enough to fit into a corner of almost *any* backyard. The loop could be made smaller, but I wanted to collect enough signal energy so that

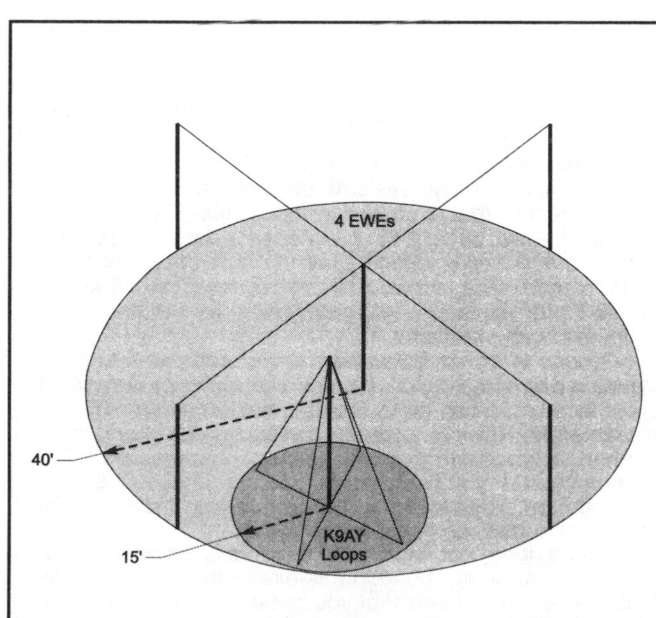

Figure 1—A comparison of the real estate needed for four EWEs and the K9AY Loop system shows that the loops need only 1/7 the area of the EWEs, yet they provide the same directional patterns.

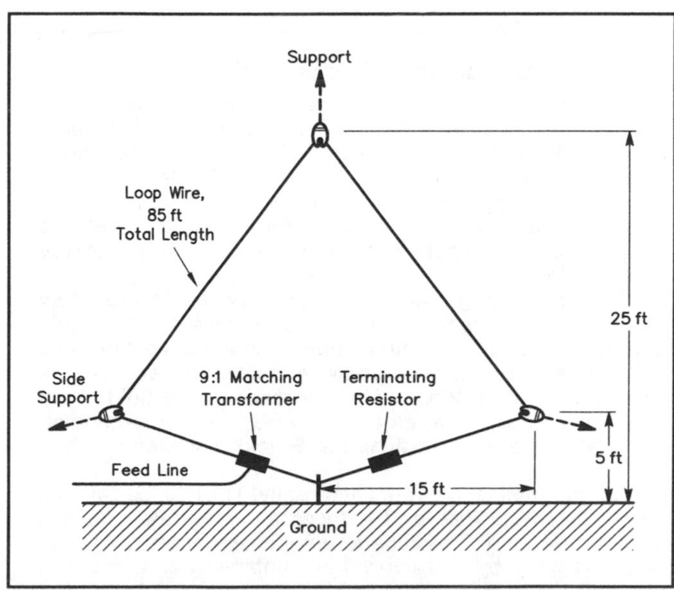

Figure 2—The basic design of a single loop in an easy-to-construct quasi-delta loop configuration. Exchanging the feed point and termination reverses the pattern. A four-direction system uses two of these loops installed at right angles to one another (as shown in Figure 1) and a relay controlled switching system.

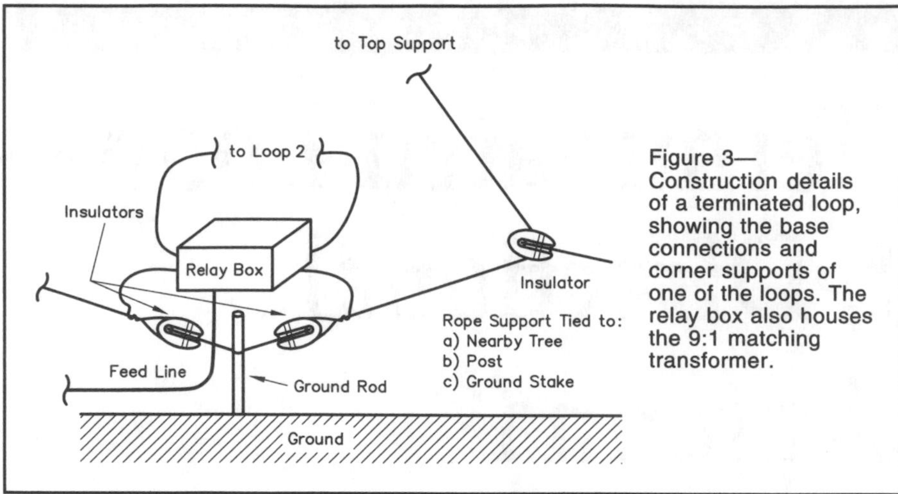

Figure 3—Construction details of a terminated loop, showing the base connections and corner supports of one of the loops. The relay box also houses the 9:1 matching transformer.

Figure 4—A photo of the central connection point at the base of a two-loop system, as installed at K9AY.

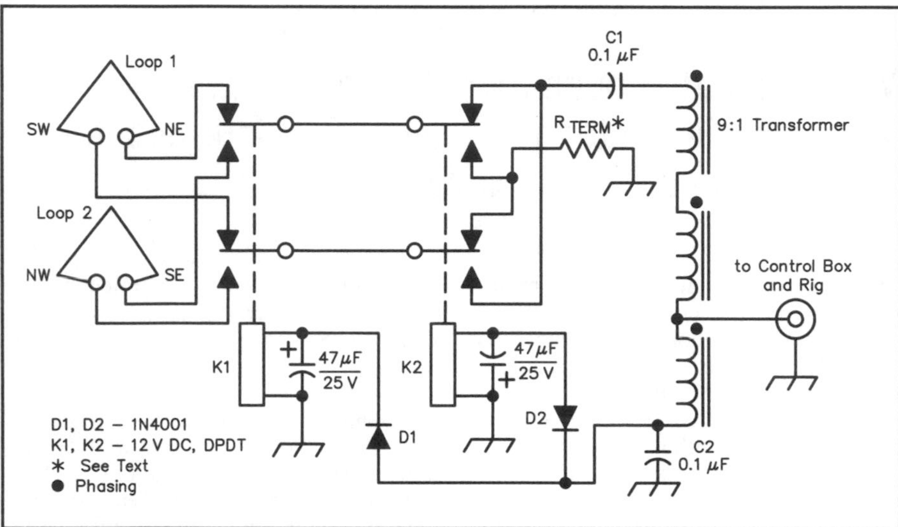

Figure 5—Schematic of the relay box located at the base of the system. The relays switch between the two loops, reversing the feed point and termination connections.

even a modest preamp (such as those included in most HF rigs) can be used. More information on the antenna's operation can be found in the sidebars "How Does the Terminated Loop Work?" and "Summary of Characteristics."

To cover all directions, two loops using the same support are oriented at right angles to each other. Each loop provides reception from two directions when the feed point and termination are reversed, for a total of four separate patterns. Here are all the details you need to build the K9AY Loop antenna system, including a relay-controlled pattern-switching system that uses the coaxial feed line to carry the switching system control voltage.

How Does the Terminated Loop Work?

The terminated loop was developed after examining the behavior of the EWE. In theory, the EWE is a *terminated half-loop.* High-frequency directional couplers such as those used in the well-known Bird Thruline wattmeters are constructed similarly, just much smaller.* My analysis determined that a *terminated full loop,* fed and terminated to a single ground point at bottom center, behaves the same way as a half loop.

How does a single loop achieve a directional pattern? By the way it responds to the electric (E) and magnetic (H) field components of the arriving electromagnetic wave. Let's say we have a signal arriving at the loop from one end. As the wave passes, the loop's wire intercepts the E-field like a short vertical antenna, creating a *voltage* at the feed point. As expected from a vertical, the E-field response is omnidirectional.

The magnetic field works differently. The H field is at right angles to the E field and induces a *current* as it passes through the loop. The voltage developed across the terminating resistor by that current is combined with the E-field voltage. If the wave arrives from the feed point end, the voltages add in phase. If the wave arrives from the opposite direction, the voltage is 180° out of phase with (and subtracted from) the E-field voltage. To maximize the front-to-back ratio, the terminating resistor must have a value that balances the voltages created by the two field components so that their sum is close to zero.

The resulting pattern is a cardioid with a single null. With an optimum terminating resistor, the null can reach 40 dB or more in depth—that's more than six S units! The null is not at ground level, but at 20 to 55° elevation, depending on the shape of the loop and local ground conditions. Unless you build a short, wide loop or a tall, skinny loop, the null will be at 30 to 40° elevation, very convenient for reducing QRM from in-country stations.

Ground is an essential part of the antenna—its resistance is part of the circuit. If ground conductivity is known at your location, it can be included in the computer modeling parameters. However, ground conditions can change over a short distance and vary with seasonal changes and moisture content. If you find that the antenna does not perform as modeled, adjustment of the terminating resistor value may be needed, as noted in the article.

Ground does not need to be lossy, as is the case with a Beverage antenna. The loop has directivity even with perfect ground. This means that you can install it over *any* type of ground, including a buried radial system (as long as it's not too close to another structure).—*Gary Breed, K9AY*

**Instruction Book*, Model 43 Wattmeter, Bird Electronic Corp, Solon, OH.

Let's Build It!

First, gather the following materials to construct the loops:

- Two lengths of **wire** about 85 feet each. Although almost any wire size will do, #14 copper is probably the best for a long-lasting installation.
- Ten simple **insulators**—anything from old toothbrush handles to fancy porcelain insulators will do.
- A three or four-foot length of **copper pipe**—the ground rod doesn't need to be driven deeply—a depth of three to four feet is sufficient. Because copper-plated steel rods eventually rust, I prefer to use ½ or ¾-inch-diameter copper water pipe.
- One **support** 25 feet above ground—my antenna hangs from a limb of a Georgia Pine, but anything that gets the top of the loops up about 25 feet will work. The wooden **A** frame support described in *The ARRL Handbook* and *Antenna Book* is an excellent choice. A metal mast can be used, but make it only as long as necessary and insulate it from ground. Large metal objects affect the antenna's performance, so install it in the clear.
- You'll also need some **rope**—nothing fancy, just something strong enough to keep things in place.

Now, grab your tools and head for the backyard. Attach the midpoint of each wire length to an insulator that will be positioned at the top of the loop. The two loops must not touch each other, so leave some room between their insulators. I separated my insulators with about a foot of rope.

Aligned directly under the tops of the loops, drive a ground rod, leaving a foot or so above ground as an attachment point. Using short ropes or wires, connect four insulators a few inches away from this post to receive the lower ends of the loops. Leave a foot of wire for a pigtail after twisting the ends of the loop wires around the insulators. Figure 3 sketches the important installation details, and the photo of Figure 4 shows how my system is installed.

The lower corners of the loops are supported by insulators and ropes tied off to nearby trees, fence posts, or stakes driven into the ground. Pull out the corners with enough tension to maintain the loop shapes. At this point, the major mechanical work is finished and you should have something resembling an eggbeater, as shown in Figure 1.

The Relay Box and Controller

Figure 5 shows the relay circuit used to switch the feed and termination points of the loop wires. House the components in a weatherproof box having external connection points for the four ends of the loops, a coax connector for the feed line and a ground-wire attachment.

K1 is a DPDT relay that switches between the two loops. K2, another DPDT relay, swaps the connections of the terminating resistor and matching transformer, thereby reversing the pattern. Relay power is supplied via the coaxial feed line. C1 keeps the control voltage from reaching the antenna, and C2 provides an RF ground for the transformer.

Most builders will likely choose to orient the loops for northeast/southwest and northwest/southeast directions. The relay box has four switching modes, one for each direction: 1) northeast, when *neither* relay is energized; 2) southeast, when K1 *only* is energized; 3) southwest, when K2 *only* is energized and 4) northwest, when *both* relays are energized. Switching is accomplished using a single power connection through steering diodes D1 and D2. D1 allows K1 to operate when +12 V is applied, while D2 blocks the voltage from reaching K2. When –12 V is applied, K2 operates, but not K1. When 12 V ac is applied, the diodes rectify it to operate *both* relays.

The matching transformer is a 9:1 impedance, 3:1 turns ratio type that should be familiar to many readers. Five trifilar turns of ordinary hookup wire are wound on a ¾-inch-diameter (0.825 inch) 43 material toroid. The terminating resistor (R_{TERM}) value will be between 390 to 560 Ω, depending on your band preference. With average ground conductivity, a value of 390 Ω provides the optimum F/B at 160 meters, while 560 Ω optimizes the loops for 80 meters. A value of 470 Ω splits the difference for "pretty good" performance on both bands. I chose to optimize the antenna for 160 meter operation, so used a 390 Ω resistor. Use an R_{TERM} power rating of at least 1 W in case some transmitter power ends up being coupled to the loops. I use a 2-W carbon resistor, but two parallel ½-W or four ¼-W units of appropriate ohmic value can also be substituted.

Figure 6 is a diagram of the control box that is located in the shack. One 12 V ac transformer provides the ac relay power and feeds two half-wave rectifiers and filter capacitors to generate ±12 V. One SP4T switch selects the proper voltage to apply to the coax. The RF choke and 0.1 µF capacitor keep the RF and control voltage separated at the shack end.

Once the control unit and relay box are built and operating properly, mount the relay box at the ground rod and connect the four ends of the loops to their proper terminals. A length of 50 Ω coax carries the signal and power between the antenna and the switch box in the shack. I highly recommend keep-

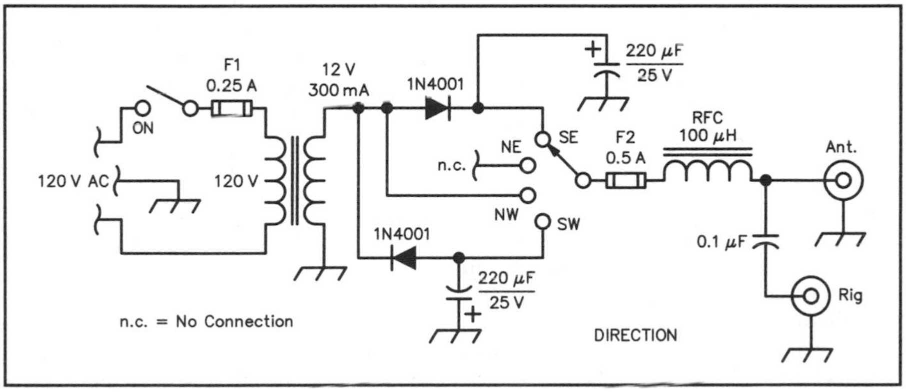

Figure 6—Schematic of the control box located in the shack. The control voltage to operate the relays is delivered via the coaxial line.

Summary of Characteristics

When choosing a specific configuration or location for your K9AY Loop system, consider this list of characteristics:

- Like all antennas, this one can be affected by nearby metallic objects or structures. Do your best to keep the antenna in the clear, away from your tower, house and power lines.
- Noise reduction in this antenna is achieved mainly by the directional pattern, although being grounded offers some reduction of wind, rain and snow static. Compared to the typical omnidirectional vertical or low inverted **V**, the reduction in overall noise and interference can be dramatic.
- The maximum circumference of the loop is a little over ¼ λ at the highest frequency of operation. If the loop is larger, the E and H-field responses of the antenna can no longer be balanced. Smaller loops (or same-size loops at lower frequencies) retain the directional pattern, which makes this an excellent antenna for AM broadcast reception. Unfortunately, the received signal voltage is proportional to the area enclosed by the loop, so sensitivity decreases rapidly as the antenna becomes smaller. Unless you have a very good preamp, keep the loop sizes near the maximum.
- The T-Loop is not just a low-frequency antenna! It can be used at high frequencies for shortwave listening, or to provide improved reception over an omnidirectional antenna such as a vertical. Just scale the size to the desired frequency using the guidelines presented in the accompanying article.—*Gary Breed, K9AY*

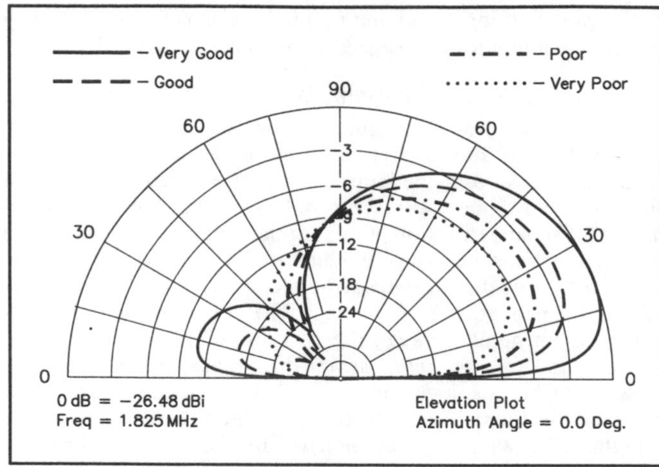

Figure 7—Vertical radiation pattern of the loops along the plane of the loop.

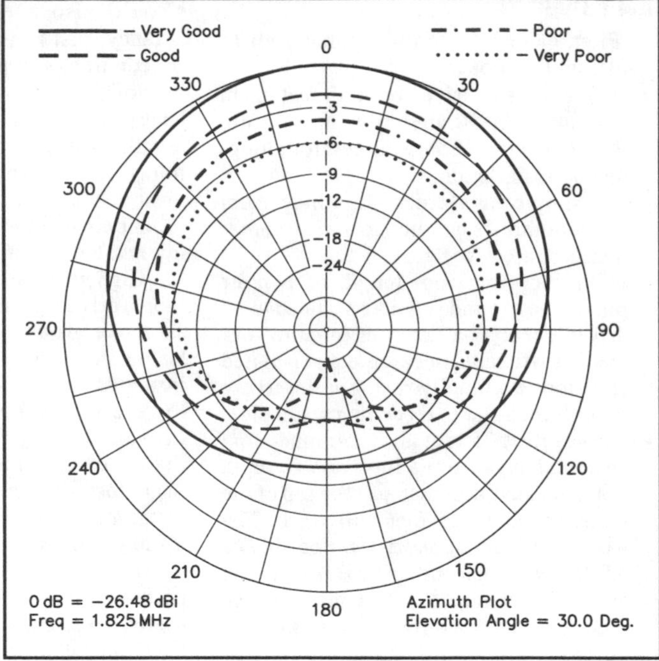

Figure 8—Horizontal radiation pattern of the loops at 30° elevation.

ing the coax on the ground (buried is better) to minimize pickup of noise or energy coupled from your transmit antenna. A receiving preamplifier is almost certainly needed, either your rig's internal preamp or an external preamp.

Evaluation and Adjustment

Next, verify that the antenna is working correctly. Some listening is probably what you'll do first, but fading makes it almost impossible to determine the antenna's actual performance. At best, you will be able to confirm that the antenna has reasonable directivity.

More accurate listening tests can be performed several ways. The best way is to enlist the aid of a nearby ham whose station is very close to being in line with one of the two loops. If such help is not available, the next-best option is to identify a local AM radio station high in the band (1400-1600 kHz) and use that as your test signal. Switch the loops to their various directions. If the test station is directly in line with one loop, you should see a front-to-back (F/B) ratio of about 2 to 3 S units as the antenna is switched toward and away from the station. You won't see a *huge* F/B because the deepest null is up at 30 to 40°.

Local ground conductivity can affect performance. You might not get an optimum pattern at your particular location with the "normal-value" terminating resistor. If you aren't getting the expected performance, substitute a 1 kΩ potentiometer for the terminating resistor and adjust it for best F/B while listening to your test station. Then, measure the pot's resistance and install a fixed-value resistor of the same value.

Antenna Performance

The vertical pattern of the antenna, in line with the loop, is shown in Figure 7. Figure 8 is the azimuth pattern at 30° elevation. Modeling was done using W7EL's *EZNEC* program,[3] which uses *NEC-2* to evaluate the antenna over "real" ground.[4] Listening tests confirm that these modeled patterns are close to the as-built antenna performance. Much of the on-air pattern evaluation was done by listening to AM broadcast stations and to WWV on 2.5 MHz, because the pattern of the antenna changes little over this frequency range. Stations at exactly the right distance and directions have verified the deep null off the back. For example, New York City area radio stations are exactly in line with my northeast/southwest loop, and show 40 dB F/B ratio when the arrival angle is just right!

The front of the pattern is quite broad, with maybe one S unit F/S. The advantage of this antenna is its rearward null, which reduces local noise and distant QRM. I installed my system just 2½ weeks before the 1996 ARRL 160 Meter Contest, so it got a thorough evaluation in a short time. During the contest, changing the pattern direction often made the difference between Q5 copy and a busted contact. For example, pointing southeast to hear Caribbean stations reduced stateside QRM by 2 or 3 S units, enough to easily hear the DX through unruly pileups.

Higher-Performance Ideas

The compact size of the K9AY Loop makes it easy to build more than one for a multielement array. Installing two of them in a broadside/endfire combination is one option, but a really ambitious approach would be to install a four-square! This array would have a substantial F/B over a wide angle, along with a much narrower front lobe than a single loop. The space required for this high-performance array is far less than what is needed for Beverages. If you try this, remember to switch the individual loops to the desired direction as you switch the feed system.

Summary

If you want improved reception on the low bands and don't have a lot of room, the terminated loop is an excellent choice. It is small, easy to build and its directional pattern makes DX much easier to hear.

Notes

[1]Floyd Koontz, WA2WVL, "Is This Ewe for You?" *QST*, Feb 1995, pp 31-33.

[2]Floyd Koontz, WA2WVL, "More EWEs for You," *QST*, Jan 1996, pp 32-34

[3]*EZNEC* 1.0 by Roy Lewallen, W7EL, PO Box 6658, Beaverton, OR 97007; tel: 503-646-2885, fax: 503-671-9046.

[4]The modeling accuracy of antennas with a direct connection to ground is somewhat uncertain, even with the power of *NEC-2*. The high accuracy ground characterization of the Sommerfield method gives answers that vary with the number of segments chosen for each wire. The *MININEC*-type ground model gives more consistent results, but on-air tests show that the deepest null is obtained with a lower value terminating resistor than this model indicates. The plots presented here are the "best fit" between modeled and observed performance.

Photos by the author.

From *QST*, May 1998 (Technical Correspondence)

Hum Problems When Switching the K9AY Loops

By Gary Breed, K9AY

I am pleased to report that many hams have successfully built the receiving antenna described in the September 1997 issue of *QST*.[1] I had some concern that variations in local ground conditions and nearby structures could reduce the antenna's performance, but the performance I obtained seems to be readily duplicated.

[1]Gary Breed, K9AY, "The K9AY Terminated Loop—A Compact, Directional Receiving Antenna," *QST*, Sep 1997, pp 43-46.

There is one problem that needs to be addressed. When the loops are switched to the northwest direction, an ac voltage is sent down the coax to the relay box. A few hams have reported hum or distortion when the antenna is switched in this direction. Two explanations are possible: a ground loop due to widely separated antenna and station grounds, or modulation of the core of the matching transformer. The presence of hum was not evident in my prototype, but to avoid either cause of the problem, I recommend using a *separate* three conductor control wire to operate the relays. The power is solely 12 V dc, and no current flows through the transformer. The modifications are shown in the accompanying figures, which can replace Figures 5 and 6 in the original article.

Secondary benefits of this arrangement include the option of using an existing 12 V power supply to provide the operating voltage, and the ability to use an additional conductor to carry power to an antenna-mounted preamplifier.

Figure 1—A revision of the schematic shown in Figure 5 of the September 1997 *QST* article.

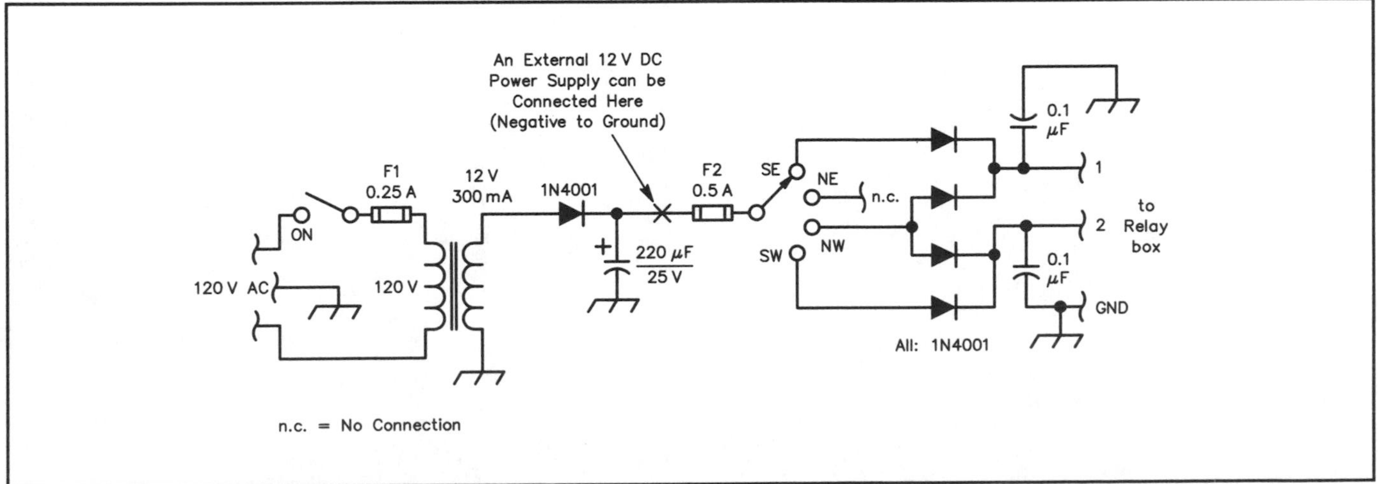

Figure 2—Changes to Figure 6 of the original article.

CHAPTER NINE

MISCELLANEA

A Skyhook for the '90s

Is your group looking to put up a BIG vertical antenna for Field Day? The crew at N4ZC started their own air corps. Their full-size 160-meter quarter-wavelength vertical is *not* supported by hot air!

Some History, as Introduction

By 1936, the Zeppelin company was without rival in the design, construction and operation of rigid airships. Flying successfully around the world, providing regular passenger service between Germany and South America, the *Graf Zeppelin* airship was nonetheless unsuited for the hardship of crossing the North Atlantic. The *Hindenburg* was designed and built for this route. This huge airship was one of the largest flying machines the world has ever known. Her crash and destruction at Lakehurst on May 6, 1937, ended the era of passenger transport by airship.[1]

Later, growing up on a farm near Akron, I sometimes saw the Goodyear airships passing silently overhead. Twenty-odd

PHOTO BY DON DASO, K4ZA

Inspiration, on a grand scale. What if we had one of *these* to play with?

PHOTO BY SCOTT DOUGLASS, K2SD

Figure 1—K4ZA fastens the antenna and tether to the miniature blimp prior to launch.

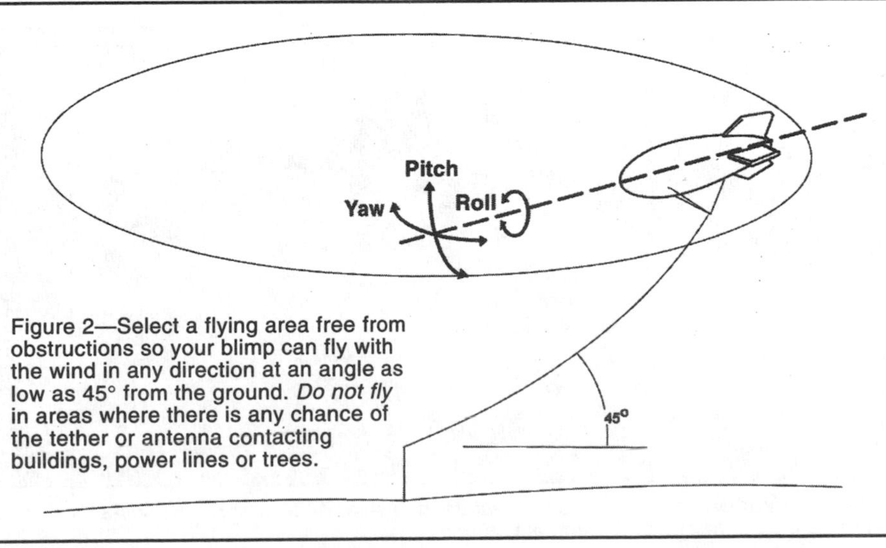

Figure 2—Select a flying area free from obstructions so your blimp can fly with the wind in any direction at an angle as low as 45° from the ground. *Do not fly* in areas where there is any chance of the tether or antenna contacting buildings, power lines or trees.

[1]The terms concerning lighter-than-air (LTA) craft are often misapplied. *Webster's Ninth Collegiate Dictionary* indicates that *dirigible*, *airship* and *blimp* all denote an LTA with *steering and propulsion* capabilities. Hence, the terms do not strictly apply to the kytoon, which has neither steering nor propusion systems. Nonetheless, *blimp* is so commonly used by manufacturers, sellers and users to describe the kytoon that we continue that usage here, as well. —*Ed.*

PHOTO BY DON DASO, K4ZA

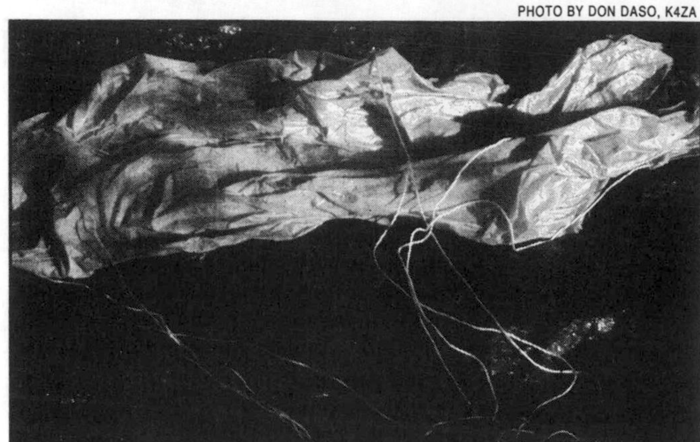

Figure 3—Here's our miniature version of the real thing—fresh out of the box and ready for inflation.

years after the *Hindenburg* disaster, I began to experiment with balloons myself—originally using surplus weather balloons to support wires, and more recently, with small blimps holding contest antennas. Many hams have considered using balloons or kites for antenna supports. Indeed, they come closer than anything else to that "skyhook" we all jokingly refer to on occasion. Yet, each one has limitations.

The Kytoon

Balloon-supported antennas tend to lose altitude in the wind, however slight. Kites cannot fly without wind. Satisfactory service from either support requires a limited range of wind and weather conditions—something that's not too common. Combining features from each—the balloon and kite—solves several of these limitations. Indeed, the Kytoon is a small, inflatable blimp-shaped balloon with kite-like surfaces that cause it to fly into the wind. An early Kytoon—with a rigid skeleton—is registered with the US Patent Office and enjoyed some success as a skyhook. (See Bibliography for further information.)

Today, the early Kytoon's bladder-and-frame construction has been replaced by completely inflatable balloons, made from 2 to 3-mil polyurethane film (see Figure 1). Brilliant skin colors are common because these miniature blimps are intended for advertising purposes. These small blimps are available from various sources, with prices varying according to size—typically from 10 to 30 feet long and starting at $300. [Beware: Many suppliers specialize in advertising Kytoons and sell them for twice the price of those sold for scientific uses.—*Ed.*]

The plastics used in their construction have been specially formulated for high elasticity, helium retention and paint application. Helium (a safe, inert gas) is available from party suppliers or welding supply companies. One tank will easily support an antenna for the typical contest weekend. If you have some room, but maybe not the resources to put up a full-sized radiator, consider a tethered Kytoon. As a modern antenna support, these miniature blimps are the skyhook of the '90s.

Tethered Flight

Most hams considering these small blimps as antenna supports concentrate on their antennas. This natural inclination can create problems. Most hams have little knowledge or experience with the basic principles of LTA flight, knots and rigging rules, with FAA regulations, with weather and especially with how these all relate to tethered flight. It's a complex process and serious business to lift an antenna aloft with a small balloon or blimp.

These blimps consist of a closed envelope of gas—usually helium—that is lighter than the atmosphere. Gas balloons can only descend by losing lift (that is, by venting gas) or when some force overcomes their lift (retrieval by tether or powered flight). As a gas balloon rises, it will reach an equilibrium altitude, where it will remain until the lift-to-weight ratio changes. These principles—which are mostly associated with larger, cargo-carrying blimps—are important, especially because our balloon will be tethered. As with any tower or antenna project, use common sense and make safety your highest priority!

Because space is three-dimensional, any "skyhook" can turn in the air—rotating in three axes: a longitudinal roll, a lateral yaw and a vertical pitch. The shape of these blimps allows them to "fly" in the wind, countering these motions, and the tether allows the blimp to move easily enough within those three axes. Generally, you will experience something like Figure 2.

A 10 to 12-foot blimp is sufficient for most antennas suitable for low-band use. Most blimp applications will be for 160 meters, because the sheer size of effective antennas for this band makes them difficult (and costly) to erect. A 10 to 12-foot-long blimp of 3 to 4 feet in diameter holds about 70 cubic feet of helium and is capable of lifting two pounds. This is sufficient for an effective $\lambda/4$ wire vertical for 160 meters.

Setup, Antenna and Ground

You'll need a working area on the ground to inflate your blimp. A 20×20-foot area should suffice. It's important to protect the thin skin of the balloon. I usually lay the balloon out on a tarp to prevent any stubs or twigs from poking the plastic skin (see Figure 3). While the plastic seems sturdy and elastic, I'm always careful and treat the balloon gently. I hold the helium tank upright, and I usually lash it securely to a two-wheeled dolly, so I can move it to the field easily. Follow the manufacturer's inflation directions carefully (each will be slightly different).

Inflation temperature is important: If possible, inflate the balloon at the same temperature it will experience in flight. Helium comes out of the tank cold. With rising temperatures, it expands quickly inside the balloon. Sunlight and warm air cause further expansion. (In very hot weather, a balloon loses some of its lifting power, but this is usually not a problem.) Do not overinflate the balloon. Finger pressure will barely dent the surface of a properly inflated polyurethane blimp. Pay special attention while closing the neck of the blimp. With many blimps, you can simply fold the neck back on itself several times and then fasten it in place with heavy rub-

PHOTO BY SCOTT DOUGLASS, K2SD

Figure 4—Vertical antennas are real-estate intensive. K4ZA and KF4HK install radials in the horse pasture at N4ZC's contest station.

PHOTO BY SCOTT DOUGLASS, K2SD

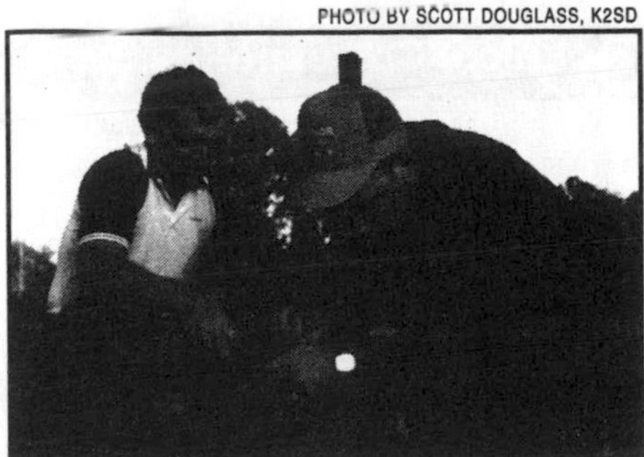

Figure 5—WA4UNZ and K4ZA work inside the large copper ground ring installing the feed system.

ber bands (folding them, if necessary) to provide a tight, leak-free fit. Refills are much easier when rubber bands, rather than knotted rope, seal the balloon.

Use a quality ball-bearing swivel between the end of the tether and the balloon's attachment line. This allows the blimp to easily turn into the wind. Fishing line works well as a tether; so do some kite lines and other synthetic ropes. I've used 100-pound fishing line and 1/8-inch nylon rope.

Stranded wire is a lightweight antenna conductor. Aluminum welding wire is light and strong enough to use. (I use phosphor-bronze wire that was originally intended for use with the "Gibson Girl" rescue unit's kite- or balloon-supported antenna. It comes on a small reel. The irony, weight and low cost appeal to me.) To save time, measure and mark the antenna length before construction begins. I have found no instances where the addition of a tether created a problem, in launch, retrieval or during flight. The antenna has never broken; and our balloon has never broken free. Obviously, it's simple insurance, and you don't want your antenna wire touching power lines or other hazards, which might happen with a runaway blimp.

The original installations at N4ZC were all within a working pasture (see Figure 4). On-ground radials were easier to install than elevated radials, albeit more labor-intensive (easier because they need no supports, labor-intensive because it requires more radials—4 versus 60). A large diameter ring of copper water pipe served as the radial connection point and made soldering easier. The ring also created a work space to stand in during set up and launch (see Figure 5). (Future installations will use a high, central wooden post, with attachment points for the antenna, tether and four elevated radials.) Our installation is over 150 feet from the nearest tower (the 20-meter array). RG-213 runs from the feed point to the base of that tower, where it connects to Hardline running to the shack.

Hints

Always wear gloves when launching and retrieving the blimp. Let the blimp rise slowly; retrieve it slowly. The lift can be terrific, and the antenna wire and tether can easily cut your hands. Always have helpers coil up the antenna and tether on spools as you walk the balloon down. (I recommend reeling in the blimp during daylight hours. It's a tempting target, and we hear gunshots in the woods all the time at N4ZC's somewhat country location. Make certain winds cannot push the blimp into nearby objects, even when it's near the ground.)

FAA regulations for flying tethered balloons—those under five pounds in weight—are clear: No person may operate a kite or balloon in a manner that creates a hazard to persons, property or other aircraft.

Avoid all dangerous situations: operation near airports, in wet or stormy weather, near electric power lines, over public streets or areas congested with people or at extreme heights (which is probably the first thing a ham will think of). Use common sense and play it safe. Using less than the legal limit with a single 1/4 vertical supported by a 10-foot blimp, I've been able to work any station heard on 160 meters.

Sources

Here are names and contact information for two suppliers of these inflatable blimps:

Toy-Tex Novelty Company, 7315 N Linder, Skokie IL 60077.

The Blimp Works, 156 Barnes Airship Dr, Statesville NC 28677.

Annotated Bibliography

There's a wide variety of applicable information and material available, but not much has appeared in Amateur Radio publications. This is new territory, in many ways, that's worthy of further experimentation. I recommend the following books and articles for further reference and study—before buying, building and flying:

R. Carleton Greene, W8PWU, "More On Balloon Supported Antennas," *QST*, Nov 1940, pp 38, 39 and 82. One of the earliest articles on using meteorological balloons as antenna supports, complete with lifting charts.

David T. Ferrier, W1LLX and William G. Baird, W9RCQ, "A New Kind of Skyhook," *QST*, Oct 1946, pp 24-25. The original use, as best I can determine, of the Kytoon as an antenna support. Although focused on emergency or Field Day use, the final line of the article presents this challenge: "Perhaps long-wire vertical antennas can now be exploited with outstanding results."

Stan Gibilisco, W1GV, "Balloons as Antenna Supports," *The ARRL Antenna Compendium*, Volume 2, (Newington: ARRL, 1989). A brief overview of the topic, with good descriptions and several safety tips.

Maxwell Eden, *Kiteworks*, (New York: Sterling Publishing Company, 1989). An excellent book about kites of all kinds—including building and flying them. It's filled with tips and techniques.

Will Hayes, *The Complete Ballooning Book*, (Mountain View, California: World Publications, 1977). Although focused on large hot-air balloons, the history, basics of flight and safety chapters are well worth your time. It also includes the FAA regulations.

John Belrose, VE2CV, "A Kite-Supported 160- (or 80-) Meter Antenna," *QST*, Mar 1981, pp 40-42. A unique application of the Parafoil—a special type of kite. Interesting tips and techniques applicable to blimp-supported antennas.

"160-Contest Results," *QST*, Jun 1976, pp 71-74. Pictures and text from W8LT, the Ohio State University Amateur Radio Club, using an original Kytoon to support a vertical antenna.

Balloon Skyhooks

By Jack M. Hughes, WB6SOI

I appreciated and enjoyed "A Skyhook for the '90s."[1] The company I work for, TCOM, L.P., has been doing this sort of thing commercially since 1981, but on a much larger scale. Since 1972, TCOM, L.P. has been an authority on and builder of aerostats.[2] We manufacture and market an aerostat system called the Tethered Aerostat Antenna Platform (TAAP).

VLF and LF transmitters and receivers are used by the military and governments worldwide for high-reliability communication of strategic information. A typical antenna system for VLF/LF communication is a complex array of wires suspended on insulated towers that take months to erect and place into operation. A unique scheme for the rapid deployment of a complete VLF/LF (up to 100 kW) communications system uses the TAAP.

An aerostat is employed to elevate an encapsulated vertical antenna/tether to a height approaching one quarter wavelength for the frequency in use. At VLF, this technique enables the system to operate with a high antenna-radiation efficiency unobtainable by other land or sea-based antennas.

The radiated-power efficiency is further enhanced by the installation of a simple groundplane consisting of eight radial wires emanating from beneath the mooring-system trailer. The radial ends are secured by ground rods and copper wire. The entire system—antenna/tether, aerostat, mooring system, transmitter and equipment shelter—is designed to be transportable by land, air or sea and to be rapidly erected at a selected site.

The aerostat is a helium-filled, aerodynamically stable tethered balloon typically 32 meters long. Its antenna/tether is composed of Kevlar with optional fiber optics and two concentric outer sheaths of aluminum braid. The Kevlar construction provides up to five times the strength of a similar-weight steel cable. A fiber-optic core provides for secure communication between the ground and the aerostat, and the double-aluminum sheaths act as the radiating elements—as well as a path for the 400 Hz electrical power to the aerostat.

The aerostat is launched and retrieved at a rate of 200 feet per minute and it does not vent helium. In addition to the helium chamber, it uses an air-filled ballonet system that expands and contracts depending on the outside air pressure to keep the aerostat rigid at all times. It can operate in winds of up to 50 knots and will survive in winds of greater than 70 knots. The system is designed to fly continuously for up to 21 days before replenishing it with helium.

I commend author Don Daso for his creativity and resourcefulness. He seems to have just as much of a thrill as we do in handling LTA vehicles.

[1] Don Daso, K4ZA, "A Skyhook for the '90s," *QST*, May 1997, pp 31-33.

[2] Aerostats are essentially helium-filled, aerodynamically shaped balloons that are tethered to ground. They're not cheap—a 15-meter-long Aerostat system costs about $175,000.

The Clothesline Antenna

Dry your laundry or work DX. Could this be the first dual-purpose antenna?

Every once in a while, you run across an idea that seems so simple, so obvious, you can't believe it hasn't been done before. Surely (you say to yourself), you're not the first person in the universe to have thought of this...

That's the case with the Clothesline antenna and me. While mulling over a variety of ideas for an antenna suited to my apartment, I started drawing some graphs of sine waves at various frequencies. One thing led to another and I wound up with a terrific solution for my antenna needs—and it was one that I haven't found any references to anywhere. As far as I'm concerned, *I've* invented the Clothesline. Still, I'm not going to be surprised if someone shows me that it's been done before!

But even if this antenna design has been around since hydrogen, it may still be new to you. And even if you have seen something similar to the Clothesline, this design may be worth a second look. For hams in a variety of situations, this could be just what the doctor ordered to cure your DX dilemma.

What It Is ... And Isn't

I will tell you what the Clothesline is, but first I'll tell you what it is not. The Clothesline isn't some dubious trick for loading up an actual clothesline with 12 cubits of RG-213 looped four turns to the foot around your washing machine. This is an antenna, it works all the HF bands, and it gets great results. It doesn't need a tuner because it's dead-on resonant on the 160, 80, 40, 20, 15 and 10 meter bands. A tiny tweak will bring in 12 and 17 meters, too.

It's remarkably easy to build. There are no traps, no stubs, no loading coils, no variable or fixed capacitors, no screws, no clamps, and you don't have to drill anything. This antenna is so simple to put up, it hurts. The Clothesline consists of little more than a piece of wire, a center insulator, some feed line, and a couple of $2 hardware-store fittings.

Too good to be true, you say? What's the catch, you ask?

Well, you *do* have to adjust this antenna for most band changes. But before you get too excited by the word "adjust," I'm not talking about unwrapping yards of black tape to get at a loading coil, or tweaking the bare shaft (whoops, hot side—sorry!) of a 50-year-old fleamarket variable capacitor. No, the Clothesline makes band changes a piece of cake. When you see how easy it is, you're going to chuckle!

The Concept

The easiest way to explain how the Clothesline works is to lead you through the simple reasoning I used to come up with the thing. I started with a drawing like the one in Figure 1. It shows a simple dipole for 80 meters. It's 132 feet long and fed in the middle with coax. Next I drew the graph of the voltage distribution along the antenna (at its fundamental resonant frequency). Note the voltage is at a maximum (with reference to ground) at the antenna ends. That's important for any dipole radiator. The ends of the antenna have to correspond to the high voltage points of the curve.

I already knew that the curve the voltage wave describes is a quarter of a full cycle—a quarter wavelength—on each half of the antenna. And that the distance between the two places where the voltage is highest, or in other words, from one end of the antenna to the other, is $^1/_4$ plus $^1/_4$—a half wavelength. That's why they call it a half-wave dipole.

I also understood why the feed point is in the middle. It's located precisely where the voltage curve is lowest with respect to ground. To me, an antenna is just like a transformer—it transforms the low impedance coming out of the coax up to the impedance needed to couple to the cosmos. Come in low, go out high. So much for the plain dipole.

THAT LAST PAIR OF LONG JOHNS YOU REELED OUT MOVED MY FEEDPOINT

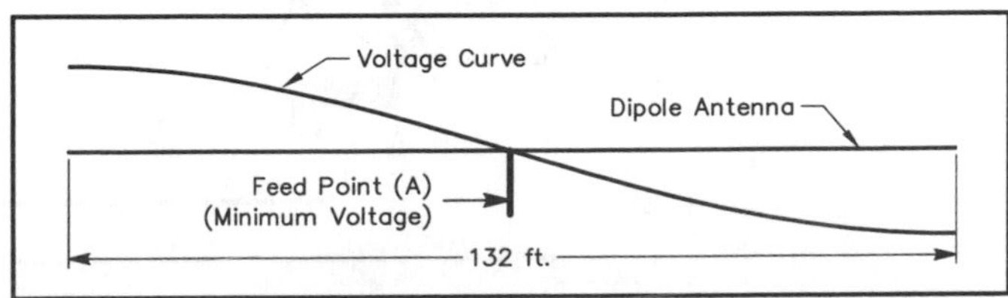

Figure 1—The voltage distribution along an 80-meter dipole antenna. Note that the voltage is at maximum (with respect to ground) at the ends of the antenna.

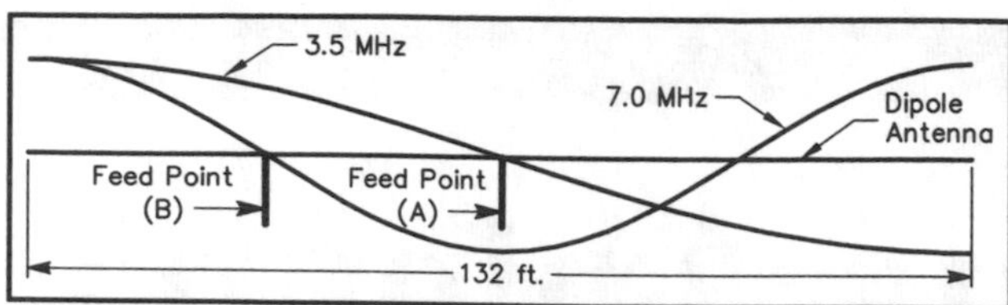

Figure 2—Take the 80-meter dipole shown in Figure 6 and overlay the voltage distribution for a 7-MHz signal. Note how the 7-MHz feed point (B) has shifted to the left of the 3.5-MHz feed point (A).

I then dropped a 7-MHz voltage curve over the original 3.5 MHz length (see Figure 2). Because the new frequency is a whole multiple of our original frequency (3.5 MHz × 2 = 7 MHz), there is still a whole number of half wavelengths from one end to the other, so the voltage peaks are still at the ends. This means this same length of wire will resonate at 7 MHz, which means it will radiate perfectly well at this frequency too.

But I noticed right away that something had changed from our earlier situation—the point where the voltage minimum crossed ground was no longer in the middle. There were now two such points, neither of them anywhere near the middle. So the antenna was resonant—the voltage peaks were at the ends—but my original feed point was now useless! Alright, I reasoned. If I shift the feed line to the new minimum I'll have the perfect setup—high voltage at the ends, low voltage at the feed point. Look out DX, here I come!

I thought about how far I'd moved the feed point. Looking at the curve, you can see that from (A) to (B) is a quarter wavelength at the new frequency, right? We doubled the frequency, which means we cut the old wavelength in half. So what used to be a quarter wave at the old frequency—from the middle to one end—will be twice that, or one-half wavelength, at the new frequency. And I wanted to get the feed point over to the new location at point (B), so I move it half that half—a quarter wavelength—over to the new minimum at (B).

I also realized that I'd added a quarter wave to the right half of the antenna, making the right-hand side 1/2 plus 1/4 wavelengths, or a total of 3/4 wavelengths long. Since 3/4 on the right plus 1/4 on the left equals 1, I now had a full-wavelength antenna with some gain compared to a dipole, fed one-quarter wavelength from one end.

I kept going. I doubled the frequency again, and the same thing happened. The voltage peaks would stay at the ends, but I'd see the feed point moving further to the left by one-quarter wavelength at the new frequency for each time I doubled frequency.

I tried it for other multiples—by 3, 5, 6, 7 and so on. Just as I expected, I always wound up with some whole number of half waves, and voltage peaks at the ends. All I ever had to do was to move the feed point to the new quarter-wave point of the antenna. And for that matter, for any band other than 80, I had a choice of feed points. Each higher-band antenna had more than one voltage minimum point in its length, and I could feed at any one!

Now I drew a bunch of these curves like the ones shown in Figure 2 for 80, 40, 20, 15 and 10 meters. I dropped reference lines down from each voltage minimum to show where the feed points would lie along this single piece of wire, depending on the frequency. Because there are multiple feed points for each of the higher bands, I could see that any time I want to change frequency, I would just shift to the closest feed point for the next frequency, and away we go!

But so what? How was I going to make a simple antenna with all these feed points? For that matter, what was simple about an antenna with all these feed points anyway? The answer was, "nothing." What about a *single* feed point that somehow travels or slides along the antenna to anywhere I want it to go? Could I make that happen? I was looking out the window at my 40-meter folded dipole when I came up with the answer.

Other than the fact that it's fed with 300-Ω line rather than coax, the folded dipole is almost identical to the standard dipole. It resonates over the same lengths at the same frequencies, so my marvelous (but yet to be invented) multi-band, sliding feed-point antenna would be just as comfortable with a folded dipole 132 feet long as it would be with a plain dipole. But what about the sliding feed point?

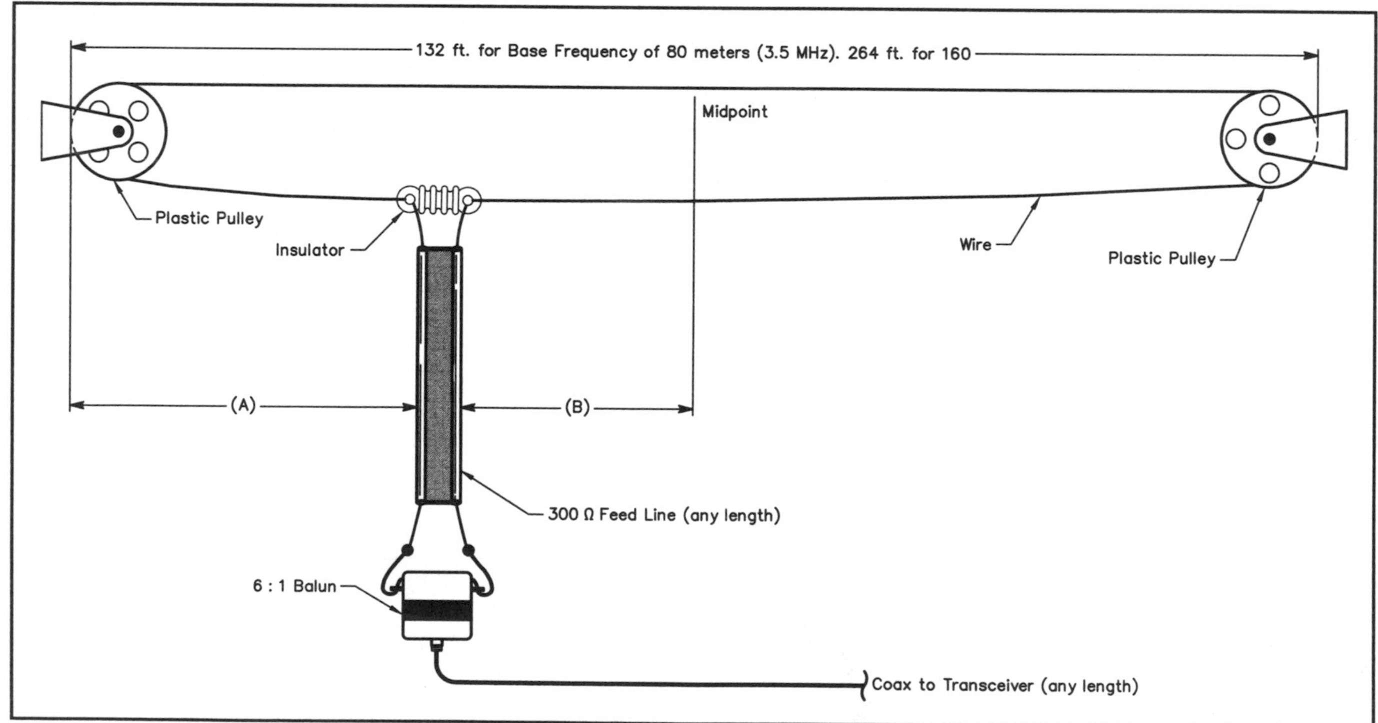

Figure 3—The Clothesline antenna. Just set (A) or (B) to equal a quarter wavelength at your operating frequency. As long as it is a multiple of your base frequency (most HF ham bands are), your antenna is tuned. You can measure out your favorite bands, peak them by observing the SWR, and put color marks or tags on the bottom wire. Roll up to the mark and you're ready to couple to the cosmos!

What would happen with my folded dipole if I just grabbed the feed line and tried to slide it along toward one end? It hit me: If the endpoints were run over pulleys the feed point would effectively slide! It would just run along the bottom of the antenna to anywhere you cared to put it. As far as the electrical nature of the antenna is concerned, absolutely nothing has changed—except the position of the feed point! And there it was—the birth of the multi-band, sliding feed-point antenna, otherwise known as the Clothesline! See Figure 3.

But wait a minute, I thought. It's a clothesline. It can't possibly work. It's too simple!

Yet no amount of poking about on paper could find fault with it. Still very skeptical, I strung up a miniature version (for 10 and 20 meters) running down my hallway. It worked! It loaded like a breeze, with the feed about 8 feet from the end on 10 meters and about 16 feet from the end on 20 meters, just where those graphs said it should. I bought some decent wire, a couple of plastic pulleys, 75 feet of 300-Ω twin lead, and threw one up on the roof to cover 40 through 10 (66 feet is all the horizontal space I have up there). Within a couple of days I had racked up dozens of DX contacts on every band. Most remarkable was the apparent gain on the higher bands. I consistently surprised other stations (and myself) with my signal strength, usually trading equal signal reports and getting lush praise from guys running far greater power and fancier antennas. My 100 W and a Clothesline are reaching the world!

How to Build and Use the Clothesline

Measure out enough wire for twice the lowest frequency. If this happens to be 80 meters, you'll need 2 × 132 feet (plus a bit for trimming). String it up like a clothesline, using nylon or rope leaders to attach the plastic pulleys to their supports. Just make sure that all the wire is used, so that the antenna is full-length. You can use long leaders if you want to hang one end off a distant support. Run each end of the wire through the pulleys top-to-bottom, attach the two ends to a center insulator and hook up your feed line. That's it!

If you're building a version for 80 or 160 meters, you'll want to install a third little pulley to ride along the top run, attached by a short piece of cord to the center insulator, to help keep the top and bottom runs of the antenna roughly parallel. Build it for 40 as the lowest frequency and you won't even need this little extra.

There is a full set of feed points between the middle of the antenna and half-way out to either end, so if you want to keep the feed line short, you can cut it to provide just enough slack to move within this range. Alternatively, because different feed points for the same frequency can affect the gain pattern, you may want the flexibility of having a choice of feed points for each frequency, and decide to leave more line slack. Losses in twin lead are so much lower than in regular coax, you can use all you want (within reason). Do be careful to stand it off from any metal it encounters on the way into the shack.

Inside, the simplest way to match the twin lead to your rig is with a balun. I use a 6:1 balun to get the impedance down to the 50 Ω my radio likes. A 4:1 balun works fine too, though the 6:1 usually presents a better match. Both baluns are sold by a number of *QST* advertisers. Ditto for the 300-Ω twin lead.

Speaking of the shack, if yours is on an upper level at your home, consider setting the Clothesline up with one end near a shack window. You'll be able to reach out, retune (that is, haul in the line), then go back to your rig to check your SWR meter.

If you don't have this luxury, don't sweat it. I calibrated my Clothesline by setting my rig on tune-up power for each band, and going up to the roof of my apartment building to peak the antenna with a field-strength meter. I confirmed each setting by checking the SWR back in the shack. On every band I tuned for, it was well below 1.5:1. As I found each feed-point setting, I marked them with various colored indelible markers for each band. Now when I want to change bands, I just go up to the apartment roof, run the Clothesline out to the right color, and I'm tuned! It takes seconds.

I can imagine other settings where you'd run the Clothesline right outside a window, out to some convenient support. Or vertical, up or down the side of a building, a tree, or a flagpole (tie a flag to it, if you like). If you're camped out in suburbia with difficult, anti-antenna neighbors, a detachable feed line using alligator clips makes it the perfect disguised antenna. Who's going to suspect a clothesline? You could even use it during the day to... wait for it... *dry clothes!* You could even make it out of your standard, garden-variety hardware-store clothesline kit—plastic pulleys, vinyl-covered wire, the works. This makes the Clothesline the only ham antenna kit I know of that you can buy complete (minus the feed line) from your local hardware store. (In a perfect world, your local hardware store would stock twin lead.) Yes, I know I said at the beginning that this wasn't a trick for loading up a real clothesline and I've been true to my word. This is an antenna that happens to *look* like a clothesline.

I mentioned covering 17 and 12 meters. If you cut the original length to resonate just a little up from the bottom of 80, at 3.615 MHz, the fifth harmonic is smack dab on 17 meters. Cut it to 3.55 MHz and harmonic number seven is on 12 meters. In fact, these are such minor variations that you can surely find a convenient center frequency on 80 that'll put you right where you want to be on all bands. Though you don't need it, a tuner (either in or out-board) can get you right down to zero reflected power. Remember, feeding with twin lead or other balanced lines keeps your losses way down, so a little elevated SWR at the band edges is no big deal.

I'm now thinking about some kind of motor drive that'll permit me to tune up from inside the shack. My first reaction to this idea was, if I'm going to get into motor drives and the like, is that any simpler than a beam? I now think the answer is "yes." The drive would be simpler, in that I wouldn't need any position feedback—all I need to do is watch my reflected power meter to know when I've hit resonance. It doesn't need a whole lot of travel because a full range of feed points is available over just a quarter-length of the antenna. As well, few beams are all-band. And since the antenna you can put up is always simpler than the antenna you can't, for those of us in apartment settings where a beam or other, more complex antenna systems are out of the question, a motorized Clothesline might well be worth the effort. I've got a sneaking suspicion that, somewhere out there, there's a cheap, off-the-shelf drive unit with "Clothesline" written all over it!

A Variable-Frequency Antenna

One to Ten Meters with a Single Antenna System

Multiband transmitters and band-switching receivers make it easy for us to jump from band to band, but the antenna has not kept pace. Many of us are prevented from operating on several bands by the thoughts of the multiplicity of antennas demanded by the conventional approach to multiband work. Here, to go with the VFO, is the VFA — tunable to resonance from the operating position. It covers 11, 10, 6 and 2, and all the television and f.m. bands in between, with the optimum performance all along the line. Ideas for the lower frequencies are included, too.

It has been the writer's lifelong ambition as a radio amateur to have a universal antenna; one that would not only work on several bands, but also tune within a band, providing optimum operating conditions on any frequency. The memory of endless trips to the rooftop or out to the mast to lower the antenna and cut off or splice on a few inches of wire to hit a special spot in the band is still fresh in mind. The old Zepp was pretty good but it required spaced feeders and tuning at the transmitter end. The half-wave aerial split in the center for a 72-ohm transmission line required no tuning at the station end but necessitated a different antenna for every band. Often the resonant frequency of these antennas varied widely from the values indicated by the formulae because of conditions not always apparent, and under some circumstances it was difficult to get adequate loading over an entire band. As the years passed matching systems were introduced, and they, too, are usually one-band devices. With all their disadvantages the antenna systems were not too bad back in the days of crystal control, but now we have VFO and often operate anywhere within the band. It goes without saying that what we need to go with VFO is a good VFA!

The group, other than amateur, most in need of a VFA, is that vast population trying to receive the various television and f.m. channels on a single antenna. A good half-wave dipole will outperform most of the existing elaborate receiving antennas providing, of course, it is possible to adjust the antenna accurately for each channel.

For years I have been giving thought to ways and means of feeding out and retrieving wire to make an adjustable center-fed half-wave antenna. The increased use of the VFO and the advent of television and f.m. broadcasting have made such an antenna practically a necessity. The folded dipole makes it a possibility, with rigging less complicated than the dial drives on some broadcast receivers! It is the purpose of this discussion to describe a remotely-controlled center-fed antenna capable of continuous adjustment from band to band and within the bands, using an untuned transmission line.

The Folded Dipole as a VFA

The folded dipole lends itself admirably to rigging with cords and pulleys so that the length of the flat top may be varied simply by pulling down on the feeders (see Figure 1). If the spacing between the center pulleys is adjusted so that the characteristic impedance of the two parallel conductors is 300 ohms the feeders will match 300 ohm ribbon regardless of their length and the length of the antenna. In this way, we have a tuned antenna with a flat line for any frequency, accomplished without resorting to sliding contacts at any point in the system. Figure 2 shows a practical hand-operated rigging that will enable the experimenter to make a set-up and observe its characteristics. Antennas of this type may be made large or small but it is suggested

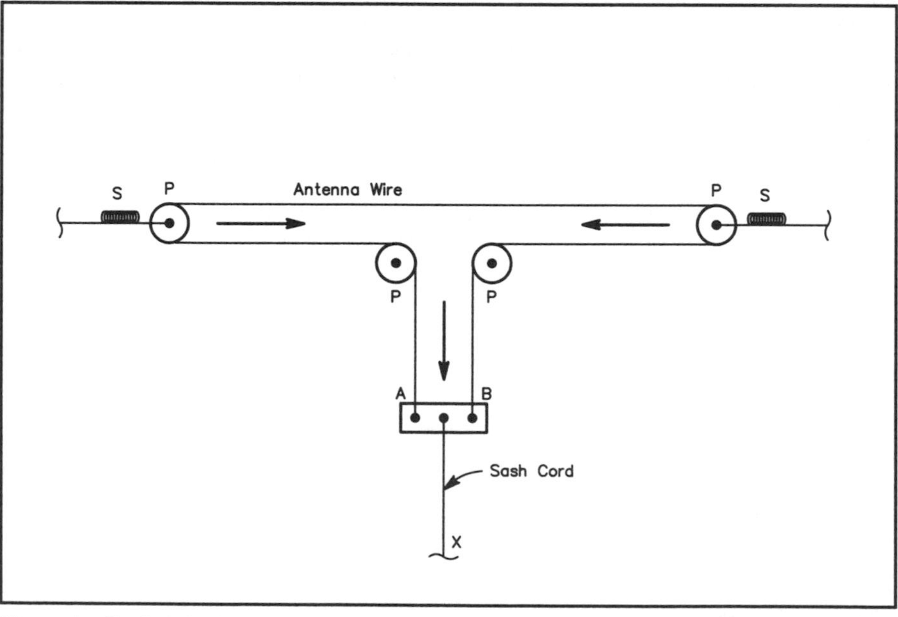

Figure 1—Basic principle of the adjustable folded dipole. Dipole and feeder section are made of one piece of flexible wire. Antenna length is changed by pulling down on the feeder at point X. Spacing of the pulleys at the center is such that the characteristic impedance of the feeder section is 300 ohms. Twin-Lead is connected at points A and B.

Practical working model of the folded-dipole VFA of Figure 3 used at W8AJC. The system is operated by the servo motor at the base of the mast, and is controllable from the operating position.

that the beginner make up small models for 2 and 6 meters, or simply for the f.m. and television bands, to prove the merit of the antenna.

The photograph and Figure 3 show the details of an experimental working model, made long enough to tune to 10 meters, but designed so it could be pulled down to a flat-top length of only a few inches. The tuning was fairly sharp and the results over conventional antennas for the reception of f.m. broadcasting were gratifying. A considerable improvement was noticed even within the 88-108 MHz f.m. band when the receiver was tuned to different stations and the antenna adjusted for maximum response. It was a real thrill to couple the feeders to the 10-meter transmitter and watch the plate milliammeter go up as the antenna came into resonance, and then pass through, and return for maximum; then without changing antennas, to switch on the f.m. broadcast receiver, run the antenna down to about four and a half feet to pick up a Detroit station. It was interesting to observe the effects on the received signal strength as the antenna was shortened from resonance at 10 meters to the proper length for the f.m. band, with the receiver tuned to a station on 98.5 MHz. Reception was possible with the long antenna and became good as the flat top hit fourteen feet (three half waves), falling off to a very sharp null at 121.5 inches (critical), after which it returned to full signal strength at 56.25 inches, approximately a half wave for the received signal. This ability to tune to an extremely critical null might find application in the elimination of an undesired, strong nearby signal under certain receiving conditions.

The uses to which a continuously-variable antenna may be put are limited only by the operator's imagination. Once the mechanical details have been worked out they may be used singly or in multiple, as antennas or reflectors, driven by common or separate servo motors and in various phase relations. Such antennas may be used for transmitting or receiving, or for special applications such as field operations covering a wide band of frequencies, signal-strength measurements, target transmitters for lining up rotary beams, and antenna studies. Their greatest commercial application will no doubt be in the f.m. and television fields where simplified versions, adjustable from the receiver, should find wide acceptance.

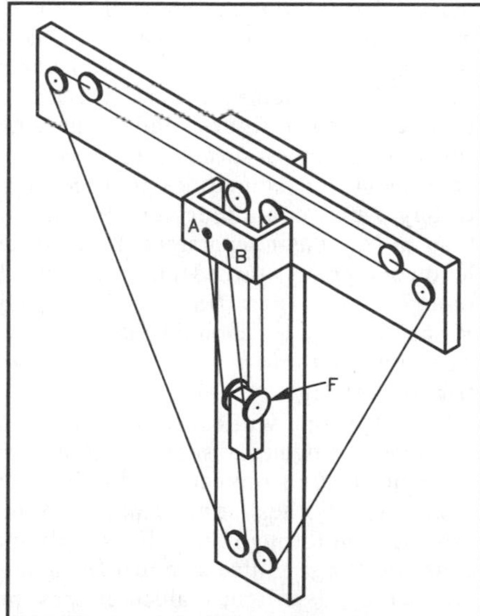

Figure 2—Hand-operated version of the variable-frequency antenna mounted on a wooden frame. To minimize mounting space, the feeders are folded back over ratio pulleys, F. The 300-ohm line connects at A and B. With this arrangement the vertical movement of pulley F is equal to the end movement of the antenna pulleys. The supporting fishline or sash cord must be taut.

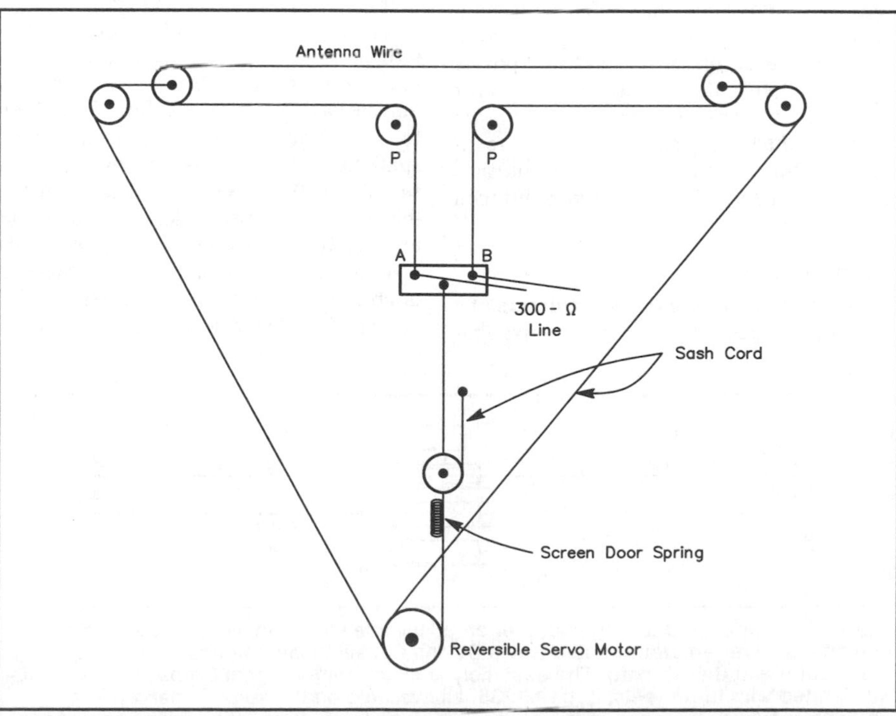

Figure 3—Diagram of the remotely-controlled antenna shown in the photograph. This system is now in use at W8AJC for 28 MHz and up. It may be reduced to a flat top of a few inches.

Other Types of Adjustable Antennas

Although the folded doublet lends itself most easily to continuous adjustment it might also be desirable to have a variable center-fed half-wave antenna suitable for use with coaxial or other low-impedance feeders. This may be done by the use of pulleys, insulators and wire, but requires a sliding contact at the center where the wind-up drums are located, and unless operated frequently it is subject to all the ills of exposed slip rings. This type of antenna is shown in Figure 4 and although at present, untried at W8AJC, it may prove worth-while on the lower frequencies, filling the long-felt need for a tunable 80-meter antenna that will work on 40 and 20 also!

Folded-Dipole Design

Useful information relative to the design of folded dipoles may be found in the ARRL *Handbook* for 1948, War Department *TM11-466*, RCA's little book *A Practical Analysis of U.H.F.* and in a paper by W. Van B. Roberts appearing in the *RCA Review* for June, 1947, page 289. The article by Roberts, in which he tells of the work done by his group at Princeton during the war under Government contract for the study of the folded dipole, is very helpful in providing the reader with a mechanism for analysis of the antenna. The literature indicates that spacing of the conductors composing the flat top should be close, in the order of 1/100 wavelength. Since our antenna is to have flexible conductors it is desirable to work with a type of wire that has a large diameter for low r.f. resistance, and at the same time has a high degree of flexibility. Such a conductor may be composed of insulated wire with a braided copper shield, the shield acting as the antenna and thus combining large area with good flexibility. The rest of the design is based upon available materials, with emphasis upon methods and mechanical devices for carrying out the function of varying the length of the flat top and handling the feeders. Refinements in both electrical and mechanical aspects will result from continued development.

Construction

When considering the construction of a tunable folded dipole many arrangements using springs or cords and pulleys will come to mind and it is up to the individual to select the method best suited to his particular use. In the beginning I tried a variety of springs, shock cords and weights to hold out the ends of the flexible antenna wire, but each of these methods had its own drawbacks and they all had the disadvantage of having to pull against a spring to shorten the antenna and depend upon the spring to pull it back out again. Metal springs come into resonance at certain frequencies. I tried metal-spring sash supports which would extend about 40 inches but they came into resonance in their extended positions and were not very smooth in operation.

It was finally decided to use cords and pulleys so arranged as to be in mechanical equilibrium and use the servo motor or other means only for the purpose of adjustment. This required less power in the servo and while it calls for more pulleys the result was smoother adjustment. I have found it convenient to support the antenna from the ends by means of insulated pulleys on a wood or other nonconducting structure. If the dipole is to be operated in a horizontal position there is no objection to using a vertical metal support pipe, but horizontal metal rods or pipes should not be used. For long antennas a center support must be provided for the feeder pulleys and the wind-up mechanism. The ends may be supported by poles, trees or buildings.

Servo Motors

The servo motor shown in the photograph is from a surplus Azon bomb tail assembly. It has plenty of power and may be reversed at will. It requires a 4-wire cable to the battery or other d.c. source. The current consumed is small and since the time of operation is also small a few dry cells will provide power for operation over a long period of time. Contained within the unit are two selenium rectifiers placed there to short circuit reverse currents to prevent sparking. They may be removed and inserted in a 30-volt a.c. line to the unit where they will provide sufficient d.c. for its operation. Reversal may be obtained at the station end by means of a double-pole double-throw toggle switch. Many other similar slow-speed servo motors are available on the surplus market, most of them reversible, and varying in size and power requirements. In some cases where d.c. is not available advantage may be taken of the gear train by connecting a universal coupling to the motor end and driving with a reversible universal fan or vacuum-cleaner motor operated from 115 volts a.c.

Sources of Materials

Antenna wire should be light, durable, flexible, of large diameter and a good conductor. For ease of adjustment it should pull around a one-inch pulley readily. Superflexible stranded copper wire of large diameter would be quite heavy whereas an insulated stranded wire, if size 20 or so and covered by a braided tinned-copper shield, would be light in weight, adequately flexible and of sufficient diameter. Belden No. 8885 shielded grid wire having an o.d. of 0. 1 inch has been found satisfactory. Too-stiff wire will make the system unwieldy.

Pulleys must be free-running for smooth operation and have as little friction as possible. A number of different kinds of pulleys normally available at hardware stores were tried and all had mechanical imperfections. Usually, though they seemed free-running when tried at the store, they turned out to have prohibitive friction when loaded and in the system. Since the number of pulleys required is fairly large and the accumulated friction may be excessive, ball-bearing pulleys are recommended. The first ones used here were homemade and turned out of fiber and used small ball bearings in the center. The ball bearings had 1/4-inch holes and a 5/8-inch o.d. and were obtained from disassembly of surplus gear trains, bomb sights, computers and other equipment so plentiful on the surplus market. Later I found a source of one-inch aircraft pulleys (AN-210-1A) with ball-bearing centers. Air Associates sells them for $1.25 each but the surplus market offers them at a lower figure. Pulleys may be found in all sorts of surplus aircraft control equipment and sometimes it is cheaper to buy a unit containing several pulleys than to buy them separately. Homemade hardwood pulleys turned out of maple and boiled in paraffin and using 1/4-inch brass axles should be satisfactory. The important thing is to have good low-friction bearings.

Twisted rope will cause the pulleys to turn over and twist and short out the aerial; therefore it is recommended that braided sash cord or clothesline be used. This, when properly fed through the pulleys, will not cause twisting. A nice size that fits available pulleys is a light braidcd clothesline 1/8-inch in diameter. Of course any flexible insulating line may be used such as dial cable, fishline or upholsterer's twine. Spring loading to prevent slipping because of stretching is advisable in some cases.

Limit Switches: If the antenna is not visible from the operating point Micro-Switches may be so placed that when the end of travel of the antenna is reached the

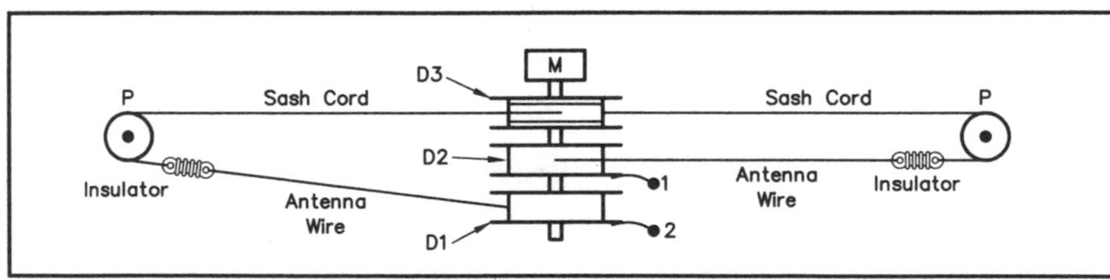

Figure 4–A suggested arrangement for an adjustable motor-driven dipole suitable for use on the lower amateur bands. Reels D_1 and D_2 reel in the antenna wire, while D_3 plays out braided sash cord. The assembly is made from a Signal Corps reel, Type RL-42-B, fitted with three reels, Type M-235, all available on the surplus market. The antenna reels are fastened to the cord reel by means of stand-off bushings secured to the center insulation. Brushes 1 and 2 connect to a 72-ohm line running to the transmitter. Success of this system depends upon maintenance of good contact at the brushes.

circuit will be opened and the motor will stop, it being possible then to reverse it and run it to the other limit where another switch will furnish protection from overtravel. The switches should not be connected in the common lead to the motor but in the circuit controlling that direction only. (Switches and associated wiring must be placed so as not to interfere with the electrical operation of the antenna.)

Conclusion

The examples shown are but a few of many possible ways of setting up remotely-controlled variable-frequency antenna systems. It is hoped that this article will serve as a basis for further development of adjustable antennae for amateur and commercial use.

By Rod Newkirk, W9BRD From *QST*, July 1993

Honey, I Shrunk the Antenna!

Think small! Communicate with multiconductor miniature loop antennas.

Great energy and ingenuity go into efforts to boost the effectiveness of compact transmitting antennas—*compact* meaning configurations with circumferences of 1/8 wavelength or less. By this definition, a 40-meter design could be a square with 4-foot sides. In the following discussion, I depart slightly from the square. I prefer 3½ feet high by 4½ feet wide, suspended in the vertical plane, for convenience in my antenna's location.

The traditional small-loop approach is to insert capacitance in series with a one-turn loop, tune it to the desired frequency, and then attempt to feed power to it as efficiently as possible–no simple matter. The antenna's Q is astronomical, current and voltage are monstrous, bandwidth is razor sharp, and its radiation resistance is ridiculously low. This lossy situation can be improved by reducing the resistance of the loop material through the use of piping, foil, etc, but the resulting plumbers' nightmares are hardly worth the pains.

A more promising avenue toward practical, simple compact antennas would appear to be in the direction of varied configuration. The goal is to "cool" the system with better power distribution, and at the same time simplify the feed. For example, vast improvement in the key parameters is possible by application of the venerable "folded dipole" technique. This can be done by adding one more turn, tuned identically, connecting the two turns *in series*, and re-resonating the antenna. Now, looking into one of the twin current nodes, the input impedance will have roughly quadrupled. The one-turn loop's skimpy 5 or 6 Ω rises to approximately 25 Ω. Adding yet another such turn doubles this to about 50 Ω–a convenient match to common coax. See Figure 1. As with the folded dipole, more than four turns are unprofitable; diminishing returns raise the input impedance. little faster than ohmic resistance. Note that the tuning capacitances, each about 40 pF, must be kept nearly equal for overall system balance.

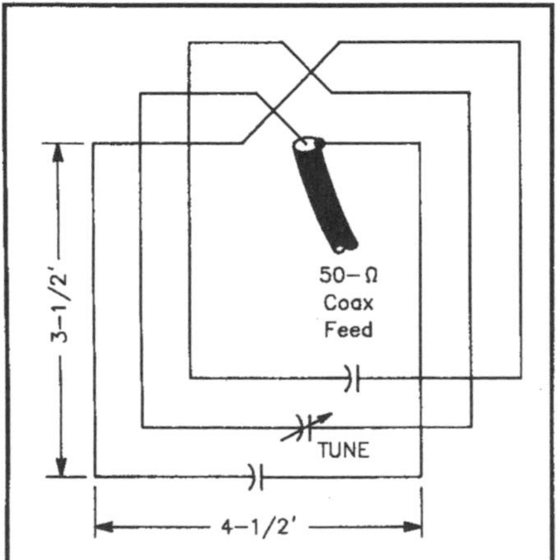

Figure 1—The three-turn loop antenna for 40 meters. Each capacitor is about 40 pF to resonate and match the antenna to a 50-Ω feed line.

> **Increasing Multiturn Loop Efficiency**
>
> Lots of hams use small loops in situations where they can't erect a larger antenna. Making contacts with these antennas largely depends on high radiation efficiency. Here are some things you can do to increase efficiency and effectiveness of such antennas:
>
> • Use large-diameter copper wire. Keep the copper surface clean. Tin-plated wire is noticeably worse than copper in this application because of skin effect–most of the RF travels in the tin, which is far lossier than copper at RF.
>
> • Use low-loss capacitors. The key here is minimizing dielectric loss, which is most favorable in vacuum- and air-dielectric capacitors. You can also parallel several capacitors to decrease losses.
>
> Increasing the wire size used in these loops raises the antenna Q, narrowing the bandwidth and increasing the voltages and currents present. For this reason, you may have to use higher-quality capacitors than those used in the experimental versions if you increase the wire size.
>
> Remember that high voltages can exist on a multiturn loop antenna's wires at power levels of a few tens of watts or more. Thus, it's best to keep the antenna well away from other objects.—*Roy Lewallen, W7EL*

Construction techniques, conductor material, insulation and components now become less critical because voltage peaks and current nodes are distributed throughout the system. I use ordinary zip cord, two lengths entwined, giving four available turns.[1] For proper continuity, the four wires on each side of the loop are clearly tagged at alligator-clip connectors, top and bottom centers. For 7 MHz, one turn is left open (at top *and* bottom), but will become useful on the lower bands as discussed later. A more classy construction might feature braided hook-up wire of contrasting colors. Small cardboard spreaders at the clips keep adjacent capacitors separated.

Midget receiver-type capacitors are adequate for CW power output up to 100 watts. One of these should be conveniently variable, preferably a double-gang or splitstator type, for tweaking the system to the desired center frequency as indicated by a 1:1 SWR. At 7 MHz, each capacitance is about 40 pF in my antenna, so the old war surplus Hammarlund APC midgets do nicely. Bandwidth can approach that of a normal linear dipole if a slightly higher center frequency SWR is accepted by stagger-tuning.

Excellent 40-meter results indicated that the same loop, dimensions unchanged, could have a shot at 80 meters. At 3.5 MHz, it becomes truly a miniloop, only 1/16 wavelength in circumference. Adding the fourth turn (Figure 2) raises the feed impedance at a slightly higher rate than ohmic resistance, so I included it. Armed with a bargain bag

The Ingredient Called *Belief*

Even though my busy family life keeps hamming time rare, the time I spend hamming has to be spent *right*. The idea isn't just to get on the air and contact anyone anywhere—that's too *easy*. The idea is to do something a hair off-the-wall and see if anyone shares the same orbit. So I'm one of those guys who calls CQ QRP for ten minutes at 10149.5 kHz even as game clogs the band's low end—someone who looks for (and *expects*) ragchews on 3560 at noon. And so I naturally was the guy to call a 5-watt, 3 × 2 × 1 CQ at 7138 kHz at around 10 PM one night last fall when the broadcasters had long since muscled straight-thinking NovTechs off to safer terrain. And so I naturally was *of course* immediately answered by W9BRD.

It figures. We've worked on schedule many times before, but he and I—he's the father; I'm the son—have never had to prearrange each new "first contact" after changes of station setup or locale. *Chance* is too facile a word for this, but then I haven't looked too hard for a better word, either. I don't need to; we just do our individual quirky things and click.

This time, we *both* had new setups: I'd just relocated the armful of gear constituting my "shack" to the bedroom and was talking to a low dipole through 100+ feet of mongrel coax; *he* was trying out an indoor, largely below-ground loop antenna—one of the versions shrunken in this article—with just a few tens of watts, tunneling past those steamroller broadcast signals as soundly as any medium-scale ham signal from Chicago.

Enjoying the magic of yet another first contact, I hardly gave his setup a second thought. But you may wonder if such antennas really work. Gooch's Paradox explains it thus: "RF gotta go *somewhere*." Seeking to soothe the linearless, towerless masses who toil with "*only* 100 watts," Newkirk's Tiresome Chant puts it, "Successful MF/HF radio communication may proceed at astoundingly low received-signal strengths." Reduced to generic essentials, it's this: The most important single factor in success with *whatever* you use is overcoming your own disbelief.—*David Newkirk, WJ1Z*

of small 1-kV ceramic capacitors, I reached 3.6-MHz resonance with a value of 100 pF for each of the four capacitances. One capacitor was then replaced with an old broadcast-style, two-gang 300-300-pF variable for tweaking. I measured the feed impedance, at the center of one of the four turns, at about 18 Ω–well within the range of simple gamma matching. So another junk-box broadcast variable, three 350-pF gangs paralleled, was suspended at the

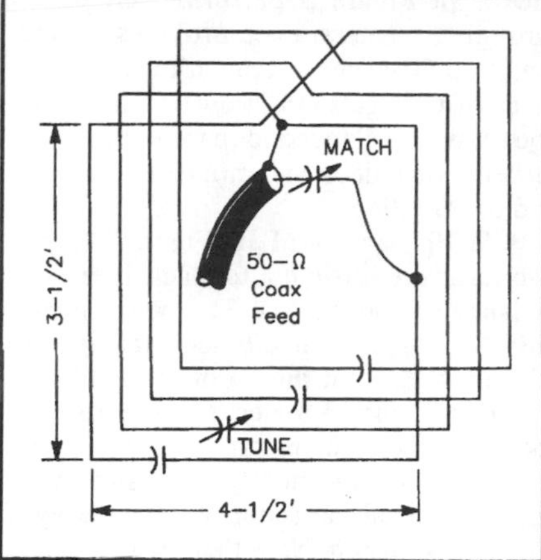

Figure 2—Extending the Fig 1 concept to 80 and 160 meters requires adding another turn to the antenna and increasing the capacitances. See the text for capacitor values.

loop's top center. Only about 650 pF was needed for a 1:1 match to coax after a 30-inch gamma lead was dangled to a tap at the center of one side of the feed turn. Then the system was tweakable over the entire 80-meter CW range. The remaining ceramic capacitors get slightly warm at 80 watts, which indicates significant power loss. These capacitors should be replaced with low-loss units for best performance, but I left them alone for my experiments. I easily contacted the East Coast and as far west as Arizona in the following few nights on the air.

My next inclination, as you may surmise, was to try the little gem on 160 meters. A four-foot-square transmitting loop for 1.8 MHz? I wasn't overly optimistic. Years of tinkering at W9BRD had failed to produce a decent indoor compact antenna, even of much larger size, for top-band work. But I dug into my bag of ceramics and gave it a go. The four-turn loop, now only 1/32 wavelength in circumference, resonated at about 1.8 MHz with four 350-pF capacitors. One of these was replaced with an old broadcast-type 500-pF variable for tweaking the center frequency between 1.8 and 1.85 MHz. I double- and triple-checked the measured feed impedance—16 Ω. Now I knew it would work. The gamma-match SWR dipped to 1:1 at about 800 pF using the same lead, tap and broadcast capacitor as on 80 meters. Immediate solid 40-watt CW QSOs with WN9W, KC4WWV, WKØB and W4VZB were most gratifying.

Variations and the Higher Bands

If such a 1/16-wavelength loop can function at 80 meters, the same should be true at 20 meters. So I wound a 1-foot-square, three-turn model on a cardboard box, hung it on a wooden bulkhead, and resonated it to 14.050 MHz with three 50-pF ceramic capacitors. This time, instead of the usual fudging capacitor, I

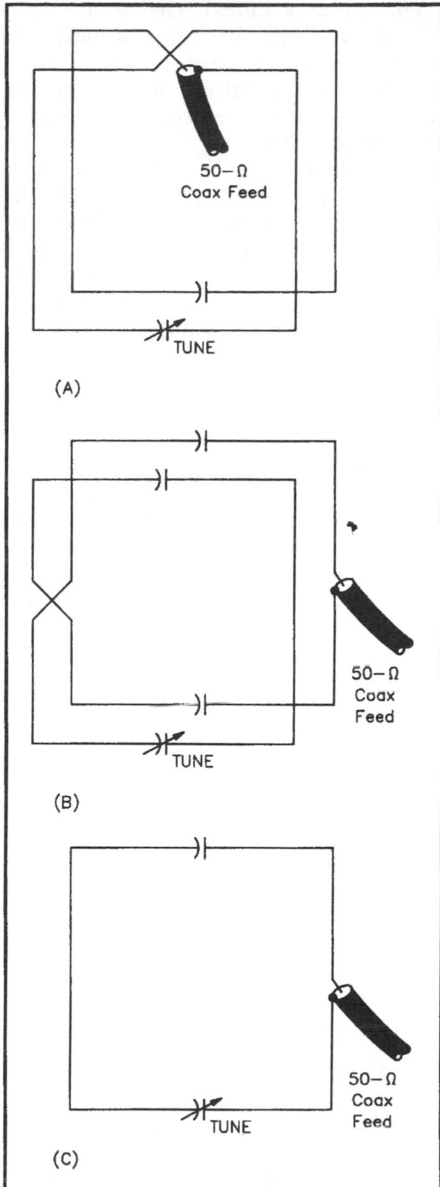

Figure 3—At A, the 30/20-meter loop; at B, the 15/17-meter version, and at C, the 10-meter variant.

spaced the turns 1½ inches apart. Center-frequency tweaking thus could be done by uniformly edging the turns away from or toward each other. This spacing approach works fine for box-wound loops, but for the lower bands, I prefer closewound zip cord. Input impedance turned out to be about 11 Ω—I should have used a fourth turn—and gamma coupling worked okay, as on 80 and 160. The 1:1 feed-line dip occurred with a 120-pF series capacitor, with a gamma lead of 13 inches. In my installation, this antenna is no low-angle DX radiator, to be sure, but I've received S9 reports during many stateside QSOs. Anyone for a 3-inch-square 2-meter version?

Antennas smaller than necessary are interesting stunts that help prove a design. However, for the lowest HF bands I recommend the largest practicable circumference.

For operators limited to indoor installations, floor-to-ceiling height is usually the limiting size factor. Thus, a cooler 7-foot by 9-foot job, requiring much smaller resonating capacitances, should be a natural for 80 and 160.

Incidentally, the configurations described are quite well balanced. Although in my tests the antennas were placed within tweaking distance of the operator, no shack RFI showed up on any band at the 80-watt level. Careful symmetry helps, notwithstanding the dangling gamma lead on 80 and 160. The coax feeder should run down and away from the top-center feed point, equidistant from the vertical sides. A balun or coaxial RF choke might seem applicable, but I found such isolation unnecessary. The directivity pattern for such a loop is the usual figure-8, with most radiation in the plane of the loop. However, there will be sharp nulls perpendicular to the loop's plane, which you should take into account when placing your antenna.

My prime purpose here is to treat loops of 1/8-wavelength circumference or smaller, but you may desire to use the same 3 1/2-foot by 4 1/2 foot dimension on the higher HF bands, as I do. The hook-ups shown in Figure 3 perform well on 30 through 10 meters. Each provides a direct match to 50-Ω coax. True, almost any old piece of wire gets QSOs on 10.1 MHz and above, but a balanced loop radiator is hard to beat for indoor hamming. A few hints and observations:

- At 10. 1 and 14 MHz (Figure 3A), just two turns are sufficient to reach a feed impedance of about 50 Ω. The two approximately 20-pF capacitance values are slightly higher for the 30-meter band.
- For 17 and 15 meters (Figure 3B), the loop must be made electrically smaller by splitting it symmetrically. Two split turns and four equal capacitances then bring a close 50-Ω match. Note that the feed point is moved to the center of one of the four vertical sides. Capacitances are in the 15-pF range.
- On 12 and 10 meters (Figure 3C), one split turn will suffice for a near-50-Ω match, with two capacitors of about 12 pF. Here, splitstator midgets can ease adjustment by minimizing hand capacitance, but it's a fancy junk box that includes them. The old APC midgets, whose rotors have little more mass than their stators, will maintain enough system balance.

Cautions and Conclusions

Like any indoor antenna, the loops I describe here generate substantial electromagnetic fields in operation. Thus, they have considerable potential to generate RFI. For this reason and to prudently avoid placing yourself or others in large RF fields, you should keep all antennas as far from consumer electronic devices and people as possible, and use the least RF power necessary to conduct the desired communications. The current editions of *The ARRL Handbook* and *The ARRL Antenna Book* cover this subject in more detail.[2]

A final comment: For overall results with compact antennas, like almost any skyhook, the higher above ground the better. In my case, radiation and reception strongly favor higher skywave propagation angles. All antenna configurations I describe here were tested and operated in a cellar ham shack, where half of the system is below ground level!

Notes

[1]See the sidebar, "Increasing Multiturn Loop Efficiency," for some ideas on optimizing such an antenna's performance.

[2]See Chapter 9 of the *Handbook* and Chapter 1 of the *Antenna Book.*

About the ARRL

The seed for Amateur Radio was planted in the 1890s, when Guglielmo Marconi began his experiments in wireless telegraphy. Soon he was joined by dozens, then hundreds, of others who were enthusiastic about sending and receiving messages through the air—some with a commercial interest, but others solely out of a love for this new communications medium. The United States government began licensing Amateur Radio operators in 1912.

By 1914, there were thousands of Amateur Radio operators—hams—in the United States. Hiram Percy Maxim, a leading Hartford, Connecticut, inventor and industrialist saw the need for an organization to band together this fledgling group of radio experimenters. In May 1914 he founded the American Radio Relay League (ARRL) to meet that need.

Today ARRL, with approximately 170,000 members, is the largest organization of radio amateurs in the United States. The ARRL is a not-for-profit organization that:

- promotes interest in Amateur Radio communications and experimentation
- represents US radio amateurs in legislative matters, and
- maintains fraternalism and a high standard of conduct among Amateur Radio operators.

At ARRL headquarters in the Hartford suburb of Newington, the staff helps serve the needs of members. ARRL is also International Secretariat for the International Amateur Radio Union, which is made up of similar societies in 150 countries around the world.

ARRL publishes the monthly journal *QST*, as well as newsletters and many publications covering all aspects of Amateur Radio. Its headquarters station, W1AW, transmits bulletins of interest to radio amateurs and Morse code practice sessions. The ARRL also coordinates an extensive field organization, which includes volunteers who provide technical information and other support for radio amateurs as well as communications for public-service activities. ARRL also represents US amateurs with the Federal Communications Commission and other government agencies in the US and abroad.

Membership in ARRL means much more than receiving *QST* each month. In addition to the services already described, ARRL offers membership services on a personal level, such as the ARRL Volunteer Examiner Coordinator Program and a QSL bureau.

Full ARRL membership (available only to licensed radio amateurs) gives you a voice in how the affairs of the organization are governed. ARRL policy is set by a Board of Directors (one from each of 15 Divisions). Each year, one-third of the ARRL Board of Directors stands for election by the full members they represent. The day-to-day operation of ARRL HQ is managed by an Executive Vice President and a Chief Financial Officer.

No matter what aspect of Amateur Radio attracts you, ARRL membership is relevant and important. There would be no Amateur Radio as we know it today were it not for the ARRL. We would be happy to welcome you as a member! (An Amateur Radio license is not required for Associate Membership.) For more information about ARRL and answers to any questions you may have about Amateur Radio, write or call:

ARRL—The national association for Amateur Radio
225 Main Street
Newington CT 06111-1494
(860) 594-0200
Prospective new amateurs call:
800-32-NEW HAM (800-326-3942)
You can also contact us via e-mail at **ead@arrl.org** or check out *ARRLWeb* at **http://www.arrl.org/**

FEEDBACK

Please use this form to give us your comments on this book and what you'd like to see in future editions, or e-mail us at **pubsfdbk@arrl.org** (publications feedback). If you use e-mail, please include your name, call, e-mail address and the book title, edition and printing in the body of your message. Also indicate whether or not you are an ARRL member.

here did you purchase this book?

☐ From ARRL directly ☐ From an ARRL dealer

there a dealer who carries ARRL publications within:

☐ 5 miles ☐ 15 miles ☐ 30 miles of your location? ☐ Not sure.

cense class:

☐ Novice ☐ Technician ☐ Technician Plus ☐ General ☐ Advanced ☐ Amateur Extra

---

ame ARRL member? ☐ Yes ☐ No

________________ Call Sign ________

ddress ________________

ity, State/Province, ZIP/Postal Code ________________

aytime Phone ( ) ________________ Age ________

licensed, how long? ____________ e-mail address: ____________

ther hobbies ________________

____

ccupation ________________

For ARRL use only	MORE WIRE
Edition	1 2 3 4 5 6 7 8 9 10 11 12
Printing	2 3 4 5 6 7 8 9 10 11 12

From ______________________________

____

____

Please affix postage. Post Office will not deliver without postage.

EDITOR, ARRL'S MORE WIRE ANTENNA CLASSICS
ARRL–THE NATIONAL ASSOCIATION FOR AMATEUR RADIO
225 MAIN STREET
NEWINGTON CT 06111-1494

– – – – – – – – – – – – – please fold and tape – – – – – – – – – – – – –